ENVIRONMENTAL SCIENCE AND TECHNOLOGY HANDBOOK

Ayers, Deb, Fisher, Hattemer-Frey,
Kelly, Kester, Knowles, Krieger,
Little, Middel, Puszcz, Silka, Slavin, Tumulty, Vajda, Young

 Government Institutes, Inc.

Government Institutes, Inc., Rockville, Maryland 20850

Copyright © 1994 by Government Institutes. All rights reserved

99 98 97 96 95 94   5 4 3 2

Library of Congress Cataloging-in-Publication Data

Environmental science & technology handbook/
  Kenneth W. Ayers...(et al.).
    p. cm.
    Includes index.
    ISBN 0-86587-362-3
    1. Environmental sciences. 2. Environmental
engineering. I. Ayers, Kenneth W.
GE105.E59 1993
628--dc20
                                                    93-27558
                                                         CIP

No part of this work may be reproduced or transmitted in any form or by any means, electronic or mechanical, including photocopying, recording, or any information storage and retrieval system, without permission in writing from the publisher. All requests for permission to reproduce material from this work should be directed to Government Institutes, Inc., 4 Research Place, Suite 200, Rockville, Maryland 20850.

The authors and publisher make no representation or warranty, express or implied, as to the completeness, correctness or utility of the information in this publication. In addition, the authors and publisher assume no liability of any kind whatsoever resulting from the use of or reliance upon the contents of this book.

Printed in the United States of America

SUMMARY CONTENTS

	DETAILED CONTENTS iv
	BIOGRAPHIES xvii
1	ENVIRONMENTAL PROCESSES .... 1 Lyle R. Silka, Hydrosystems, Inc.
2	HUMAN HEALTH RISK ASSESSMENT .... 37 Janet E. Kester, Holly A. Hattemer-Frey, and Gary R. Krieger, Dames & Moore
3	ECOLOGICAL RISK ASSESSMENT .... 75 Janet E. Kester, Holly A. Hattemer-Frey, and Gary R. Krieger, Dames & Moore
4	ENVIRONMENTAL CHEMISTRY AND ANALYSIS OF REGULATED COMPOUNDS .... 97 Joann M. Slavin, Ursula R. Middel, and Ellen R. Kelly, H2M Labs, Inc.
5	AIR QUALITY .... 125 James W. Little, Dames & Moore
6	AIR POLLUTION CONTROL TECHNOLOGIES .... 155 Perry W. Fisher, Kaushik Deb, Dames & Moore
7	SOLID AND HAZARDOUS WASTE TREATMENT AND DISPOSAL .... 177 Kenneth W. Ayers, Willis Coroon
8	UNDERGROUND AND ABOVEGROUND STORAGE TANK TECHNOLOGY .... 209 A.D. Young, Jr., Consultant
9	GEOLOGY AND GROUNDWATER HYDROLOGY .... 249 Porter-C. Knowles, Ground-Water Hydrologist
10	GROUNDWATER POLLUTION CONTROL TECHNOLOGIES .... 317 Stanley G. Puszcz and Michael V. Tumulty, H2M Group
11	POLLUTION PREVENTION THROUGH TOTAL QUALITY MANAGEMENT .... 347 Gary Vajda, Dames & Moore
	INDEX 376

DETAILED CONTENTS

BIOGRAPHIES .... **xvii**

1   ENVIRONMENTAL PROCESSES 1
OVERVIEW 1
BASIC TRANSPORT PROCESSES 1
THE HYDROLOGIC CYCLE 3
    Residence Time 4
PROCESSES CONTROLLING MIGRATION IN AIR 4
    Equilibrium Partitioning of Chemical between Liquid and Air . 5
    Evaporation of a Liquid Chemical Spill into the Air 5
PROCESSES CONTROLLING MIGRATION THROUGH SURFACE WATER 7
    Uptake or Partitioning of a Chemical into Surface Water 7
    Colloidal Transport in Surface Water 7
    Transport by Adsorption on Suspended Sediment 7
    Fate of Chemicals in Surface Water 8
PROCESSES CONTROLLING MIGRATION THROUGH THE SUBSURFACE 9
    Transport of Volatile Organics in Soil 9
    Partitioning of VOC between Pure Liquid and Soil Gas 10
    Partitioning of VOC between Soil Gas and Soil Moisture 10
    Partitioning of VOC between Soil Moisture and Soil Solids 12
    Effects of Temperature and Soil Moisture Content on Partitioning 14
    Transport of VOC Vapor through Soil Gas 15
    Transport of Contaminants in the Saturated Zone
    Recharge and Discharge 19
    Groundwater Basins 19
    Aquifers and Aquitards 19
    Artesian Groundwater 20
    Flow of Groundwater and Darcy's Law 20
    Calculating True Groundwater Velocity 21
    Determining Flow Path 22
    Transport of Chemicals through Groundwater
    Advection 22
    Mechanical Dispersion 23
    Molecular Diffusion 24
    Movement of Immiscible Liquids 24
BEHAVIOR OF CHEMICALS IN THE WATER-SOIL SYSTEM 26
    Processes That Retard Chemical Movement 26

Adsorption	27
Ion Exchange	27
Retardation Factor	29
Chemical Precipitation	29
Biodegradation	29
HYDROGEOLOGIC CONTROLS OF MIGRATION THROUGH GROUNDWATER	31
Unconsolidated Sands and Gravels	31
Fractured Hard Rock	32
Karst and Highly Fractured Rock Regions	33
SUMMARY	34

2 HUMAN HEALTH RISK ASSESSMENT 37

OVERVIEW	37
Summary	39
DATA EVALUATION	39
Review of Site Information	39
Summary	41
Evaluation of Available Data for Use in Risk Assessment	41
Statistical Evaluation of Data	43
Selection of Chemicals of Potential Concern	44
Summary	45
TOXICITY ASSESSMENT	45
Carcinogenic Health Effects	46
Non-Carcinogenic Health Effects	48
Chemicals Without Toxicity Values	50
Summary	51
EXPOSURE ASSESSMENT	52
Identification of Potential Fate and Transport Mechanisms	52
Identification of Potentially Exposed Populations	55
Selection of Complete Exposure Pathways	55
Estimating Exposure Point Concentrations	58
Estimating Chemical Intakes	59
Modeling Exposures from Soil-Related Pathways	59
Incidental Ingestion of Soil	59
Dermal Contact with Soil	60
Modeling Exposures via Inhalation	60
Ingestion of Groundwater	61
Modeling Food Chain Intakes	62
Summary	62
RISK CHARACTERIZATION	63
Carcinogenic Health Effects	63
Non-carcinogenic Health Effects	64
Summary	65

Uncertainty Analysis 65
    Qualitative Uncertainty Analyses 66
    Quantitative Uncertainty Analyses 67
    Summary 69
REFERENCES 69

3   ECOLOGICAL RISK ASSESSMENT 75
OVERVIEW 75
CURRENT APPROACHES FOR ERA 76
EXPOSURE ASSESSMENT 77
    Characterization of the Site 78
    Identification of Relevant Exposure Pathways 78
    Selection of Ecological Receptors of Concern 79
    Identifying Chemicals of Potential Concern 80
    Selection of Ecological Endpoints 82
    Fate and Transport Analysis 83
    Estimating Exposure Point Concentrations 84
    Estimating Chemical Intakes 85
    Summary 86
TOXICITY ASSESSMENT 87
    Toxicity Criteria for Aquatic Organisms 88
    Toxicity Criteria for Terrestrial Organisms 90
    Summary 90
ENVIRONMENTAL RISK CHARACTERIZATION 91
    The TQ Approach 91
    Summary 92
UNCERTAINTY ANALYSIS 92
REFERENCES 93

4   ENVIRONMENTAL CHEMISTRY AND ANALYSIS OF REGULATED COMPOUNDS 97
OVERVIEW 97
ENVIRONMENTAL CHEMISTRY AND ENVIRONMENTAL
REGULATION 98
DRINKING WATER ANALYSIS 99
HAZARDOUS WASTE ANALYSIS 100
WATER POLLUTION 102
AIR QUALITY 102
LABORATORY OPERATIONS FOR THE ANALYSIS OF
REGULATED SUBSTANCES 103
METHODS 104
INSTRUMENTATION FOR ORGANIC ANALYSIS 105
GAS CHROMATOGRAPHY 106
DETECTORS 106

SAMPLE PREPARATION 107
IMPROVEMENTS IN GC METHODS AND TECHNOLOGY 108
GAS CHROMATOGRAPHY/MASS SPECTROMETRY 109
HIGH PRESSURE LIQUID CHROMATOGRAPHY 110
INSTRUMENTATION FOR INORGANIC CHEMICAL ANALYSIS . 111
QUALITY ASSURANCE/QUALITY CONTROL IN THE
ANALYTICAL LABORATORY 114
STANDARD OPERATING PROCEDURES 115
"IT DIDN'T HAPPEN IF IT'S NOT DOCUMENTED" 115
FIELD MEASURES OF QA/QC 115
DOCUMENTATION CONTROL 116
INTERNAL QUALITY CONTROL 117
CALIBRATION 118
HOLDING TIMES 119
DELIVERABLES 119
DATA VALIDATION: COMPLIANCE V. USEABILITY 120
CONCLUSION 120
REFERENCES 121

5 AIR QUALITY 125
OVERVIEW 125
TYPES OF AIR POLLUTANTS 126
TYPES OF AIR POLLUTION SOURCES 127
   Mobile Emission Sources 127
   Non-Industrial Stationary Emission Sources 128
   Industrial Stationary Emission Sources 128
ESTIMATION OF EMISSION RATES FOR STATIONARY
SOURCES 129
ATMOSPHERIC DISPERSION, TRANSFORMATION, AND
DEPLETION 131
   Dispersion 131
   Transformation 133
   Depletion 133
EMISSION CONTROL METHODS 133
   Stationary Emission Sources 133
   Mobile Emission Sources 135
AMBIENT AIR QUALITY EVALUATION METHODS 135
   Ambient Air Quality Monitoring 135
      Monitoring Methods 136
      Requirements for a Monitoring Program 136
   Air Quality Modeling 137
   Types of Data Required for Modeling 137
      Available Models 138
      Good Engineering Practice Stack Height Concept 139

viii / Environmental Science and Technology Handbook

STRATOSPHERIC OZONE DEPLETION 139
INDOOR AIR QUALITY 140
REGULATORY APPROACH TO AIR QUALITY PROTECTION 141
    Overview 141
        Legislation 141
        Types of Air Quality Regulations 143
        Types of Pollutants from a Regulatory Perspective 143
        Attainment and Nonattainment Areas 144
        State and Local Programs 145
    Prevention of Significant Deterioration Requirements 145
        PSD Area Categories 146
        Emission Sources Subject to PSD Review 146
        Pollutants Subject to PSD Review 146
        Potential PSD Permitting Requirements 146
    Emission Offsets 147
        Offset Pollutants Related to Ozone 148
        Offset Ratio 148
        Offset Credibility Requirements 148
    General Air Quality Permitting Process 149
        Application Preparation Phase 149
        Application Review Phase 149
        Schedule for Construction Permits 150
    Clean Air Act Operating Permit Provisions 150
    Hazardous Air Pollutant Control Requirements 151
    Clean Air Act Enforcement Provisions 152
REFERENCES 152

6   AIR POLLUTION CONTROL TECHNOLOGIES 155

OVERVIEW 155
AIR POLLUTANT CHARACTERIZATION 155
IN-PROCESS AIR POLLUTION CONTROL 157
ADD-ON AIR POLLUTION CONTROL 161
SUMMARY 174
ENDNOTES 174
ADDITIONAL REFERENCES: 175

7   SOLID AND HAZARDOUS WASTE TREATMENT AND DISPOSAL TECHNOLOGIES 177

OVERVIEW 177
OVERVIEW OF WASTE TREATMENT TECHNOLOGIES 178
THERMAL PROCESSES 187
    Incineration 187
    Thermal Desorption 189

Contents / ix

- Pyrolysis 189
- Plasma Torch 190
- Vitrification 190
- BIOLOGICAL TREATMENT 191
  - Bioreactors 191
  - Solid Phase Bioremediation 192
  - Soil Heaping 192
  - Composting 192
  - In Situ Bioremediation 193
- CHEMICAL/PHYSICAL PROCESSES 193
  - Traditional Physical/Chemical Treatment Processes 193
    - Separation 193
    - Dewatering 194
    - Densification 195
    - Neutralization 195
    - Oxidation/Reduction 195
    - Ion Exchange 196
    - Activated Carbon Adsorption 196
  - Soil Washing 196
  - Soil Flushing 197
  - Soil Vapor Extraction 198
    - Vacuum Extraction 198
    - Steam Stripping 198
    - Radio Frequency Heating 198
  - Dechlorinization 199
  - Solidification/Stabilization 199
- TREATMENT TRAINS 201
- MATERIALS HANDLING 201
- CONTAINMENT/DISPOSAL 202
  - Simple Soil Cover 202
  - RCRA Caps 202
  - RCRA Landfill 204
- NATURAL ATTENUATION AND EXCAVATION WITH OFF-SITE DISPOSAL 204
- RADIOACTIVE AND MIXED WASTES 205
- GROUNDWATER TREATMENT 205
- RECYCLING AND REUSE 206
- SOLID WASTES 206
- PROCESS WASTE STREAMS 207
- SUMMARY 207
- ADDITIONAL REFERENCES 207

8   UNDERGROUND AND ABOVEGROUND STORAGE TANK TECHNOLOGY 209

    OVERVIEW 209
    UNDERGROUND STORAGE TANKS 210
        Causes and Solutions for UST Tank Failure 210
        Regulations 210
        Technical Developments 211
        UST Tank Design 211
        UST Piping 214
        Overfill and Spill Protection 215
        Secondary Containment 216
        Monitoring and Leak Detection 217
        UST System Testing for Liquid-tight Integrity 219
        Upgrading Existing UST Systems 222
        UST System Closure 224
        Installation of UST Systems 225
        UST System Management and Operations Planning 228
            Operating Practices 229
            Regulatory Requirements 230
            Engineering and Construction Program 231
    ABOVEGROUND STORAGE TANK TECHNOLOGY 231
        AST Regulations and Standards 232
        AST System Designs 233
        SPCC Plans and AST Operations 240
        Comparison of Benefits of AST vs. UST Configurations 242
    SUMMARY 246

9   GEOLOGY AND GROUNDWATER HYDROLOGY 249

    OVERVIEW 249
    GEOLOGY 250
    THE EARTH 251
        Metamorphic Rocks 268
    GEOLOGIC TIME 270
    STRUCTURAL GEOLOGY 276
    GEOLOGIC MAPPING AND INTERPRETATION 281
        Geophysical Surveys 282
    GROUNDWATER HYDROLOGY 284
    GROUNDWATER FLOW VELOCITY 297
    AQUIFER SYSTEMS 298
        Igneous and Metamorphic Rock Aquifers 300
        Glacial Deposits 303
    GROUND WATER CLASSIFICATION 304
    GROUNDWATER REGULATIONS 306

CONTAMINATION 306
WELL HYDRAULICS 310
GROUNDWATER MODELING 314
SUMMARY 315
REFERENCES 316

10 GROUNDWATER POLLUTION CONTROL TECHNOLOGIES 317

OVERVIEW 317
EVALUATION CRITERIA 318
REMEDIAL TECHNOLOGIES 318
DEVELOPMENT OF A REMEDIAL PROGRAM 319
   Theis Equation 320
DESIGN FLOW AND CONTAMINANT LOADING 323
   Contaminant Migration And Characteristics 324
BIOLOGICAL TREATMENT 327
AIR STRIPPING 328
CARBON ADSORPTION 334
AIR STRIPPING WITH CARBON ADSORPTION 335
ULTRAVIOLET (UV) OXIDATION 336
LIQUID PHASE GRANULAR ACTIVATED CARBON (GAC) TREATMENT 337
VAPOR PHASE GRANULAR ACTIVATED CARBON (GAC) TREATMENT 338
INORGANICS REMOVAL 339
SYSTEM OF UNIT OPERATIONS (TYPICAL GROUNDWATER TREATMENT SCENARIO) 340
   Gravity Differential Separation 342
   Surge Tank/Surge Pump 342
   Solids Filtration 343
   Counter-Current Packed Bed Air Stripping 343
   Vapor Phase GAC Adsorption 343
SUMMARY 344
REFERENCES 344

11 POLLUTION PREVENTION THROUGH TOTAL QUALITY MANAGEMENT 347

ROLE OF POLLUTION PREVENTION 347
DEFINITIONS 348
HIERARCHY OF WASTE MANAGEMENT 350
IMPLEMENTING A PROGRAM 354
   Economic Concerns 354
   Political Pressures 355
REGULATORY 355

POTENTIAL OBSTACLES 358
INTEGRATING MANUFACTURING AND ENVIRONMENTAL
ISSUES 359
SETTING THE STAGE 360
BASELINE CHARACTERIZATION 361
ALTERNATIVES DEVELOPMENT 363
CASE STUDIES 371
   Plastic Container Manufacturing Facility 371
   Aircraft Component Manufacturing Complex 372
   Metal Product Manufacturing Facility 373
REFERENCES: 374

INDEX 376

Contents / xiii

TABLES

1-1. Values of Henry's Law Constant, Vapor Pressure, and Solubility at 25°C for Selected Chlorinated Solvents   p. 11

1-2. Reported Values of Octanol: Water Partition Coefficient and Calculated Values of Partition Coefficient for Selected Chlorinated Solvents   p. 14

2-1. EPA's Weight-of-Evidence Carcinogenicity Classification Scheme   p. 47

2-2. Potential Exposure Pathways That Could Be Evaluated During a Human Health Risk Assessment   p. 56

2-3. General Uncertainty Factors   p. 68

6-1. Combustion Modification Schemes for NO_x and CO Control   p. 159

6-2. Post-Combustion Strategies for NO_x and CO Control   p. 160

6-3. Typical Characteristics of Particular Control Equipment   p. 166

6-4. Advantages and Disadvantages of Particulate Control Equipment   p. 166

6-5. Advantages and Disadvantages of Gaseous Pollutant Control Processes   p. 173

7-1. Representative Volatile Organic Compounds (VOCs)   p. 180

7-2. Representative Semi-Volatile Organic Compounds (SVOCs)   p. 181

7-3. Representative Metals   p. 183

10-1. Volatile Organic Pollutant Characteristics   p. 325

10-2. Liquid Film Resistance Percentages for Volatile Organic Pollutants   p. 332

FIGURES

4-1. Semi-Volatiles Sample Preparation Record   p. 108

4-2. Chain of Custody Form   p. 117

6-1. Dust Cyclone   p. 163

6-2. Pulse-Jet Fabric Filter System   p. 164

6-3. Principle of Electrostatic Precipitation   p. 165

6-4. Carbon Adsorption System   p. 170

6-5. Catalytic Incineration System   p. 172

6-6. Thermal Incineration System   p. 172

7-1. Technology Decision Flow Chart   p. 184

9-1. Earth   p. 253

9-2. Bowen's Reaction Series   p. 261

9-3. Porosity and Permeability   p. 265

9-4. Transportation of Sediments   p. 267

9-5. Glaciers & Glacial Sediments   p. 269

9-6. The Metamorphic Cycle   p. 271

9-7. Geologic Column   p. 273

9-8. Typical Cross Section of Transgressive and Transgressive-Regressive Ocean   p. 273

9-9. Law of Superposition and Law of Cross-Cutting Relationships   p. 275

9-10. Rock Relationship at Erosional Surface   p. 277

9-11. Rock Cross Section   p. 279

Contents / xv

9-12. Fault Cross Section   p. 280

9-13. Strike and Dip of Formation   p. 283

9-14. The Hydrologic Cycle   p. 285

9-15. Confined, Unconfined, and Perched Water   p. 288

9-16. Coefficient of Permeabiliy and Transmissibility   p. 292

9-17. Unconfined Aquifer   p. 294

9-18. Confined Aquifer   p. 295

9-19. Typical Arrangement of a Pumped Well and Observation Well   p. 296

9-20. Determination of Groundwater Gradient   p. 299

9-21. Typical Groundwater Contamination Sources   p. 307

9-22. Stiff Diagram Example (Aquifer "Signatures")   p. 308

9-23. Cone of Depression for Different Transmissivity   p. 312

9-24. Hydraulic Barrier   p. 313

10-1. Cross Section of Remediation Well   p. 321

10-2. Cross Section Showing Removal of Floating Petroleum   p. 326

10-3. Air Stripping Components   p. 329

10-4. Treatment Process Schematic   p. 341

11-1. Pollution Prevention Components   p. 349

11-2. Pollution Control Hierarchy   p. 351

11-3. Functional Analysis Diagram   p. 364

11-4. Components of Product Cost   p. 365

11-5. "True Cost" Analysis   p. 366

11-6. Evaluating Potential for Environmental Improvement   p. 368

11-7. Evaluating Potential for Operational Improvement   p. 369

11-8. Structured Decisionmaking   p. 370

BIOGRAPHIES

Kenneth W. Ayers

Kenneth W. Ayers, is a corporate resource available to Willis Corroon account executives who are designing and implementing risk management programs for clients who have environmental loss exposures. These clients include potentially responsible parties under Superfund, parties complying with RCRA, waste transporters, and hazardous waste treatment storage and disposal operations.

Mr. Ayers has 18 years of experience in the environmental engineering disciplines of water and wastewater treatment, hazardous and solid waste treatment, and industrial hygiene. Prior to joining Willis Corroon, Mr. Ayers directed all design and construction activities for the EPA's Superfund program.

He received his B.S. in civil engineering with highest honors from the University of Texas. He has an M.S. in environmental engineering from the University of Minnesota, and a Master of Public Administration from the University of Oklahoma.

Kaushik Deb

Kaushik Deb is an environmental engineer in Dames & Moore's Chicago office. He joined Dames & Moore in 1989. Mr. Deb specializes in air pollution control (APC) technology evaluations, equipment selection, and air permit applications for both criteria and toxic air pollutants for a wide variety of industrial facilities.

He has wide-ranging experience in source sampling, top-down BACT/LAER (Best Available Control Technology/Lowest Achievable Emission Rate) analyses, APC system design, air emissions inventory development, and air dispersion modeling.

As principal investigator on all of the toxic air pollutant projects performed by Dames & Moore for 13 Wisconsin pulp and paper companies, and on two projects for a domestic and foreign automaker in Wisconsin and Kentucky, Mr. Deb has considerable experience in air emissions estimating, permitting, and compliance in the Midwest.

Perry W. Fisher

Dr. Perry W. Fisher joined Dames & Moore in 1973, and is presently a Principal and Certified Consulting Meteorologist in their Chicago office. Dr. Fisher specializes in air quality regulations and permitting, as well as toxic air pollutant assessments. His experience is particularly extensive with the pulp, paper, and automotive industries.

During his career with Dames & Moore, he has directed more than 50 projects for pulp and paper companies and more than 30 Prevention of Significant Deterioration (PSD) permit applications for a variety of industrial facilities, primarily in the Midwest. He has also directed 75 projects for automotive companies related to air quality.

Since the passage of the 1990 Clean Air Act Amendments (CAAA), he has taken the lead for Dames & Moore in speaking at their seminars and other presentations across the nation for a wide variety of industries. He has also spoken in Japan, summarizing these amendments and recommending how industries can be proactive in responding to them. He has served in a variety of capacities at the local and national level of the Air and Waste Management Association (AWMA).

Dr. Fisher has published over 25 publications relating to air quality.

Holly A. Hattemer-Frey

Holly A. Hattemer-Frey is a senior risk assessment scientist with Dames & Moore. She provides technical support in the areas of human health risk analysis and exposure assessment, has authored more then 35 open literature publications and technical reports, and has been invited to give presentations on diverse risk assessment topics.

In addition, she was a member of the Health Risk Assessment Panel for the proposed Clean Harbors hazardous waste incinerator, Braintree, Massachusetts (1988). As an Environmental Scientist with Oak Ridge National Laboratory's Office of Risk Analysis, she gained experience in assessing human exposure to organics through the food chain, evaluating human health risks associated with municipal solid waste incineration, evaluating the potential human impacts of genetically altered organisms, and in using pharmacokinetics to improve the risk assessment process.

Ms. Hattemer-Frey graduated *cum laude* from Ohio Wesleyan University and received an M.S. in environmental sciences from the Institute of Environmental Sciences at Miami University.

Ellen Kelly

Ellen Kelly currently serves as marketing director for H2M Group.

She has over ten years of experience in the communication of environmental information, and has coordinated a nationwide seminar program on underground storage tank regulation. In addition, she has served as technical editor for the first edition of the book *Underground Storage Tank Management: A Practical Guide*, published by Government Institutes.

Ms. Kelly holds an M.A. in government from Georgetown University.

Janet E. Kester

Janet E. Kester is a senior toxicologist with the international environmental consulting firm of Dames & Moore. She provides expert technical coordination and support company-wide in matters related to toxicology and ecological and human health exposure and risk analysis.

Dr. Kester participated in the development of innovative approaches to ecological risk assessment for both aquatic and terrestrial ecosystems using GIS technology to develop spatially weighted site-specific biomagnification factors and hazard indices for a variety of receptors. A systematic methodology for identification of toxicity reference values for a variety of chemicals and receptors, including threatened and endangered species was also developed.

She has developed and taught graduate courses in basic toxicology, published research papers in peer-reviewed literature, and co-authored a symposium paper regarding the bioavailability of metals associated with mining wastes.

Dr. Kester received her undergraduate degree in biology from Cornell University, with distinction, and completed her M.S. and Ph.D. at the University of Rochester. She served as an adjunct professor of toxicology at the Rochester Institute of Technology and as an associate of toxicology at the University of Rochester School of Medicine.

Porter-C. Knowles

Porter-C. Knowles, formerly a partner and officer with Dames & Moore, is a ground-water hydrologist and consulting engineer in private practice in Overland Park, Kansas.

Mr. Knowles has over 25 years of experience involving a wide variety of multi-discipline projects, including hazardous waste, ground-water, water resources, civil engineering, and contamination. In more recent years, his extensive background in project management and his ability to provide expert testimony have enabled him to focus on problems related to hazardous and solid waste regulations, such as RCRA and CERCLA.

Mr. Knowles attended Yale University and the Colorado School of Mines (M.S.). He is a member of numerous professional societies and regularly speaks and publishes on technical topics. He was an elected member of the Dames & Moore Board of Directors from 1987 to 1991.

Gary R. Krieger

Gary R. Krieger manages Dames & Moore's firm-wide Health Systems Program and Toxicology and Risk Assessment Group. Dr. Krieger is board-certified in internal medicine, occupational medicine, and toxicology.

In addition to performing medical surveillance, Dr. Krieger has assisted in the development of health risk assessments, medical case management, health and safety training programs, medico-legal consultation, and has also been a consultant to private industry on the health effects of bio-medical waste incineration. Dr. Krieger has also been involved in multiple human health risk assessments, public evaluations, and siting studies for a variety of industrial facilities. He is either Medical Director or Chief Medical Consultant to the following organizations: U.S. Department of Commerce/Boulder, National Bureau of Standards, National Weather Service, National Oceanic Atmospheric Agency, and National Renewable Energy Laboratory, Kodak Colorado Division.

Dr. Krieger obtained his undergraduate medical degrees from the University of North Carolina, completed residency training at the Mayo Clinic, and holds a Masters of Public Health from JohnsHopkins University.

James W. Little

James W. Little is a senior air quality consultant with Dames & Moore. His responsibilities include air quality permitting, air quality modeling, toxic air pollutant risk evaluations, environmental audits, project management, and regulations assessment.

He has 24 years of experience in air quality and atmospheric sciences, and he is also a certified consulting meteorologist.

His educational background includes an M.S.P.H. in air and industrial hygiene from the University of North Carolina.

Ursula Middel

Ursula Middel is a technical manager with H2M Labs., Inc. She is responsible for implementing new protocols and methodologies and keeping the company at the forefront of technical advances.

After a brief career as a service/application engineer for gas chromatographs, she switched to laboratory work. After years of working as a radiochemist in her own laboratory, she joined H2M in 1977 to establish an organics program. She has followed the development of organics analysis in environmental samples as organics supervisor and later as technical manager.

Ms. Middel graduated in chemical engineering at the Ohm Polytechnikum in Nurnberg, Germany. She earned her M.B.A at Dowling College, New York.

Stanley G. Puszcz

Stanley G. Puszcz is a project manager with H2M Associates, Inc., where he has managed remedial investigation and design projects which incorporate a wide variety of specialties.

Mr. Puszcz is a member of the American Institute of Chemical Engineers and the Association of Groundwater Scientists and Engineers.

He is a graduate of Manhattan College with a bachelors degree in chemical engineering.

Lyle R. Silka

Lyle R. Silka is president and co-founder of Hydrosystems, Inc., headquartered in Sterling, Virginia. Mr. Silka has been providing environmental management consultation to industry, financial institutions, and government for over 20 years.

Mr. Silka has written extensively on various aspects of environmental management, including site investigation techniques, computer simulation of contaminant migration through soil and groundwater, and remediation technologies. He has been an invited speaker in numerous environmental forums, both in the United States and abroad.

Mr. Silka received his undergraduate degree in geology fromthe University of Northern Iowa and his M.S. in geology from Oklahoma State University, where he was awarded the U.S. Environmental Protection Agency traineeship in water quality. He is a certified professional geologist in several states.

Joann McJenkins Slavin

Joann M. Slavin is a principal of H2M Labs., Inc., and has been the quality assurance manager of the environmental laboratory since 1985. Her responsibilities include the implementation and maintenance of the laboratory quality control program and CLP.

Ms. Slavin has extensive experience in data validation of both inorganic and organic CLP protocols, as well as in all facets of organic analyses by gas chromatography and mass spectrometry. As the supervisor of the GC/MS department at H2M Labs, her responsibilities also included interpretation of mass spectral data.

Ms. Slavin holds a B.S. degree in toxicology/chemistry, as well as an M.S. in toxicology.

Michael V. Tumulty

Michael V. Tumulty is a vice president and general manager of H2M Associates, Inc. In addition to being a licensed professional engineer, he is a certified groundwater professional (CGWP) with the Association of Groundwater Scientists and Engineers. Mr. Tumulty is a gradate of Hofstra University and the State University of New York, with an M.S. in environmental engineering.

Gary F. Vajda

Gary F. Vajda is the managing principal of the St. Louis, Missouri office of Dames & Moore. Mr. Vajda directs environmental audits, regulatory consultation and strategies, site investigations and remediations (RI/FS), and pollution prevention evaluations for a number of corporate and facility clients in the midwest and elsewhere in the United States.

Mr. Vajda has over 18 years of experience in environmental engineering. His previous responsibilities have included process and engineering design, project management, and serving as a regulatory liaison in the engineering/construction and manufacturing industries for projects dealing with air, water, and hazardous waste. He has worked on projects in the chemical, petrochemical, iron and steel, and general manufacturing industries, and has worked in the United States, Great Britain, and Poland.

Mr. Vajda received his M.S. in environmental engineering from the Illinois Institute of Technology in Chicago. He is a member of the Water Pollution Control Federation and the Air and Waste Management Association. Mr. Vajda is a registered professional engineer in Missouri, Illinois, Indiana, Ohio, Kentucky, Michigan, and Louisiana.

Albert D. Young, Jr.

Albert D. Young, Jr., a consultant in storage systems and liquid handling, is associated with Hart Environmental Management Corporation, and has provided tank management counsel and guidance to a wide variety of industrial and governmental tank owners.

He was previously employed by Exxon Company, USA for over 28 years in marketing engineering and management. He supervised both underground and aboveground tank design, installation, operation, retrofit, and remedial response in operating areas containing over 15,000 UST systems, with related bulk storage and distribution facilities. He has developed and supervised corrective action programs for over 600 petroleum release incidents. Mr. Young has also represented the oil industry in UST regulatory hearings and before legislative committees.

He has published numerous articles and papers on UST technology, and is the principal author of a definitive work entitled *Underground Storage Tank Management: A Practical Guide*, published by Government Institutes, Inc. Since 1984, he has presented seminars on UST technology in 29 states, reaching over 4,800 industry and regulatory specialists.

1

ENVIRONMENTAL PROCESSES

Lyle R. Silka
Hydrosystems, Inc.

OVERVIEW

The fate of contaminants in the environment are controlled by physical, chemical, and biological processes. These processes act to cause the contaminants to be transported along migration routes, or pathways, through air and water. Important to the consideration of the fate of contaminants in the environment are the processes which affect the concentration and form in which the contaminants occur, because it is the level and form of the contaminants that determine the potential risks to human health and the environment from that exposure.

This chapter provides a beginning discussion of the important processes that determine the migration routes of contaminants along the principal pathways of exposure through environmental media: air, surface water, groundwater, and soil.

BASIC TRANSPORT PROCESSES

A molecule of contaminant can be transported through environmental media as a vaporized molecule in air, as a dissolved molecule in water, or as a molecule attached to a noncontaminant, such as a soil particle. The mechanisms operating to transport the contaminant molecules under these situations are:

Diffusion — The movement of molecules in air or water from higher to lower concentration regions.

Advection — The transport of a contaminant as it is carried along by the mass movement of water or air from higher to lower regions of pressure.

In relative terms, transport by advection is much more important than that by molecular diffusion. For example, in water, advective transport of a molecule can be more than 1,000,000 times faster than diffusive transport.

As the contaminant molecules are transported, their concentration is controlled by several basic processes that are similar for all media. These basic processes are:

Partitioning —The separation of the contaminant between two co existing phases, such as air-water, water-soil, chemical-water, chemical-air, which is controlled by concentration of the contaminant in one phase, and an equilibrium constant, which in general terms can be written as:

$$\frac{C_1}{C_2} = K_{1\text{-}2} \qquad (1)$$

where C_1 is the equilibrium concentration in Phase 1,
C_2 is the equilibrium concentration in Phase 2, and
$K_{1\text{-}2}$ is the partition coefficient between Phases 1 and 2.

Dispersion — The spreading and mixing of a contaminant plume with the surrounding uncontaminated air or water due to heterogeneities in the movement of the air or water, such as caused by turbulence. Many authors place dispersion with the discussion of the transport mechanisms of advection and diffusion. However, dispersion is an artifact of turbulence and heterogeneities in the medium, not a physical mechanism of transport.

Dispersion can occur in three directions, x, parallel with flow direction, y, perpendicular to flow, but in the same plane, and z, vertical to the x-y plane.

Dispersion in air and surface water depends on the turbulence of the flowing air or water. In perfectly laminar flow, there will be no dispersion. In groundwater, dispersion is a result of the heterogeneities of the aquifer matrix.

In all cases, dispersion is a function of the size of the problem. Dispersion in the x or longitudinal direction is generally one-tenth the magnitude of the distance of the problem. Dispersion in the y or transverse direction is approximately one-tenth the magnitude of the longitudinal dispersion, and dispersion in the z direction is one-tenth the magnitude of the transverse dispersion.

Adsorption — The fixing of a molecule onto the surface of a mineral grain or natural organic matter, which generally should be considered reversible. Adsorption is discussed in more detail in the section on transport through the subsurface.

Degradation — Any of several processes that cause the breakdown or conversion of a contaminant molecule into one or more different molecules that may or may not be less toxic, such as hydrolysis, photolysis, and biodegradation (which is discussed later).

In the following discussions of the transport and fate of contaminants in the environment, many of the above basic processes will be important factors and described in more detail.

THE HYDROLOGIC CYCLE

Water and air are dominating factors in life processes on Earth, equally important to bacteria as to humans. Water and air are also the primary pathways for the exposure of humans and other life to chemicals released into the environment.

A fundamental understanding of the transport and fate of contaminants in the environment begins with the concept of the hydrologic cycle. This is the general model of the transport of water through the atmosphere, oceans, surface water, and groundwater. Movement of water is the result of the forces of solar energy and gravity. The movement of water as a vapor, liquid, or solid occurs through the following five processes:

1. Evaporation of water into the atmosphere,
2. Precipitation of atmospheric vapor to the Earth's surface,
3. Transpiration of water vapor from plants back to the atmosphere,
4. Surface water flow through streams, lakes, and oceans, and
5. Flow of water through the subsurface.

Residence Time

An important aspect of the hydrologic cycle is the residence time of water in each phase of the cycle. Residence time is the time needed for water to travel through a particular element of the cycle. It varies considerably. For instance, the average residence time of water in oceans is measured in terms of thousands to hundreds of thousands of years, while the average residence time of water in the atmosphere is on the order of several years or less. For surface water, residence time in streams is measured in days or weeks; in lakes it is months or years. For groundwater, residence time varies from days to hundreds of thousands of years. The variation in residence time for groundwater depends on the velocity of groundwater movement and the length of the flow path.

Residence time is important to the fate and transport of contaminants. Shorter residence times for water means greater flushing action to dilute and clean out the contaminants faster. Longer residence times may mean lower transport rates and restricted migration. Depending upon the mobility, toxicity, and persistence of the chemical, residence time can be an important factor in the potential risks presented by the contamination.

PROCESSES CONTROLLING MIGRATION IN AIR

In its most simplistic form, transport of contaminants in air can be described as **Fickian,** where transport is governed by molecular diffusion under a concentration gradient in a stagnant body of air. However, the assumption that a body of air is stagnant is not often met in the real world. Stability of the air mass, thus, becomes a very important factor in the uptake and transport of chemicals in air.

Modeling of air contaminant plumes from point sources, such as stacks, became an important field after the passage of the Clean Air Act in 1972. In this discussion, such plume modeling will be omitted. Instead, we will evaluate the problem of the uptake and transport of a chemical from a spill on the ground.

Equilibrium Partitioning of Chemical between Liquid and Air

At a specified temperature, the partitioning of a chemical between the chemical liquid phase and air is a function of the chemical's partial pressure. At equilibrium, the concentration of the chemical in the air next to the chemical liquid is:

$$m_a = \frac{P_a}{P_T} \tag{2}$$

where m_a is the mole fraction of chemical a in air,
P_a is the vapor pressure of the chemical liquid at the temperature of concern, and
P_T is the constant total pressure of the air space (1.0 atmospheres at sea level).

As an example, at 25° centigrade (C) and 1.0 atmosphere (atm), one mole of dry air occupies 24.5 liters (l) of volume and has a mass of 28.97 grams (gm). Thus, one liter of dry air would have a mass of about 1.2 gm. If pure heptane liquid reaches equilibrium with stagnant dry air with a vapor pressure of 0.06 atm at 25° C, the mole fraction of heptane (100 grams per mole) in the air would be 0.06 and would have a mass of 0.24 grams per liter of air (100 grams per mole times 0.06 mole fraction divided by 24.5 liters per mole of air). This 0.24 grams per liter of air is equivalent to 60,000 parts heptane per million parts air (ppm).

Evaporation of a Liquid Chemical Spill into the Air

The loss of a chemical from a free-liquid surface such as a spill will be directly proportional to the temperature, vapor pressure, and wind speed.

Temperature of the spilled liquid can often be the most important factor affecting the evaporation of the liquid into the air. Summer time road surface temperatures can be quite high, especially on asphaltic roadways. A simplified model of the loss of liquid chemical from the spill and its mixing into the near-surface atmosphere (approximately the lower 10 meters of air) solves an upward evaporative loss equation and a mass mixing equation with air flow across the spill site, estimated by:

$$E = \rho_e \, v_x^{0.78} \, L^{0.89} \tag{3}$$

where E is the evaporative loss from the spilled liquid surface per unit length perpendicular to the wind,
ρ_e is the equilibrium concentration of the chemical in air at the ambient temperature of the liquid spill from Equation 2,
v_x is the estimated average wind velocity at a 1.0 meter height above the spill, and
L is the length of the spill in the direction of the wind.

For example, a spill of heptane covering a 10 meter by 10 meter by 0.01 meter thick spill, which is about 800 kilograms, would take less than one minute to evaporate in a wind of 1.0 meters per second, and the downwind concentration could be on the order of 45,000 ppm.

Equation 3 provides a conservative concentration in air for a simple spill consisting of a single chemical. For multi-chemical spills, such as petroleum products or creosote, evaporation proceeds with the more volatile components leaving the surface of the spilled oil, enriching the surface of the oil with lower volatile components. This creates a resistance within the liquid to the upward diffusion of the volatile components toward the air-oil interface, which reduces the evaporation rate. As the lighter fractions of the multi-chemical liquid evaporate, the liquid becomes more viscous, reducing the internal mechanical mixing of the liquid, which further reduces the evaporation of volatile components remaining within the liquid.

These processes that slow the evaporation of the lighter components in multicomponent spills create a longer time for those lighter components to infiltrate into the subsurface, along with the heavier components, thus creating greater potential soil and groundwater clean-up problems. The lighter components generally have higher solubilities in water and are more mobile in water than the heavier, less volatile components.

PROCESSES CONTROLLING MIGRATION THROUGH SURFACE WATER

Transport of chemicals via surface water occurs by the uptake of a chemical as dissolved molecules in water, by the suspension of very small droplets of a chemical in the water (as a **colloid**), or by the adsorption of the chemical on suspended sediment carried by the water.

Uptake or Partitioning of a Chemical into Surface Water

When precipitation hits the ground and comes in contact with contaminated soil or with a pool of liquid or solid chemical, the opportunity exists for some fraction of the chemical to become dissolved into the water. Published solubility of chemicals in water most often are based on idealized equilibrium studies under laboratory conditions. Relating these laboratory-derived solubilities to the real world presents problems.

If uptake of the chemical by the surface water is the only process governing the chemical transport, then the equilibrium solubilities generally will over predict the concentrations in nature. Only in isolated, stagnant pools or ponds of water could equilibrium concentrations be expected. Where running water is present, lower nonequilibrium solubilities generally will be observed in the bulk water.

Colloidal Transport in Surface Water

On many occasions, the equilibrium solubility can be exceeded by the entrainment of particles of the chemical, or a microscopic suspension of chemical colloids. Turbulent or agitated flow of surface water over a contaminated soil or spill site can cause the formation of small particulates of the contaminant which become waterborne. This can occur for both solid and liquid contaminants and contaminated soil. The formation of a colloidal suspension of the contaminant can cause the maximum total concentration of the contaminant in the water to be 10 or more times greater than the true solubility of the dissolved chemical.

Transport by Adsorption on Suspended Sediment

The normal process of erosion of soil by running water can cause the suspension of soil particles, both mineral and organic, which have chemicals adsorbed to their surfaces. The Soil Conservation Service developed the **Universal Soil Loss Equation** for estimation of the loss of soil due to sheet and rill erosion of farm land. The equation has also provided useful estimates of the magnitude of contaminants transported via adsorption onto suspended soil.

The Universal Soil Loss Equation has the form of:

$$X = 1.29 \times E \times K \times 1_s \times C \times P \tag{4}$$

where X is the annual soil mass eroded per area,
E is the rainfall/runoff erosivity index, which is related to rainfall intensity,
K is the soil erodibility factor, which is a function of soil texture and organic matter content,
1_s is the topographic factor, which is related to the degree of slope of the land and the distance to the next runoff collection point (ditch, stream, terrace, etc.),
C is the cover management factor, which is a function of the vegetative cover, and
P is the supporting practice factor, which relates to soil conservation practices (mulching, silt fencing, etc.).

Fate of Chemicals in Surface Water

The factors affecting the concentration of chemicals in surface water include the following.

Settling of Suspended Sediment. Adsorption of metals and organic contaminants onto suspended sediment is an important transport mechanism as well as a mechanism for removal of contaminants when the suspended sediment settles out. However, this process only transfers the contamination problem from the water column to the bottom sediment.

Precipitation and Coprecipitation of Metals. The fate of many metals is dominated by direct precipitation as insoluble oxides or oxyhydroxides. Coprecipitation of many heavy metals along with precipitating iron is an effective mechanism for removing those heavy metals from surface water. However, this process only transfers the contamination problem from the water column to the bottom sediment.

Volatilization of Organic Chemicals. Volatilization of organics from surface water is important for those chemicals with high vapor pressures (P_a). For comparison, the half-lives determined for a 1.0 meter thick water column is approximately 5.0 hours for benzene (P_a is 0.125 atm) versus 100 hours for napthalene (P_a is 0.0001 atm).

Photolysis of Organic Chemicals. The striking of many organic molecules by certain wavelengths of light can cause the breakage of molecular bonds resulting in transformation into another compound or breakup into smaller fragments. For example, polynuclear aromatics, with carbon ring structure, are susceptible to ultraviolet wavelengths. For example, at 35° north latitude during the typical summer day in the Mid-Atlantic states, anthracene can be transformed photolytically with a half-life of 1.6 hours. In the winter, the photolysis of anthracene can proceed with a half-life of 4.8 hours. Within four half-lives, over 94 percent of the compound has been transformed.

Thus, photolysis can be an important determinant of the fate of organics in surface water. The best conditions for photolysis are low turbidity in the water, bright sunny days, and slower moving streams.

PROCESSES CONTROLLING MIGRATION THROUGH THE SUBSURFACE

The infiltration of a chemical into the subsurface can occur as the pure liquid chemical migrating downward or as dissolved chemical in infiltrating water moving downward. From the ground surface downward to the water table, the chemical is migrating through an unsaturated zone that contains voids that are partially filled with water and partially filled with air. Below the water table, by definition, the earth material is saturated with water.

The transport and fate of a chemical in the unsaturated zone is different from its transport and fate in the groundwater below the water table.

Transport of Volatile Organics in Soil

When a liquid **volatile organic compound** (VOC) is spilled on the soil or leaks from a tank into the soil, the VOC partitions among the liquid and vapor phases and becomes dissolved in soil moisture and adsorbed onto the surfaces of soil minerals and organic matter. The degree of partitioning of the VOC among these four components will depend on the volatility and water solubility of the VOC, the soil moisture content, and the type and amount of soil solids, i.e., the minerals and organic matter.

Partitioning of VOC between Pure Liquid and Soil Gas

Partitioning between the pure liquid and soil gas is controlled by the vapor pressure of the VOC and the temperature. At equilibrium, the mole fraction of a VOC in the air space above the pure liquid VOC at a specified temperature is expressed as:

$$m_a = \frac{p_a}{p_T} \tag{5}$$

where m_a is the mole fraction of chemical a,
p_a is the vapor pressure of chemical a, and
p_T is the total pressure in the air space.

Vapor pressures for many VOCs at ambient temperatures are available in the literature. In contrast to the case of a liquid spilled on the ground surface evaporating into the atmosphere, the evaporation of a liquid chemical into the soil gas can be expected to approach equilibrium concentrations according to Equation 5.

Partitioning of VOC between Soil Gas and Soil Moisture

Partitioning between the VOC vapor in the soil gas and VOC dissolved in soil moisture may be expressed as the ratio of its concentration in each of the two phases. At equilibrium, this ratio is constant for constant temperature and is referred to as **Henry's Law constant:**

$$K_H = \frac{C_G}{C_L} \tag{6}$$

where K_H is Henry's Law constant for the VOC at a specified temperature,
C_G is the concentration of the VOC in soil gas, and
C_L is the dissolved concentration of the VOC in the water phase.

Henry's Law constant may also be expressed as a function of the VOC vapor pressure, the concentration of the VOC in water, and temperature as:

$$K_H = \frac{16.04 \, p_a M_a}{TC_L} \qquad (7)$$

where M_a is the gram molecular weight of the VOC,
T is the temperature (in degrees Kelvin), and
the other parameters are as previously defined.

Values of K_H for selected VOCs are presented in Table 1. Empirically derived values of Henry's Law constants are in reasonable agreement with the calculated values of K_H, keeping in mind the temperature dependence of K_H.

Empirical tests also have shown that sorption of VOCs at low vapor concentration (<10 mg/l) into soil moisture can be accounted for by the linear partitioning process as described by Henry's Law. However, at very low moisture contents (equivalent to less than eight monolayers of water), the sorption isotherm becomes

TABLE 1

**Values of Henry's Law Constant,
Vapor Pressure, and Solubility at 25° C
for Selected Chlorinated Solvents**

Chemical	Solubility in Water (ppm)	Vapor Pressure (mm Hg)	Henry's Law Constant (dimensionless)	
			Calculated	Tested
1,1,2,2-Tetrachloroethane	3000	6.5	0.019	NA
1,1,2-Trichloroethane	4420	23	0.038	NA
1,1-Dichloroethane	8700	82	0.050	0.04
Tetrachloroethylene	140	18.6	1.2	0.50 0.43
Trichloroethylene	1100	74	0.49	0.33
trans-Dichloroethylene	6300	326	0.27	NA
cis-Dichloroethylene	3500	206	0.31	NA

NA is Not Available

nonlinear, with the observed partitioning into the water phase being higher than that predicted by Henry's Law. The increased sorption at low moisture content may be due to sorption onto the bound water on mineral surfaces, as opposed to being solely dissolved. To expand on this theory, the observed increase in VOC sorption by soil moisture over that predicted by Henry's Law may be attributable to a greatly expanding surface area of water across which partitioning may occur as soil moisture decreases below a critical value, i.e., a nonlinear relationship may exist between soil water surface area and soil moisture content. Mineral morphology may also play an important role in the observed increased sorption rates. It has been indicated that the critical soil moisture content below which vapor-water partitioning becomes nonlinear varies according to soil type. However, these low moisture contents will be limited principally to arid regions and, in humid regions, limited to shallow soil during dry months.

Partitioning of VOC between Soil Moisture and Soil Solids

Sorption of VOCs onto soil solids from soil vapor has been described as a two-step linear process for soil moisture above a critical value. Above that soil moisture content, soil solids will be coated with layers of water molecules. VOC vapor will partition from the vapor phase into the liquid water phase. Once in the liquid water, some of the VOC will be adsorbed onto the soil mineral and organic matter. At equilibrium, the degree of partitioning between the soil solids and the soil moisture has been expressed as the linear isotherm:

$$K_D = \frac{S}{C_L} \tag{8}$$

where K_D is the partition coefficient or distribution coefficient (with units of length^3/mass),
S is the mass of chemical adsorbed per unit dry mass of soil solids, and
C_L is the concentration of the chemical in the soil moisture.

It has been observed that strongly hydrophobic organic chemicals tend to adsorb more strongly onto the soil solids. Empirical studies have found that K_D is proportional to the organic carbon content of the soil as well as the octanol:water partition coefficient (K_{ow}), which is a measure of the **hydrophobicity** (the degree to which the chemical is immiscible with water) of an organic chemical. For the equilibrium condition, this relationship has been expressed by the following equation:

$$K_D = 0.63 \, K_{ow} f_{oc} \tag{9}$$

where K_D is the distribution coefficient of Equation (4),
f_{oc} is the soil organic carbon content, and
K_{ow} is the octanol:water partition coefficient.

The amount of carbonaceous matter in the soil is the dominant factor controlling the extent of adsorption of organic chemicals. The particle size of the mineral fraction is also important. For example, the partition coefficients for semivolatile compounds, such as pyrene and methoxychlor, for the sand-sized fraction are approximately 100 times less than their partition coefficients for the silt- and clay-sized fraction, due primarily to the lower organic carbon content of the sand. The following table presents data for K_{ow} and calculated values of K_D using the above equation for selected VOCs. From the example calculations of partition coefficients, these VOCs are not strongly adsorbed onto the soil solids due to their relatively low octanol:water partition coefficients. In comparison, compounds with low volatility and low solubility, such as pentachlorophenol with a log K_{ow} of 4.74, has a K_D of 35, i.e., pentachlorophenol will be preferentially adsorbed to the soil solids by a factor of 100 to 1,000 times greater than the chlorinated solvents.

TABLE 2

Reported Values of Octanol:Water Partition Coefficient and Calculated Values of Partition Coefficient For Selected Chlorinated Solvents

Chemical	Octanol:Water Partition Coefficient	Calculated Partition Coefficient	
		Fraction Organic Carbon	
		0.001	00.1
1,1,2,2-Tetrachloroethane	2.56	0.23	2.3
1,1-Dichloroethane	1.79	0.04	0.4
Tetrachloroethylene	2.88	0.48	4.8
Trichloroethylene	2.29	0.12	1.2
trans-Dichloroethylene	1.48	0.02	0.2
cis-Dichloroethylene	1.48	0.02	0.2

Effects of Temperature and Soil Moisture Content on Partitioning

For compounds with a Henry's Law constant of 0.05 (for example, the chloroethanes), effects of temperature on equilibrium partitioning are negligible up to about 40°C, above which partitioning of the VOC to the soil-gas phase increases. For VOCs with a Henry's Law constant of 1.0 (such as tetrachloroethene), a relatively constant increase in partitioning to the gas-phase is predicted with increasing temperature. TCE adsorption has been found to decrease as soil moisture content increases from zero to the saturation soil moisture content (the soil moisture content that is in equilibrium with 100% humidity). The ratio of the concentration of TCE adsorbed on the vadose-zone soil to its concentration in the soil gas was one to three orders of magnitude greater than the ratio predicted by equilibrium models. This apparent disequilibrium probably results from the slow desorption of TCE from the organic matter in the soil relative to the faster volatilization loss of TCE from the soil system. Therefore, the assumption of Henry's Law constant and equilibrium adsorption isotherms may not hold in the field.

Transport of VOC Vapor through Soil Gas

Transport of VOC vapor tough soil gas may be under density gradients, thermal gradients, pressure gradients, or concentration gradients.

Density Gradient. Density-induced convective flow of VOC vapor may be important where concentrations of denser VOCs are high, especially near the source. When the density of the VOC contaminated soil gas is more than several times the ambient soil-gas density, density gradient induced transport can be important.

The importance of density gradients to the transport of VOC vapor is related to permeability. Density-driven transport can be neglected for permeabilities of less than 10^{-11} m^2 (10 darcy), i.e., equivalent to a clean sand. Since the effective permeability of the material to gas is affected by the water content, density-driven transport also should be expected to become less important as the moisture content increases.

Aside from consideration of permeability, density-driven transport of VOC vapor through the unsaturated zone will be important for regions close to a source of dense volatile compounds and will be important early in the history of a spill when free-phase liquids are present. As the source dissipates and as distance from the source increases, density-driven transport will become less important.

Thermal Gradient. Thermally-induced convective transport of VOC vapors is generally limited to the near surface due to the rapid attenuation of temperature fluctuation with depth. While no research on the thermally-driven transport of VOC vapors through soil has been reported, research on transport of water vapor has been. For example, the warming of soil lowers the suction pressure and raises the vapor pressure of soil water. The resulting thermal gradient induces water vapor migration from warmer to cooler regions in the soil. Therefore, the effect of warming of the soil surface during the day would be a decrease in the upward flux of water vapor.

Thermal gradients in soil that may be significant to soil-gas transport are generally limited to the upper several feet of the soil column.

The amplitude of thermal fluctuations in soil decreases exponentially with depth below the surface. Diurnal soil temperature fluctuations caused by the rising and setting of the sun are damped out within a short distance below the soil surface. Under conditions that would be common to a large portion of humid temperate climatic regions with vegetated soil or partial shading of the soil surface, the diurnal soil temperature variation is damped out to less than 5% of the average soil temperature at a depth of 0.2 m (0.66 ft). Thus, in general, diurnal soil temperature fluctuations are not an important factor in VOC transport through the unsaturated zone.

For seasonal soil temperature changes, under the same conditions as above, the fluctuation in the average soil temperature is not damped out until much deeper, i.e., the largest fluctuation is less than 10% of the average soil temperature at a depth of 3.0 m (9.8 ft).

Soil temperature fluctuations may be an important factor on the partitioning of the VOC between the liquid, gaseous, and adsorbed phases, however. The vertical soil temperature profile below a depth of 0.2 to 1.0 m (0.66 to 3.3 ft) can be considered constant over the short term of several months. Laterally, where soil surface conditions may change from bare soil, to vegetated soil, to shaded soil, to pavement, localized lateral thermal gradients may exist at these lateral boundaries for short periods at shallow depths due to differences in the depth to which the thermal fluctuation penetrates the soil. These localized regions experiencing thermal gradients will produce insignificant effects on the soil-gas transport with respect to the typical scale involved.

Pressure Gradients. Barometric pressure changes due to the weight of the atmospheric column overlying the ground surface at any instant will have spatially uniform effects over large areas, especially for open ground surfaces. Barometric pressure changes due to the passage of atmospheric highs and lows will act to depress and raise the water table slightly. This barometric pressure fluctuation will cause a slight compression of soil gas during the passage of a high atmospheric pressure system and a slight expansion of soil gas during the passage of a low atmospheric pressure system. These pressure effects will be insignificant on the transport of soil vapor.

Significant pressure gradients can be established in the unsaturated zone due to the effects of manmade structures. Increased flow of gaseous contaminants, gasoline vapors, and radon can occur toward basements. This effect is attributed to wind causing a lower pressure in the basement, which generally will produce a small pressure differential. It has been found that the artificial venting, or depressurization, of houses, such as for radon gas mitigation, can increase the pressure differentials by 10-20 times the natural differentials, and the velocity of radon gas migration through the soil toward the basement can approach 1.0 meter per hour.

Concentration Gradients. In general, except for the region affected by density-induced transport of concentrated DNAPL vapors, the primary transport mechanism for VOCs in the unsaturated soil is by diffusion through the soil gas under concentration gradients. The distribution of VOC concentration in the soil gas can be modeled by **Fick's second law** which in one dimension is expressed by the following equation:

$$\frac{C}{t} = \frac{D^2 C}{z^2} \tag{10}$$

where C is concentration of the VOC in air,
D is the diffusion coefficient, and
z is the distance traveled.

Assuming the outer boundary condition is zero concentration, Equation (10) can also be expressed as:

$$\frac{C_{(z,t)}}{C_{(z=0, t=0)}} = erfc \left[\frac{z}{4Dt^{0.5}} \right] \tag{11}$$

where $C_{(z,t)}$ is the concentration (as mole fraction) at a distance z and time t,
$C_{(z=0,t=0)}$ is the initial concentration, and
erfc is the complimentary error function.

Diffusion Coefficient in Soil Gas. The estimated diffusion coefficient for VOC vapor in air is 0.43 m²/d. This single value for D has been used for intermediate molecular weight VOCs based on the similar values obtained in measurements of the gas diffusion coefficient for numerous organic chemicals of intermediate molecular weight.

However, the diffusion coefficient in soil gas is reduced from that in air by a tortuosity factor that accounts for a decreased cross-sectional area for flow and an increased length of the flow path through soil pores. The **Millington-Quirk tortuosity formula** adequately defines the soil-gas diffusion coefficient:

$$D_G = \frac{D a^{\frac{10}{3}}}{n^2} \tag{12}$$

where D_G is the diffusion coefficient in soil gas,
D is the diffusion coefficient in air, e.g., 0.43 m²/d,
a is the volumetric air content of the soil, and
n is the total soil porosity.

Since the VOC vapor may partition between the gas, liquid, and solid phases, an effective diffusion coefficient incorporating that partitioning can be formulated. The removal of VOCs from the soil gas by partitioning into the soil moisture and soil organic matter results in a reduction of the apparent diffusion rate, and consequently, the apparent, or effective, diffusion coefficient. The following relationship between the diffusion coefficient in soil gas has been developed:

$$D_E = \frac{D_G}{\frac{bK_D}{K_H} + \frac{w}{K_H} + a} \tag{13}$$

where D_E is the effective diffusion coefficient in soil gas corrected for effects of partitioning,
D_G is the diffusion coefficient,
b is the bulk dry density of the model for the effective diffusion coefficient,
b is the dry bulk density of the soil,
K_D is the soil partition coefficient,
K_H is Henry's Law constant,
w is the volumetric soil moisture content, and
a is the volumetric air content, where n, total porosity, is equal to a+w.

Transport of Contaminants in the Saturated Zone
Recharge and Discharge

Since the groundwater flow path is measured from the point of entrance into the subsurface to the point of exit, it is reasonable to conclude that the location of these recharge and discharge points partly controls the flow of groundwater. **Recharge** occurs as rain falls on the surface, infiltrates into the subsurface, and percolates to the groundwater. Recharge occurs more easily where the surface soil or rock is more permeable and the land surface is relatively flat, so that the time available for infiltration is greatest.

Considerable recharge can occur as surface water seeps through stream and lake beds. When water levels in a stream or lake are higher than the water table, surface water will seep into the ground. A stream of this type is referred to as a *losing* stream. Where the stream level is lower than that of the groundwater next to the stream, the groundwater flows into the stream, and the stream is referred to as a *gaining* stream. At any one time, different portions of the same stream may be gaining or losing. From season to season, certain reaches of a stream may change from losing (fall and winter) to gaining (spring and early summer).

Groundwater Basins

Perennial streams flow all year because they receive groundwater; they are gaining streams. These streams delineate major surface water drainage basins. In conjunction with upland recharge areas and topographic highs, these streams delineate boundaries of groundwater flow and groundwater basins.

Aquifers and Aquitards

Groundwater occurs in the pores of the earth material. A geologic formation that can supply significant quantities of water to a well or spring is defined as an *aquifer*. Aquifers consist of material such as sand and gravel or fractured rock. At an undefined lower permeability, the earth material becomes so tight that groundwater does not readily flow through it. Materials that do not allow the passage of significant quantities of water to a well or spring are called *aquitards* or confining layers. Aquitards generally consist of clay and silt or unfractured, dense rock.

Artesian Groundwater

At the water table, groundwater is subject to atmospheric pressure, and is said to be **unconfined**. If the groundwater is under pressure greater than atmospheric pressure, the groundwater is confined, or **artesian**.

Artesian groundwater conditions develop when an aquifer is overlain by an aquitard. The aquitard acts as a cap on the groundwater reservoir. Therefore, when a well is completed in an artesian aquifer, the water will rise in the well to a height above the top of the aquifer. This height above the aquifer top is a measure of the artesian pressure. If a number of wells were completed in the artesian aquifer, a contour map connecting the elevations of the water in each well would describe the potentiometric surface for the groundwater in the artesian aquifer. In comparison, water in a well completed in an unconfined or water-table aquifer will rise to the height of the water table outside the well.

Direct recharge to an artesian aquifer can occur only where the aquifer is exposed at the surface, in its unconfined outcrop region. Elsewhere, recharge to an artesian aquifer occurs indirectly through leakage across the confining layers.

Flow of Groundwater and Darcy's Law

A knowledge of the controls exerted by the properties of the earth material and the water is also necessary to understand chemical transport processes. One of the first descriptions of groundwater flow was made by the Frenchman D'Arcy (altered to Darcy) in the nineteenth century. While studying water flow through sand filters at the water treatment plant for Paris, he observed a simple relationship between the volume of water flowing through the sand and properties of the sand. Darcy devolved the following mathematical statement:

$$\frac{Q}{t} = KA \frac{dH}{dL} \qquad (14)$$

where Q is the volume of flow per given time interval, t, through a given cross sectional area of flow, A, under a given pressure gradient defined as dH/dL, the change in water level over a given length, L.

K is the proportionality constant termed the hydraulic conductivity.

The difference, or change, in elevation of the water table, dH or (h_2-h_1) over the length (L) is the slope of the water table, or the **hydraulic gradient**. In practice, the hydraulic gradient is the difference in elevation between two points on the same flow path divided by the distance between those two points.

Also important to understanding the meaning of what has come to be known as Darcy's Law are the measurement units used. If English units are applied to equation (14), then the units of K are gallons per square foot per day, which can be converted to cubic feet per square foot per day, and, in turn, reduced to feet per day. Because K can be described in units of length per time, the hydraulic conductivity is sometimes erroneously thought of as a measure of velocity. When volume per time **(Q/t)** is determined for a unit cross-sectional area (A=1.0), the rate of flow per unit cross-sectional area per time has the units of feet per day. Thus, Darcy's Law has also been incorrectly used to calculate velocity of water movement. Darcy's Law calculates the volumetric flow rate through a unit cross section of the aquifer, not the actual velocity of water movement.

Calculating True Groundwater Velocity

Calculating the true velocity of a molecule of water flowing through the aquifer requires consideration of the actual area through which flow occurs. Since flow occurs only through the pore spaces, the cross-sectional area must be reduced by its **porosity**, the ratio of voids to total volume. Also, many of the pore spaces are blocked and do not allow flow. Therefore, the total porosity must be reduced. The actual pore space that allows flow of groundwater through the aquifer is called the **effective porosity**. The average velocity of the groundwater, then, is:

$$v = \frac{Ki}{n_e} \quad (15)$$

where v is the actual average velocity of groundwater movement,
K is the hydraulic conductivity,
i is the hydraulic gradient, and
n_e is the effective porosity.

The effective porosity can be determined by calculating the velocity of the groundwater directly from measurements in the field using a tracer to follow the groundwater movement. By knowing v, K, and the hydraulic gradient, n_e can be determined. This is a very expensive method. A simpler and less expensive, but less accurate, method is to estimate n_e from another easily measured parameter. It is commonly accepted that n_e can be approximated from the specific yield of the material. The **specific yield** is the fraction of water drained under gravity from a saturated sample and can be estimated from aquifer tests and laboratory studies that are much less costly than tracer studies.

Determining Flow Path

The path of groundwater flow is dependent on the type of material through which the groundwater flows. For example, the pore spaces between sand grains are generally continuously connected. Flow of groundwater through sand or similar porous material is referred to as porous-media flow. At the other extreme, in a hard rock such as granite, the amount of pore space created when granite was formed is very small, and the pores are not interconnected. Flow cannot occur through the pores of granite. If the granite is cracked, or fractured, groundwater flow can occur through the cracks. This type of medium influences the flow paths of the groundwater. In general, porous media do not restrict flow, while fractured media place extreme constraints on flow paths.

The measurement of the flow path in a porous medium can be accomplished by measuring the water table or potentiometric surface. Groundwater flow is down-slope and at right angles to the water-table contours. In a fractured medium, this approach is not valid. Only an average flow direction over a large area can be obtained, although even an average flow direction thus obtained may be erroneous. The actual flow of groundwater in a fractured medium can be very complex. The complexity of the flow depends on the degree of fracturing, the degree of interconnection between fractures, and the distance between fractures: The greater the fracturing and the greater interconnection, the more the flow is like porous-media flow.

Transport of Chemicals through Groundwater
Advection

The primary mechanism of chemical migration in groundwater is the movement of the chemical as a dissolved or suspended constituent of the groundwater. This transport process is called **advection**, or convection, and is governed by Darcy's Law. It is often desirable to assume that the chemical is transported solely by advection with no retardation effects. Under such conditions, referred to as **conservative transport**, the chemical will travel the fastest and farthest. This assumption yields the most pessimistic predictions.

Mechanical Dispersion

Mechanical dispersion causes the chemical dissolved in the groundwater to spread out as it moves through the ground. Predictions based on dispersion plus advection will show the chemical to arrive at a specified point faster than by advection alone. Predictions based on advection only will show the chemical to have higher predicted concentrations.

At some point in time and at some distance from the source of the contamination, dispersion and other attenuation mechanisms may decrease the concentration of the slug to comply with a water quality criterion. However, the time and distance required to produce a decrease in concentration adequate to render the groundwater nonhazardous may be too great. Commonly, groundwater will discharge to pumping wells or streams before the degradation is alleviated naturally.

Since some pores are blocked, preventing passage of water, the actual flow of the groundwater and chemical can be very tortuous or winding. As the flow paths hit these blockages, the flow branches out, tending to spread the chemical in the groundwater. This process causes a mechanical mixing of the contaminated groundwater with uncontaminated groundwater, resulting in a decrease of the peak concentration. Since aquifer heterogeneity causes dispersion, it should be no surprise that dispersion increases as the aquifer media become more heterogeneous.

Dispersion is difficult to estimate. Tracers, such as chloride, can be used to track the groundwater, and dispersion can be determined from the concentrations observed in a number of monitoring wells over time. This method is expensive and contains inherent errors. As will be discussed later, the size of an experiment (scale effect) will affect measurement of many of the hydrogeologic processes. For example, measurements based on well spacing of tens of feet will produce dispersions on the order of a few feet, while measurements based on wells spaced 100 feet apart will produce dispersion estimates of tens of feet.

This paradox is caused by the fact that the larger the volume sampled, the greater the heterogeneity of the sampled volume.

Molecular Diffusion

Molecular diffusion causes migration of a chemical from a region of high concentration to a region of low concentration. It is a very slow process that, in typical situations, will account for a few feet of migration over thousands of years. This is insignificant unless the groundwater velocity is of the same order of magnitude, such as in clay.

Movement of Immiscible Liquids

Many of the chemicals that may find their way into groundwater have properties that differ from those of the groundwater. Many of the organic chemicals are not very soluble in water and remain separate fluids in the ground. For an immiscible organic liquid, the separate liquid will float on top of the water table if the organic liquid is less dense than water and will sink to the bottom of the aquifer if it is more dense than groundwater.

Calculating the movement of these immiscible liquids in the subsurface requires a modification of the hydraulic conductivity equation. Hydraulic conductivity depends on the properties of the 1medium and the properties of the fluid. The equation relating hydraulic conductivity to the basic properties of the fluid is:

$$K_a = k \frac{\rho_a}{\mu_a} \tag{16}$$

where K_a is the hydraulic conductivity for the immiscible fluid,
k is the intrinsic permeability of the medium,
ρ_a is the density of the immiscible fluid, and
μ_a is the dynamic viscosity of the immiscible fluid.

The **intrinsic permeability** is a measure of the permeability of the medium independent of the fluid type or fluid properties. Intrinsic permeability is dependent on the geometry of the pores, that is, the diameter of the pores and the tortuousness or crookedness of the flow paths.

Thus, when a fluid with density and viscosity different from water is considered, the hydraulic conductivity (K) must be corrected. If the K for water (K_w) is known, then the hydraulic conductivity for the immiscible fluid can be determined. For example, heptane, a component of gasoline, has a density of 0.68 grams per cubic centimeter and dynamic viscosity of 0.4 centipoise at normal temperature, the $K_{heptane}$ is about 1.7 times larger than K_w. This means that when a liquid of pure heptane saturates the medium to a sufficient thickness to become mobile, it can move 1.7 times faster than the groundwater.

Another complication of analyzing contamination by immiscible fluids is that the two fluids will tend to intermix in the porous medium; that is, in the case of gasoline, there will be a region that is 100 percent saturated with gasoline, a transition region in which the gasoline and the water share pore spaces (mixed-flow region), and a region 100 percent saturated with water.

Soil highly saturated with gasoline has most pores filled with gasoline, and only a few pores filled with water. Gasoline is referred to as the continuous phase and will be mobile and flow under such conditions. In this region, water is a discontinuous phase and will have considerable difficulty flowing.

At a moderate degree of saturation for each liquid, both gasoline and water are continuous. Therefore, both the gasoline and the groundwater can flow, but at reduced rates. Finally, where the water content is sufficiently greater than the gasoline content, the water is the continuous, mobile phase, while the gasoline is the discontinuous, immobile phase.

The importance of this phenomenon is that in the mixed region and at lower degrees of gasoline saturation, pure product recovery will cease, since the gasoline is discontinuous and immobile. Although gasoline is immiscible, it is very slightly soluble, to about 50 to 500 milligrams per liter (mg/l, essentially the same as parts per million). Since the threshold concentration of gasoline that results in unpalatable taste and odor in water is about 0.01 to 0.1 mg/l, the small amounts of gasoline that cannot be recovered from a spill may still render the groundwater unusable for drinking purposes. This problem is present with all the immiscible organic chemicals, and any corrective action must be able to address it.

BEHAVIOR OF CHEMICALS IN THE WATER-SOIL SYSTEM

After a spill or leak has occurred, what happens to the chemicals? The answer to this problem is complex and is based on the interactions among and between the chemicals, the groundwater, the soil, and the rock. The properties of the chemicalS that are important are:

1. Vapor pressure or volatility,
2. Miscibility with groundwater,
3. Solubility in groundwater,
4. Density or specific gravity,
5. Dynamic viscosity,
6. Reactivity, and
7. Susceptibility to biodegradation.

Properties of the environment into which the chemical is released also need to be determined. With respect to the fate of the chemicals in the subsurface, important properties include:

1. Infiltration capacity of the soil,
2. Natural organic content of the soil and rock,
3. Unsaturated and saturated hydraulic conductivities in the soil and aquifer,
4. Effective porosity,
5. Relative permeabilities for immiscible chemicals and water in the soil and aquifer,
6. Depth to the water table,
7. Groundwater flow paths,
8. Mineralogy of the soil and rock,
9. Oxygen content in the subsurface, and
10. Bacterial community present.

Processes That Retard Chemical Movement

Fortunately, there are several processes that act to control the rate and extent of migration of chemicals in the subsurface. These processes trap or delay the chemical from spreading and cause the chemical to degrade or change in chemistry, most often to a less hazardous state.

Adsorption

Adsorption of chemicals occurs as the chemical is attracted to the surfaces of minerals and natural organic matter in the soil and rock. Mineral surfaces have defects in their crystalline structure that result in imbalances of electrical charges on the mineral surface. Dissolved molecules and ions that also have charge imbalances will be attracted to a mineral surface that has the opposite charge imbalance.

Adsorption has been defined, under the assumption that it is a linear, reversible process, as:

$$K_d = \frac{C_s}{C_w} \qquad (17)$$

where K_d is the distribution coefficient, or water-soil partition coefficient,
C_s is the mass of the chemical adsorbed on the soil per unit bulk dry mass of soil, and
C_w is the mass of the chemical dissolved in a unit of water in contact with the soil.

For chemicals with a K_d of less than 1.0, the chemical is mobile in water systems. For a K_d of greater than 10, the chemical is essentially immobile in water systems, because the chemical is being preferentially removed from the dissolved phase by adsorption onto the solid phase.

Ion Exchange

Ion exchange is a further step in the adsorption process whereby a dissolved chemical substitutes itself for another chemical already adsorbed on the mineral surface. For example, the ion exchange reaction of a dissolved metal cation with a sodium cation on the clay mineral structure is:

Sodium-clay + metal cation = metal-clay + sodium cation

These ion exchange reactions will not change the overall total dissolved solids content of the groundwater, but the process can exchange a hazardous chemical for a nonhazardous chemical. On the other hand, a concern with the adsorption process is that the process can be reversed. The direction of the above exchange reaction depends on the relative concentrations of the sodium and metal cations in the water. As the dissolved metal concentration in the water rises, the reaction proceeds to the right; that is, metal increasingly replaces the sodium adsorbed on the clay. When the sodium concentration in the water reaches a certain level, the exchange reaction will slow down. Eventually, when the sodium concentration in the water has increased and the metal concentration in the water has decreased, the reaction will begin to reverse, with the metal adsorbed onto the clay being replaced by sodium.

This reversal in the exchange reaction is important in the cleanup of groundwater contamination. Although the ion exchange process causes a decrease in the concentration of the chemical in groundwater, once the contamination source has ceased, the reaction will gradually reverse. As the chemical concentration in the groundwater gradually declines, the adsorption equation will reverse direction, and the chemical will gradually desorb from the clay and become redissolved in the groundwater. The cleanup of groundwater will cause the dissolved chemical to decrease and result in a delayed release of the adsorbed chemical. If the concentration of the delayed release is above the critical concentration allowed, the cleanup may take substantially longer than expected.

Ion exchange and adsorption vary considerably, depending on the concentrations of competing dissolved chemicals, the acidity of the groundwater, the organic humus content of the soil, and the clay mineralogy. Higher levels of dissolved chemicals and lower pH result in lower ion exchange capacity. Lower clay and soil organic content also decrease ion exchange capacity.

Finally, the type of chemical affects the ion exchange capacity. For inorganic chemicals such as metals, the smaller metal atoms tend to be preferentially adsorbed over the larger metals. For organic molecules, humus and other organic matter in the soil are strongly adsorbent for heavy organic chemicals such as the polynuclear aromatic hydrocarbons. Smaller organic molecules with higher vapor pressures, such as phenol and trichloroethene (TCE) are poorly adsorbed by soil organic matter.

Retardation Factor

A measure of the retardation of a chemical due to adsorptive and ion exchange processes is the retardation factor, defined as:

$$R_f = \frac{v_w}{v_a} \qquad (18)$$

where R_f is the retardation factor,
v_w is the average interstitial groundwater flow velocity, and
v_a is the average interstitial transport velocity of chemical a.

The retardation factor also is related to the distribution coefficient, K_d, discussed above by the following equation:

$$R_f = 1 + K_a \frac{\rho_b}{n} \qquad (19)$$

where ρ_b is the bulk density of the solid, and
n is the total porosity of the solid.

Chemical Precipitation

Dissolved chemicals also are removed from the groundwater by various precipitation reactions. Most of the hazardous inorganic chemicals can be precipitated out of the groundwater to form oxides, carbonates, and other solid minerals. These reactions are controlled by the acidity and the electrical potential (Eh) of the groundwater. In part, the natural acidity and Eh will be modified by the waste stream coming from the contamination source and by biological activity. In general, the lower the pH and Eh, the more soluble are the metals.

The processes of coprecipitation and scavenging are also very important in removing hazardous metals from solution. Iron is a ubiquitous constituent of groundwater and generally occurs at concentrations much higher than those of hazardous metals. As iron hydroxide precipitates out of solution, many other metals will become trapped in the precipitating iron and be removed from solution.

Biodegradation

For many years, biological activity in groundwater was thought to be insignificant. Recently, though, microorganisms have been recognized as important agents in the attenuation of groundwater contamination. The type of organism present depends on the oxygen content of the groundwater and the availability of nutrients. The major factor in the biodegradation of chemicals, or the metabolism of chemicals by microbes, is the oxygen content. In the upper surface of groundwater, recharge by percolating water supplies oxygen to the groundwater, maintaining aerobic conditions. With increasing depth, the oxygen content becomes depleted, and anaerobic conditions prevail.

The depth at which oxygenated conditions are replaced by anaerobic conditions varies with the recharge rate, the permeability of the aquifer, and the amount of natural organic matter. Aerobic, oxygenated conditions have been observed to depths of 100 feet in sandy aquifers in southern California. Groundwater running through fractured limestone aquifers may remain oxygenated throughout. Groundwater seeping through less permeable silts and clays or aquifers containing greater content of natural organic matter may become depleted in dissolved oxygen relatively quickly at shallow depths.

Aerobic and anaerobic microorganisms will digest different chemicals at different rates. If the effectiveness of the biodegradation of a chemical is measured in terms of a half-life (the time it takes to decrease the amount of the chemical by 50 percent), some organic chemicals are rapidly degraded, while others are only slowly degraded. For example, phenol can be rapidly degraded with a half-life on the order of one month, while napthalene has a half-life on the order of four months. Therefore, an initial concentration of 30 parts per million (ppm) of phenol would be reduced to less than 0.001 ppm in 1.5 years. On the other hand, naphthalene would require 4.5 years to be decreased from 30 to 0.001 ppm. With a half-life of 10 years, dichloromethane has an even slower biodegradation rate. It would take 150 years for dichloromethane to be reduced from 30 to 0.001 ppm.

Some organic chemicals act as a poison to the microbes and interfere with the degradation process. For example, chlorophenol above 0.1 ppm concentration will prevent the biodegradation of phenol. Also, the concentration of the chemical itself can affect the rate or effectiveness of biodegradation if toxic levels are reached. For phenolic compounds, concentrations above 300 to 500 ppm will cause biodegradation to cease.

Chlorinated solvent compounds require an anaerobic biodegradation step to remove the chlorine atoms from the solvent molecules, i.e., to break the chlorine-carbon bonds. The chlorine atoms are stripped sequentially, i.e., perchloroethylene (four chlorine atoms) is converted to trichloroethylene (three chlorines), which is converted to dichloroethylene (two chlorines), and then to vinyl chloride (one chlorine).

In contrast, aerobic bacteria can efficiently break the carbon bonds of the organic molecules once the chlorine atoms have been removed. Since unsaturated zones are predominantly aerobic, even though aerobic bacteria are present, little destruction of chlorinated solvents occurs under natural conditions. Thus, solvent contamination remains in the soil a long time.

HYDROGEOLOGIC CONTROLS OF MIGRATION THROUGH GROUNDWATER

Geologic and hydrogeologic factors will affect not only the susceptibility of an aquifer to contamination but also the mobility and fate of the contaminants in the aquifer. Geologic conditions can be grouped into three divisions based on susceptibility to contamination and migration.

Unconsolidated Sands and Gravels

River-deposited sands and gravels are found throughout the United States. Along the east coast from New Jersey south to Florida, the Atlantic Coastal Plain deposits contain extensive sandy strata deposited by ancestral rivers carrying eroded material from the Appalachian Mountains. Similar sand and gravel deposits occur along the Gulf Coast from Florida to Mexico, in numerous river valleys and glacial outwash deposits, and in intermountain basins of the western United States.

Underlying the flat to gently rolling topography of the Atlantic and Gulf Costal Plains are wedged-shaped strata that are at, or near, the surface toward the west, and that thicken and deepen toward the coastline. The occurrence of groundwater in these deposits is characterized by a water-table aquifer in the near-surface strata and deeper artesian aquifers in deeper sands overlain by extensive clay beds. The Coastal Plain aquifers are relatively permeable with values of K ranging from about 10 to 100 feet per day. Effective porosities range from 10 to 30 percent. With typical gradients of about 0.002 vertical feet to horizontal feet, the interstitial pore velocity ranges from about 0.1 to 0.7 feet per day, or 37 to 256 feet per year.

With vertical hydraulic conductivities ranging from 0.001 to 0.0000001 feet per day, effective porosities of about 5 percent, and high gradients, the interstitial pore velocity through the clays can range from 2 feet per day to a less than 0.0002 feet per day. The lower permeabilities are more typical of the clay layers, and the smaller pressure differentials across the clay layers would reduce the velocity so that the typical velocity of water through a clay would be much less than 1 feet per year.

There is only minor leakage through the generally thick clay confining layers that serve to protect the deeper sand aquifers. However, small fractures and heterogeneities in the clay layers have been found to allow contaminants to penetrate through the clays at surprising high rates.

Thus, while the water-table aquifers are not protected naturally from contamination, the clay-confining layers do not guarantee protection of deeper aquifers either.

Many of the sand aquifers contain significant amounts of natural organic debris and clay minerals that increase their ability to adsorb and slow the migration of contamination. In the Coastal Plain sands, glauconite, a clay with a high ion exchange capacity, is often found. Typically, these clays will retard the migration of metals in groundwater by a factor of 1,000 to 10,000 times slower than the velocity of the groundwater.

While cleaner and coarser sands and gravels may lack the adsorptive capacity of the "dirtier" sands, they sustain aerobic conditions to greater depths, which supports higher, aerobic biodegradation rates.

Fractured Hard Rock

Hard rock, in this discussion, includes igneous and metamorphic rocks such as granite, gneiss, and schist, as well as some sedimentary rocks such as well-cemented sandstone, siltstone, and shale. These hard rocks are treated as one groundwater region in considering their contamination potential. They consist of hard, dense rock in which groundwater flow is predominantly through fractures, the cracks in the rock. Examples include the Piedmont and Blue Ridge provinces of the Appalachians, and granitic rocks in the northeastern and western United States.

The permeability of fractured hard rock depends on the width and frequency of the fractures. Overall, hard rocks have very low permeability, with values in the range of one to five feet per day. With lower effective porosities of 1 to 5 percent, flow velocity throughout the fractured rock will be 0.04 to 1.0 feet per day (using an average hydraulic gradient of 0.002 vertical feet per horizontal feet).

The groundwater contamination potential for hard rock is high. In the eastern United States, hard rock weathers to form *saprolite* (literally, "rotten rock") at the surface. The saprolite ranges in thickness from 25 to 75 feet in the uplands to less than 10 feet in the valleys. In many areas, unweathered hard rock is exposed at the surface.

Since saprolite is derived directly from the underlying rock, its character is dictated by the parent rock's characteristics. Saprolite usually has a permeability on the same order as the underlying parent rock. Saprolite does not contain any organic matter except in the plant root zone. The saprolite may not have an appreciable ion exchange capacity because clay minerals have not yet formed. Once a chemical finds its way into the fracture system of the underlying hard rock, there is essentially no attenuation of the contamination.

To confound matters further, the fractured hard rock presents difficulties in defining the extent of contamination, because it is difficult to intersect the fractures with monitoring wells. In addition, flow paths can become indirect and tortuous. Thus, it is difficult to develop adequate remedial systems to halt the spread of contamination and clean up the groundwater.

Karst and Highly Fractured Rock Regions

Highly fractured and cavernous limestone aquifers are the most vulnerable to contamination. These aquifers occur east of the Rocky Mountains predominantly as limestone formations, i.e., the Valley and Ridge provinces of Pennsylvania, West Virginia, and Virginia, the limestones of the Florida peninsula, or the cavernous limestones of the Ozarks. In the western United States, the extensive basaltic flows of the Columbia Plateau and other regions can have similar properties with respect to high contamination potential.

Karst areas underlain by limestone formations often provide principal water supply aquifers. Groundwater has dissolved the limestone in the fractures and bedding planes to form enlarged fractures and solution cavities. In many places, the solution of limestone has produced networks of caverns.

Hydraulic conductivity of the limestone is controlled by the degree of fracturing and the extent of solution-widening of the fractures. In cavernous areas, the hydraulic conductivity may be quite high, ranging from one foot per day to more that 10,000 feet per day. Porosity will vary from less than 5 percent in unfractured limestone, to 10 percent or more in highly fractured and cavernous rock.

Groundwater contamination potential is greater in Karst terrain because of high permeability and ease of entry of contaminants into the subsurface. Groundwater flow is often difficult to determine without tracers. Groundwater flow velocity can be very high, leaving little chance for responses to contain groundwater contamination. Velocities in large fractures and caverns can be well above 100 feet per day up to 1,000+ feet per day.

Chemical attenuation in limestone, as well as other fractured rock, generally is very low. The fractured flow does not allow contact with a large surface area, and the limestone has almost no adsorptive capacity. But, acidic waste is neutralized by the reaction with limestone, and some metal precipitates may result as the pH of the groundwater is altered.

In addition to the lack of natural attenuation processes, the limestone terrain is subject to catastrophic collapse. Many instances have been reported of wastewater impoundments failing because of increased solution of the limestone and development of sinkholes. When an impoundment bottom collapses into an underground cavern, the release of wastewater to groundwater is abrupt and catastrophic. Remedial actions are worthless in these situations. The contamination can spread so fast that little can be done other than to warn downstream users.

SUMMARY

It should be apparent that the development of sound and rational programs to protect and clean up groundwater must be based on a clear understanding of the hydrogeology and the transport mechanisms of potential contaminants. Groundwater protection strategies can best be implemented by considering the susceptibility of the aquifers and by using siting procedures that minimize the potential impacts on the aquifers.

Before allowing land use activities that are potentially a threat to surface water and groundwater, investigations should be conducted to determine the site's hydrogeologic characteristics. Groundwater flow directions, chemical transport processes, and heterogeneity in the system must be defined to gain as complete an understanding of the system as possible. Without this understanding, harmful land use activities may be permitted, monitoring programs may be ill-conceived, and remedial actions may be incorrectly designed.

2

HUMAN HEALTH RISK ASSESSMENT

Janet E. Kester, Holly A. Hattemer-Frey, and Gary R. Krieger
Dames & Moore

OVERVIEW

The purpose of human health risk assessment is to characterize the nature and extent of potential adverse impacts from chemicals found in soils, air and/or water at a site. The risk assessment process synthesizes available data on exposure of specified receptors and the toxicity of **chemicals of concern (COCs)** to estimate the associated risk to human health. Four basic steps are involved:

- Data Evaluation
- Exposure Assessment
- Toxicity Assessment
- Risk Characterization

The first step, **data evaluation,** involves review of available site data to characterize the site, determine potential site-related COCs, and identify exposure media and receptor populations. Key questions are identified and plans for addressing them are made at this initial stage. The objective of the **exposure assessment** is to estimate the amount, frequency, duration, and routes of human exposure to site-related chemicals. The exposure assessment should consider both current and likely future site uses. The identification of exposure routes (such as inhalation and ingestion), and receptors (*i.e.*, the person(s) who could come in contact with a chemical) is crucial to determine the validity of an exposure pathway. After potentially complete exposure pathways are identified, exposure point concentrations and chemical intakes by exposed humans are calculated.

In the **toxicity assessment,** available toxicity data are reviewed to identify the nature and degree of toxicity of the COCs. The next step is to characterize the dose-response relationship: the relationship between magnitude of exposure and magnitude of adverse health effects. **Risk characterization** is the final step of risk assessment, where exposure and toxicity information are combined to (1) determine the degree of potential risks at a site, and (2) estimate what residual levels of chemicals may pose unacceptable risks to receptors.

Qualitative and/or quantitative analysis of the sources and the magnitude of uncertainty associated with each of the key phases and the results of the risk assessment process serve to put exposure and risk estimates and resultant target criteria into appropriate perspective for decision making.

The information generated in the risk assessment can be used to:

- Determine whether remedial action is necessary from a human health standpoint (*i.e.*, if calculated risks exceed some health-based criterion);
- Determine what specific areas and what volumes of affected media may require remediation; and
- Provide a basis for establishing cleanup levels that must be achieved in order to adequately protect human health, if remediation is required.

Typically, risk assessments are done in accordance with guidelines outlined in the **Risk Assessment Guidance for Superfund, Volume I, Human Health Evaluation Manual (Part A), Part B, Part C,** the Office of Solid Waste and Emergency Response (OSWER) Directive 9285.6-03, **Supplemental Guidance: Standard Default Exposure Factors; Exposure Factors Handbook, Guidance for Data Useability in Risk Assessment,** and **Guidelines for Exposure Assessment.** The general approach described in this chapter follows these guidelines, but is not intended to describe in detail every nuance of risk assessment.

Summary

- A risk assessment characterizes the nature, extent, and potential toxicological significance of human exposure to site-related chemicals.
- The results of the human health assessment can be used to manage risks by: (1) determining whether remedial action is warranted; (2) identifying the location(s) where remediation may be necessary; and (3) providing a rationale for establishment of chemical-specific cleanup levels that satisfy health criteria.

DATA EVALUATION

Review of Site Information

A first step in the data evaluation process is to obtain and review all available site-specific data on:

- The levels of chemicals present in various environmental media;
- The chemicals and processes that were used at the site in the past;
- The physical characteristics of the site;
- Potentially exposed populations; and
- Reliable site-specific data for various human activities and behavior patterns.

The site characterization process identifies where chemicals may be present at a site and at what concentrations these chemicals are present in various media. A part of the site characterization process of particular importance in the risk assessment is the identification of chemicals present on-site that may also be present in the background (*i.e.*, unrelated to operations on-site). Since the risk assessment typically quantifies health effects from potential exposures to chemicals derived from the site (over-and-above background concentrations), the careful evaluation of background concentrations can justify the exclusion of certain chemicals from the quantitative component of the risk assessment process. This is especially important for naturally occurring entities such as metals (*e.g.*, arsenic and lead), certain ubiquitously distributed pesticides (*e.g.*, DDT) and, depending on the site location, such common anthropogenic compounds as **polycyclic aromatic hydrocarbons (PAHs),** polychlorinated biphenyls (PCBs), and polychlorinated dioxins and furans.

An understanding of historical operations at the site can help guide the site characterization process by suggesting what chemicals might be found and in what form they might exist. A characterization of the physical setting should summarize information on climate, hydrogeology, geologic setting, and soil type, as well as the location and description of surface water and groundwater bodies on or near the site.

A description of potentially exposed populations should include information on sensitive subgroups as well as activity patterns that could affect exposures. Sensitive subpopulations include children, pregnant women, and the elderly. Information on populations and activity patterns can be obtained from local agencies and site-specific survey results. This section should also include information on individuals who could be exposed to site-related chemicals even if they do not live on or near the site (*e.g.*, a nearby resident who consumes fish from a river or creek that traverses the site).

The risk assessment typically identifies potentially exposed populations associated with the following land-use categories:

- Residential
- Commercial/industrial
- Recreational
- Trespasser

To accurately define current land uses, it is often necessary to gather information on activity patterns (*e.g.*, determining the amount of time residents, workers, or trespassers spend in potentially contaminated portions of the site, and assessing how activities and exposures can change with the seasons). Information on the proximity of individuals to contaminated areas can be obtained from zoning maps, census data, and population surveys conducted on or near the site. Although future land use of the site is typically characterized by reviewing zoning plans, census projections, and making reasonable assumptions about the probable future use of the area, information on how long residents have lived on or near the site can aid in making realistic judgments about future use (*e.g.*, if the site is in the middle of an industrial area, it is not likely to be developed into a single-family residential area).

The collection of site-specific data allows risk to be more accurately assessed in its geographical context and obviates use of often overly conservative default assumptions recommended in agency guidelines. For instance, if warranted, a survey of the people living on a site or who use a site could be conducted to more accurately estimate key exposure factors. The goal of such an exposure assessment survey would be to obtain site-specific information on behavioral patterns and frequencies so that site-specific exposure parameters can be incorporated in the risk assessment. In addition, such a survey would allow the risk assessment to identify the factors that contribute most to exposure. Conducting a site-specific exposure survey can be a relatively inexpensive means of achieving more realistic risk estimates.

Summary

- A review of site-specific information is required to identify the physical characteristics of the site, COCs, and potentially exposed receptor populations.
- Current and future land uses should be defined as clearly as possible.
- The collection of site-specific survey data on behavior patterns and the likelihood of exposure can yield more accurate risk estimates.

Evaluation of Available Data for Use in Risk Assessment

A principal objective of the data evaluation task is to organize the chemical data into a form appropriate for use in the risk assessment. It includes the following steps:

- Evaluate the data sampling plan for adequacy in characterizing the amounts and distributions of site-related chemicals;
- Qualify the analytical methods with regard to their appropriateness for use in the risk assessment;
- Evaluate the quality of data with respect to sample quantitation and detection limits;
- Examine laboratory qualifiers assigned to monitoring data and evaluate potential quality assurance/quality control problems;
- Evaluate data with respect to blanks and **tentatively identified compounds (TICs)**;

- Summarize information on background concentrations of chemicals and compare background levels with site-related levels;
- Test the statistical distribution of the data; and
- Identify COCs.

It is important that the analytical methodology employed be appropriate for the medium and chemicals being tested and, in most instances, be approved by a regulatory agency.

Detection limits vary depending on the chemical analyzed, the analytical instrument used, and the characteristics of the medium being tested. Some chemicals may not be detected in samples from some sampling events due to varying detection limits. Since detection limits can vary, it is necessary to establish that the reported detection limit does not exceed levels practical for use in a quantitative risk assessment (i.e., health-based criteria). If a detection limit concentration is greater than twice the most commonly observed detection limit, then the sample with the elevated detection limit should be eliminated. Chemicals not detected in any of the samples that have appropriate detection limits can be eliminated from further consideration.

Results classified as non-detects are typically treated as equal to one-half the detection limit and included in the calculation of mean and upper-bound (95 percent upper confidence limit (UCL)) on the mean estimates. This method tends to bias estimates of the mean and standard deviation. Non-detect data should therefore be handled, if possible, using standard, EPA-approved statistical techniques for left-censored data sets. Log-probit analysis has been shown to be a robust method for estimating the geometric mean and standard deviation of samples with values in the nondetetable range. In a **log-probit analysis,** all measurements (both detects and non-detects) are assumed to be samples taken from the same lognormal probability distribution. When samples from a lognormal distribution are plotted on a probit scale, they tend to lie on a straight line. In a probit analysis involving both detects and non-detects, the non-detects are treated as unknowns, but their percentile values are taken into account. Thus, if there are 100 samples, 30 of which are non-detects, the first detectable data point would be plotted at the thirty-first percentile. If sufficient detectable values exist, they can be used to establish the line, using linear regression, that characterizes the entire data set. The geometric mean for the data set (both detects and non-detects) is the 50th percentile value.

Thus, a probit analysis allows the geometric mean to be extrapolated from a data set even if values are below detection limits (provided there are sufficient detected values to define the probability distribution.) Available probit tables and computerized programs facilitate the use of log-probit or other statistical analyses in the risk assessment process.

Statistical Evaluation of Data

It is important to define study areas that represent potential exposure units, *i.e.*, geographic areas or sets of chemical concentrations that could be contacted by receptors over the exposure period. The appropriate definition of exposure units depends on the spatial distribution of contamination as well as on site-specific land use and exposure conditions. It is recommended that: 1) surface soil across the site should be evaluated as a single exposure unit where contact with soil is spatially random, *i.e.*, in a trespasser scenario; 2) subareas of the site where a specific activity is expected to occur should be evaluated as **hot spots**, when contamination is unevenly distributed across the site; and 3) groundwater from a specific site location, and potentially from a single aquifer, should be evaluated as a possible future source of drinking water only if it is shown to be potable by other criteria (*e.g.*, total dissolved solids content is less than 10,000 mg/liter, flow rate is greater than 1,500 gallons/day).

Geostatistical methods take into account the spatial structure of the data, and should be used in defining potential exposure units. At a minimum, figures and tables should be prepared which show the spatial distribution of potential chemicals of concern.

The purpose of background comparisons in conducting baseline risk assessments is to determine if chemical concentrations detected at the site are distinguishable from concentrations originating from causes other than activities that have occurred at the site. Chemical concentrations that are indistinguishable from background may not have resulted from site activities. Also, in characterizing health risks, the risks associated with concentrations indistinguishable from background would have less significance than concentrations that are clearly associated with site activities. Statistically, this is a test of the **null hypothesis** that the mean concentration of a chemical at the study area is not significantly different from the mean value at the background location.

Site concentrations should be compared to data collected from background areas that are otherwise comparable to study areas defined for the site. Factors that determine comparability include natural features, such as soil types and mineralized zones in the subsurface. Anthropogenic influences derive from current and past land use, *e.g.*, former agriculture. Where background concentrations are unevenly distributed, background data should be grouped into geographic areas of similar size for comparison to study areas on the site.

Selection of Chemicals of Potential Concern

The process of COC selection accounts for: (1) the concentration of chemicals present in various environmental media; (2) their predicted mobility, persistence, and potential transformation/degradation in the environment; and (3) their observed toxicities.

The first step includes the screening of chemicals after detection limits, detection frequencies, and blank samples have been examined. If a chemical is detected in fewer than 5 percent of samples and is not a carcinogen, it can be eliminated. Chemicals not detected in any samples are eliminated from further consideration, while any carcinogen detected at an order of magnitude or more above the detection limit should be retained regardless of its frequency of occurrence. Chemicals considered essential human nutrients (zinc, iron, sodium, calcium, etc.) can also be eliminated unless known to be site-related and significantly elevated above background.

Chemicals not eliminated during these steps of the screening process are then compared to background levels. If local background data for a given medium or chemical are not available, U.S. or regional background data, such as those reported by Adiano, Shacklette and Boerngen, or Dragun and Chiasson may be used. If the mean concentration of a chemical in a given medium is statistically higher (at the 95% level of confidence) than its mean level in background samples as determined by a one-tailed t-test, that compound should be retained as a COC. As a result of this process, only chemicals likely to be site-related, and detected at statistically significant concentrations, are ultimately selected as COCs.

If a further reduction of COCs is required (*e.g.*, to reduce the number of chemicals that must be quantified to a more manageable number), a **toxicity-concentration screen** can be used to focus the quantitative risk assessment on those chemicals posing the greatest risk for the exposure scenarios being considered. Non-carcinogens and carcinogens identified in each medium are combined with their appropriate toxicity constants (*i.e.*, reference doses (RfDs) or cancer slope factors (CSFs)). The maximum concentration for each non-carcinogenic chemical is multiplied by the inverse of its reference dose (1/RfD), while the maximum concentration of each carcinogen is multiplied by its CSF. The resultant risk factors are summed for each group (carcinogens and non-carcinogens), and the contribution of each to the total risk is calculated on a percentage basis. Those chemicals contributing more than one percent of the total risk are usually retained. A disadvantage of this screening process is that it does not account for bioaccumulation in the food chain. Thus, it may be inappropriate to use in cases where food chain pathways (e.g., ingestion of fish or garden produce) are expected to be substantial, relative to others.

Summary

• Selection of COCs involves two phases. The first phase includes the screening of chemicals after detection limits, detection frequencies, and blank samples were examined, while the second phase involves comparing chemical concentrations in various environmental media (*e.g.*, air, soil, and water) to medium-specific background data. A concentration-toxicity screen can be used to further reduce the number of COCs to a more manageable number.

• Log-probit analysis is an alternative method of estimating the mean and standard deviation of a data set with values in the nondetectable range. Since a log-probit analysis treats non-detects as unknowns (versus as one-half the detection limit), this technique will likely yield lower mean estimates, and, consequently, lower risk estimates.

TOXICITY ASSESSMENT

Toxicity assessment evaluates the nature and potency of health effects associated with exposure to a unit dose (milligram per kilogram body weight per day, or mg/kg-day) for each site-related chemical. It consists of identifying the **intrinsic** toxicity of a chemical (*i.e.*, a hazard evaluation), and determining the frequency of an adverse health effect (*e.g.*, cancer) associated with a unit daily dose (dose-response assessment).

The hazard evaluation involves a comprehensive review of available toxicity data to identify the nature and severity of toxic properties associated with the COCs. Once the potential toxicity of a chemical has been established, the next step is to determine the amount of chemical exposure that may result in adverse human health effects.

COCs are classified into two broad groups based on their toxic end point: carcinogens and non-carcinogens. These classifications are selected because certain chemicals can have both properties, and health risks are calculated differently for carcinogenic and non-carcinogenic effects.

Carcinogenic Health Effects

Carcinogenic effects are expressed as the probability that an individual will develop cancer from a lifetime of exposure. This probability is based on projected intakes and chemical-specific dose-response data, developed by EPA, *i.e., CSFs*. Because CSFs, represent the 95th percentile UCL on the slope of the dose-response curve, risks calculated using these CSFs represent upper-bound values. Thus, there is only a 5 percent probability that the actual risk is greater than the estimated risk, and a 95 percent chance that the true risk is lower than the projected risk (it may even be zero). Evidence of chemical carcinogenicity originates primarily from two sources: (1) lifetime studies with laboratory animals, and (2) human epidemiological studies. For most chemical carcinogens, animal data from laboratory experiments represent the primary basis for the extrapolation. Major assumptions arise from the necessity of extrapolating experimental results (1) across species (from laboratory animals to humans), (2) from high doses (used in animal studies) to low doses (levels to which humans are likely to be exposed in the environment), and (3) across routes of administration (*e.g.*, extrapolation of results obtained in an oral study to inhalation or dermal exposures). Federal regulatory agencies have traditionally estimated human cancer risks associated with exposure to chemical carcinogens on the basis of administered dose according to the following approach:

- The relationship between administered dose and cancer incidence in animals is based on experimental animal bioassays.
- The relationship between administered dose and cancer incidence in the low dose range is based on mathematical models.
- The dose-response relationship is assumed to be the same for both humans and animals, if administered dose is measured in the proper units (typically, in mg/kg/day).

The uncertainty involved in extrapolating from experimental animals receiving relatively high doses to humans receiving much lower exposures is qualitatively addressed by classifying chemicals into groups based on the weight of evidence from epidemiological and animal studies that they are actually carcinogenic in humans. The EPA's weight-of-evidence classification system is shown in the Table 1.

Table 1

EPA'S Weight-Of-Evidence Carcinogenicity Classification Scheme

Group	Description
A	Human carcinogen
B1	Probable human carcinogen - limited human data are available
B2	Probable human carcinogen - sufficient evidence in animals and inadequate or no evidence in humans
C	Possible human carcinogen
D	Not classifiable as to human carcinogenicity
E	Evidence of non-carcinogenicity for humans

A quantitative risk assessment is usually performed for all Group A and B carcinogens, while Group C carcinogens may be included on a case-by-case basis.

The EPA assumes that a small number of molecular events can evoke changes in a single cell that can lead to uncontrolled cellular proliferation and tumor induction. It is assumed that there is no level of exposure to a given chemical that does not pose a small, but finite, probability of generating a carcinogenic response. Since risk at low exposure levels cannot be measured directly either in laboratory animals or in human epidemiology studies, mathematical models have been developed to extrapolate from high to low exposures. Currently, cancer potency or slope factors are estimated with the **linearized multi-stage model**, which is based on the theory that multiple events may be needed to yield tumor induction.

The linearized multi-stage model incorporates procedures for estimating the 95 percent **UCL** on the slope of the dose-response curve extrapolated to doses lower than those used in experiments. The probability that the true risk is higher than that estimated is thus only 5 percent. The animal data used for extrapolation are taken from the most sensitive species studied, based on the assumption that humans are at least as sensitive as the most sensitive animal species. The risk estimates made with this model should be regarded as conservative, representing the upper limit of risk. Actual risk is likely to be lower, and could even be zero. Cancer dose-response relationships can also be derived from human epidemiological studies, but such data are seldom adequate for regulatory purposes.

Other available low-dose extrapolation models produce quantitatively similar results in the range of observable data but yield estimates that can vary by three or four orders of magnitude at lower doses. Animal bioassay data are not adequate to determine if any of the competing models are better than the others. Moreover, there is no evidence to indicate that the precision of low-dose risk estimates increases through the use of more sophisticated models. Thus, if a carcinogenic response occurs at the exposure level studied, it is assumed that a similar response will occur at all lower doses. While the output of such models is a single point estimate of cancer potency, the magnitude of uncertainty surrounding these estimates is unquantifiable but probably very large. Thus, to enable informed risk management, this source of uncertainty must be clearly presented in the risk assessment document. It should also be noted in discussing uncertainty that the background incidence of cancer is 33 percent.

Non-Carcinogenic Health Effects

Potential effects from chronic exposure to non-carcinogenic compounds are assessed by comparing exposure levels to chronic RfDs. RfDs represent a level of intake that is not expected to produce adverse effects, even in sensitive sub-populations, over a lifetime of exposure. Unlike carcinogenic compounds, chemicals that cause toxic effects other than cancer appear to do so through mechanisms that exhibit a physiological threshold. Thus, a certain dose of the chemical must be present in the body before toxic effects are observed. The approach used to estimate the likelihood that exposed individuals will experience non-carcinogenic effects assumes that there is some level of exposure (*i.e.,* the *RfD*) that individuals can tolerate without experiencing adverse health effects. Conversely, if exposure exceeds this threshold level, there may be some concern that exposed individuals will experience non-carcinogenic health effects. RfDs represent a level of intake that is not expected to produce adverse effects, even in sensitive sub-populations, over a lifetime of exposure.

RfDs are calculated by dividing a **NOEL, NOAEL, or LOAEL dose** (in units of mg/kg/day) by an uncertainty or safety factor that typically ranges from 10 to 10,000. NOEL, NOAEL, and LOAEL are defined as follows:

NOEL:No Observed Effect Level--The dose at which there are no statistically or biologically significant increases in the frequency or severity of effects between the exposed population and the corresponding control population (*i.e.*, no measurable effects are produced at this dose).

NOAEL:No Observed Adverse Effect Level--The dose at which there are no statistically or biologically significant increases in the frequency or severity of adverse effects between the exposed population and the corresponding control population. Effects are produced at this dose, but they are not considered adverse.

LOAEL:Lowest Observed Adverse Effect Level--The lowest dose of a chemical in a study or group of studies that produces statistically or biologically significant increases in the frequency or severity of adverse effects between the exposed population and its appropriate control.

Uncertainty factors (UFs) or safety factors are conservatively applied to ensure health protective standards for all segments of the potentially affected population. RfDs are derived from the NOAEL or the LOAEL for the critical toxic effect by the consistent, conservative application of uncertainty factors (UF) and a modifying factor (MF), if necessary. UFs generally consist of multiples of 10 (although values less than 10 such as 3 are sometimes used), with each factor representing a specific area of uncertainty inherent in the extrapolation from the available data.

In general, RfDs represent the level of exposure not expected to cause adverse health effects even in sensitive sub-populations, with variability typically spanning one order of magnitude. This method is conservative since it assumes that humans are more sensitive than laboratory animals. If the laboratory animal is, in fact, more sensitive to a given chemical than humans, then the application of an UF of 10 to account for variability between humans and other animals implies that the adverse health impact estimated using that RfD could be overestimated. Hence, EPA's approach may overestimate the possibility that exposed individuals could experience adverse health effects. EPA's rationale is that the risks estimated using this approach represent the **highest** risk to which any individual living near the site is likely to be exposed.

Inhalation RfDs are derived from NOAELs by applying UFs similar to those used for oral RfDs. Inhalation toxicity data can be expressed as a **reference concentration** (RfC) (in mg/m^3) or as an inhaled intake (in mg/kg-day). **RfCs** are the concentrations of chemical in air (in mg/m^3) that will not cause adverse health effects in exposed populations assuming a continuous, 24-hour-a-day exposure. The default adult body weight and inhalation rate of 70 kg and 20 m^3/day, respectively, can be used to convert a reference concentration to an inhaled intake (and vice versa).

Oral and inhalation RfDs are typically expressed as one significant figure in units of mg/kg-day. Most RfDs developed by EPA to date are based on administered versus absorbed doses. The administered dose approach is also conservative, since it does not account for the fact that some fraction of the dose administered to the animal may not be absorbed across the lung, skin, or gastrointestinal tract, making the dose that actually affects a target organ or system lower. It is generally agreed, however, that the administered dose approach is reasonable and provides results that are protective of human health.

Chemicals Without Toxicity Values

Toxicologists are routinely confronted with the problem of attempting to derive safe levels of exposure to chemicals based on animal studies. The primary objective of analyzing available toxicological data to obtain a realistic measure of the potential toxicity of a chemical is to establish the dose-response relationship for a given chemical (specifically, to identify the critical health effect and intra- and interspecific differences). When long-term epidemiological data are not available, the most common technique for deriving health-protective benchmark toxicity values is the application of uncertainty or safety factors. It has become customary to use a multiplicative chain of factors, each of which accounts for a different source of uncertainty or is designed to add an extra margin of safety as exemplified in the following equation:

$$\text{Benchmark Value} = \frac{\text{Mean Chronic Value}}{(UF1 \times UF2 \times x \ldots \times U}$$

UF1 is the adjustment to account for the anticipated greater susceptibility in the test population than was observed during the experiment (**i.e.**, to account for intraspecies variability). UF1 may range from 1 to 10, with values of 1 to 3 being most likely when a NOAEL is derived from a state-of-the-art study. Based on a study evaluating 490 LD_{50} studies, Weil showed that a factor of 3 to 6 was adequate to adjust for intraspecies differences. A value of 10 indicates that high variability was observed or would be expected among species.

UF2 is used to extrapolate data from subchronic studies to estimate risks from lifelong exposures. UF2 may range from 1 to 10, with values of 1 to 3 being the most likely. Weil and McCollister examined ratios of LOAELs to NOAELs from subchronic and chronic studies. Combining data from all studies resulted in a subchronic-to-chronic NOAEL ratio of less than or equal to five. A value of 10 implies that there is great uncertainty associated with extrapolating a chronic value from a subchronic one.

UF3 is used to extrapolate LOAELs to NOAELs. For a LOAEL from a state-of-the-art study, a value of 2 is usually adequate. Extrapolation from a subchronic to a chronic NOAEL showed that a factor of 5 was adequate about 90% of the time, while a factor of 10 was adequate 97% of the time.

Lewis *et al.*'s results suggest that uncertainty factors need not be large to provide values protective of human health. A composite adjustment of 250 is recommended as sufficiently protective of human health versus values of 1,000 to 10,000 commonly used by EPA.

Summary

- Toxicity assessment evaluates the nature and extent of health effects from exposure to site-related chemicals.
- RfDs represent the level of exposure not expected to cause adverse health effects even in sensitive sub-populations, with variability typically spanning one order of magnitude.
- EPA CSFs represent the 95% UCL on the extrapolated slope of the carcinogenic low dose-response curve; hence, there is a 95 percent chance that the true risk is lower than the projected risk, and may even be zero.
- Since carcinogenic risks at low exposure levels cannot be measured directly either in laboratory animal or human epidemiology studies, mathematical models are used to extrapolate from high to low exposure levels. Most models produce quantitatively similar results in the range of observable data but yield estimates that can vary by three or four orders of magnitude at lower doses.

- Lewis *et al.* evaluated numerous studies regarding toxicological extrapolations from animals to humans in an effort to determine more representative uncertainty factors and concluded that a composite adjustment of 250 may be sufficiently protective of human health (versus values of 1,000 to 10,000 commonly used by EPA).

EXPOSURE ASSESSMENT

The objective of the exposure assessment is to estimate the amount, frequency, duration, and routes of human exposure to site-related chemicals. This task involves the following steps:

- Characterizing the exposure setting;
- Identifying potential human receptors;
- Characterizing pathways of exposure;
- Estimating exposure point concentrations; and
- Estimating total chemical intake by potentially exposed individuals for all relevant pathways of exposure.

The exposure assessment should consider both current and future risks to human health under land use scenarios consistent with known or likely activity.

Identification of Potential Fate and Transport Mechanisms

Environmental fate and transport modeling and/or monitoring data are used to estimate chemical concentrations in the transport medium (*i.e.*, air or groundwater) at the point of contact with the receptor. Such contact constitutes human exposure. Since the movement of chemicals within and between various environmental media can substantially influence how far the chemical can migrate and in what form it might exist at the exposure point, it is necessary to employ measurement or predictive techniques to evaluate the fate and transport of COCs. Many chemical, physical, and environmental parameters, such as water solubility, vapor pressure, octanol-water partitioning, and bioaccumulation, influence the behavior and fate of chemicals released into the environment. The importance of these factors and how they influence each other is incompletely understood. However, the examination of a few basic physicochemical properties can provide insight into the behavior and fate of chemicals released into the environment.

Water solubility is the maximum amount of a chemical that will dissolve in pure water at a specific temperature and pH. The solubility of a chemical in water affects its fate and transport in soils and could significantly influence human exposure to organics through aquatic pathways. Highly soluble compounds tend to leach rapidly from soil into ground and surface water supplies. In addition, they tend to be less volatile, more biodegradable, and more mobile than less soluble or insoluble chemicals.

Vapor pressure measures the relative volatility of a chemical in its pure state and is useful for determining the extent to which a chemical will be transported into air from soil and water surfaces. **Volatilization** is a major route for the distribution of many chemicals in the environment. The volatility of a chemical is affected by its solubility, vapor pressure, and molecular weight, as well as the nature of the air-to-water or soil-to-water interface through which the chemical must pass. Chemicals with a low vapor pressure and a high affinity for soil or water are less likely to volatilize than chemicals that have a high vapor pressure and a weak affinity for soil or water.

Assessing the environmental fate of chemicals depends largely on being able to predict the extent to which they will bioaccumulate in living organisms, including edible plants, fish, cattle, and humans. Living organisms can concentrate chemicals in their tissues at levels substantially higher than the concentration in water (in the case of aquatic organisms) or in food (in the case of terrestrial organisms). Organisms tend to accumulate chemicals in the lipid (fatty) portions of their tissues. Thus, one way to determine the bioaccumulation potential of a chemical is to measure how lipophilic it is. Since it is difficult to directly measure the lipophilicity of a chemical, researchers typically use the **octanol-water partition coefficient (K_{ow})** to predict its tendency to partition between an octanol component (a good surrogate for fat) and water.

K_{ow} is defined as the ratio of the concentration of a chemical in the octanol phase to its concentration in the aqueous phase. It is directly related to the tendency of a chemical to concentrate in biota and inversely correlated with water solubility. Hence, the octanol-water partition coefficient is used extensively to estimate the bioconcentration potential of organics in living organisms. Chemicals with large K_{ow} values (i.e., lipoplilic compounds such as DDT, dioxins, and PCBs) tend to accumulate in soil, sediment, and biota but not in water. This class of chemicals tends to become transferred to humans through the food chain. Conversely, chemicals with small K_{ow} values tend to partition mostly into air or water. For example, volatile organics such as trichloroethylene and tetrachloroethylene are widely distributed in air, with inhalation being the primary pathway of human exposure.

The migration of chemicals to groundwater and surface water is largely controlled by aqueous solubility. However, soil particles bearing sorbed chemicals may also be transported by flowing water and eventually be deposited as sediments. The release of chemicals from soil to groundwater may occur through diffusion or through leaching and percolation. Gases and vapors forming in the **vadose zone** (the unsaturated soil generally above the water table) are transported by diffusion to the surface, where they may escape into the atmosphere.

Both diffusion and percolation through the soil column may be slowed by sorption of a chemical to the surface of soil particles. Many nonionic organic compounds form a hydrophobic bond with organic matter and mineral groups present on the surface of soil particles. The extent of sorption for most organic compounds can be reasonably estimated from the organic carbon content of the soil and the **soil sorption coefficient, K_{oc}**. High K_{oc} values reflect decreased mobility in the soil due to sorption. Under equilibrium conditions, the distribution of a chemical between the soil and infiltrating groundwater can be estimated using the **partition coefficient, K_d**, which may be derived using the formula

$$K_d = f_{oc} \times K_{oc}$$

where f_{oc} is the fraction of organic carbon in the soil.

Chemicals present in various environmental media can undergo a number of physical and chemical processes that may alter their chemical makeup or their chemical properties. For example, chemicals exposed to direct sunlight may be degraded by photolysis. Chemicals sorbed to soil particles may be effectively immobilized. However, changes in soil moisture content or chemistry can serve to release adsorbed chemicals. With the exception of sorption, chemicals present in soil, surface water, and sediments are exposed to similar fate mechanisms. These mechanisms include chemical reactions, biotransformation, and bioaccumulation.

Identification of Potentially Exposed Populations

Generally, one or more of four distinct receptor groups could be exposed to COCs at a given site: residents, workers, and trespassers or recreational visitors. Residential exposures should be evaluated whenever individuals currently live on or near a site or when future use of the site is likely to include residential development. Residential receptors are assumed to come in frequent and repeated contact with site-related chemicals and generally experience the highest exposures.

Under the commercial/industrial land use scenario, employees who work on or near the site could be exposed to site-related chemicals. Exposures are typically lower than those estimated for residential receptors, since worker exposure is generally limited to eight hours a day, 250 days a year. The trespasser and recreational land use scenarios address individuals who do not live on-site but may spend a limited amount of time at or near the site playing, fishing, hunting, or engaging in other recreational activities. Recreational exposures are expected to be higher at inactive versus active sites, since there is generally less surveillance at inactive sites.

Selection of Complete Exposure Pathways

For risk assessment purposes, an important objective in evaluating the environmental behavior and fate of various chemicals is predicting the major pathways and extent of human exposure.

A complete exposure pathway consists of the following components:

- A source of chemical;
- A mechanism of chemical release;
- Retention and transport media;
- A point of potential receptor contact; and
- An exposure route at the contact point.

The relative importance of an exposure pathway depends upon the concentration of a chemical in the relevant medium and the rate of intake by exposed individuals. Each land use category (residential/industrial/recreational) is defined by a distinct set of exposure pathways. Exposure pathways that might be evaluated at a hypothetical site are shown in Table 2.

Table 2
Potential Exposure Pathways That Could Be Evaluated During a Human Health Risk Assessment[1]

Pathway[a]	Current and Future Residents	Current and Future Workers	Current and Future Trespassers	Rationale
Soil				
Dermal Absorption	Maybe	Maybe	Maybe	Only if COCs will be absorbed through the skin.
Ingestion	Yes	Yes	Yes	Incidental ingestion of outdoor soil and indoor dust is possible.
Air				
Inhalation (Vapors)	Maybe	Maybe	Maybe	Only if the COCs volatilize at ambient temperatures.
Inhalation (Particulates)	Yes	Yes	Yes	Resuspension of particulates via wind or other mechanical processes is possible.
Food				
Ingestion of Produce Grown On-Site	Yes	No	No	Trespasser and workers do not live on-site or consume food items originating from the contaminated area.
				It is possible that residents could consume produce grown in a backyard garden.
Ingestion of Meat and Milk/Dairy Products	No	No	No	Trespasser and workers do not live on-site or consume food items produced on-site.
				It is not likely that residential receptors raise their own cows.
Ingestion of Fish	Yes	No	Yes	The resident and trespasser could catch and consume fish from local water bodies.
Ingestion of Game	No	No	No	Intake of COCs from ingesting game from the contaminated area is assumed to be low relative to other exposure pathways.

[1] Some or all of these pathways may exist at a given site. This table is shown to illustrate possible exposure routes.

Table 2 (Continued)

Pathway[a]	Current and Future Residents	Current and Future Workers	Current and Future Trespassers	Rationale
Surface Water				
Dermal Contact While Swimming	Maybe	No	Maybe	Residents and trespassers could use potentially-affected surface water bodies for recreational purposes. This pathway would apply only if the COCs will readily absorb through the skin.
Ingestion While Swimming	Yes	No	Yes	Residents and trespassers could use potentially-affected surface water bodies for recreational purposes and ingest water while swimming.
Groundwater				
Ingestion	Maybe	No	No	The trespasser does not live on-site or consume local groundwater. Worker is expected to consume municipal water while at work. This pathway is applicable only if residents use or could use the groundwater now or in the future.
Inhalation During Showering	Maybe	No	No	The trespasser and worker do not live on-site or use the local groundwater for domestic purposes. This pathway is applicable for residents only if COCs are likely to volatilize during showering.
Dermal Absorption During Showering	Maybe	No	No	The trespasser and worker do not live on-site or use the local groundwater for domestic purposes. This pathway is applicable for residents only if COCs are readily absorbed through the skin.
Irrigation of Crops with Ground or Surface Water	Maybe	No	No	The trespasser and worker do not live on-site or use the local groundwater for agricultural purposes. This pathway is applicable for residents only if sufficient rain does not fall annually to support agricultural production.

[a] All pathways apply to current and future child and adult receptors except for the worker, who is assumed to be an adult.

Estimating Exposure Point Concentrations

Accurate estimates of chemical concentrations at points of potential human exposure are a prerequisite for evaluating chemical intake in potentially exposed individuals. Actual concentrations as determined from sampling data should be used whenever possible. In some cases, direct measurement of chemical concentrations and chemical contribution to another medium may not be feasible, accurate (*e.g.*, health effects may occur below limits of detection), or cost-effective.

EPA defines the exposure point concentration used to estimate **reasonable maximum exposure (RME)** as the 95% UCL on the arithmetic mean. An important complication in calculating the RME concentration is that a statistical distribution (usually either normal or lognormal) must be assumed. Supplemental guidance for risk assessment cites EPA's experience that most large environmental data sets are lognormally rather than normally distributed. In this case, the UCL can be calculated using a method presented in Gilbert. However, UCL values based on the lognormal model are often an order of magnitude (ten times) or more above the range of observed concentrations, for data sets that exhibit high variability. This is because the UCLs are extrapolations representing the average concentrations associated with extreme long-tailed (highly skewed) distributions that could have generated the observed sample data with 5% probability. Therefore, where the UCL is higher than any measured value, a more reasonable approach may be to consider the maximum concentration detected for any composite area.

Because of the uncertainty associated with estimating the extent of human exposure to site-related chemicals, recent guidance recommends developing an average as well as an RME exposure scenario. The **RME scenario** is defined as the **highest** exposure that is reasonably expected to occur at a site (*i.e.*, well above average). Since this scenario uses a mix of conservative assumptions and data, the uncertainty associated with its results is unquantifiable, but may be substantial. It should therefore be regarded as an upper bound on exposure and risk.

To provide some estimate of what more common exposures and risks might be, an average or most likely exposure (MLE) scenario using mean exposure assumptions should also be considered. Mean chemical concentrations are used to estimate exposure point concentrations.

Human Health Risk Assessment / 59

The advantage of evaluating both exposure scenarios is that they provide decision makers with a broader perspective on the true range of risks likely to be experienced by individuals exposed to site-related chemicals. The use of a point estimate of risk, especially an upper-bound estimate, does not provide adequate information to individuals who must decide if the health risks are excessive and therefore if the site warrants remediation.

Estimating Chemical Intakes

Chronic daily intakes are estimated using standard EPA exposure equations. Calculated intakes are expressed as the amount of chemical actually taken into the body versus the amount that is absorbed through the lung or gut once the chemical has been ingested or inhaled. This method of calculating exposures is appropriate if the toxicity data used in the assessment are based on administered (versus absorbed) dose, such as the numerical toxicity values recommended by EPA. If the toxicity constants used are based on absorbed versus administered dose, it is appropriate to account for the fact that not all of the chemical taken into the body will be absorbed through the lung, skin, or gut. This approach may result in less conservative toxicity data and lower risk estimates.

Modeling Exposures from Soil-Related Pathways

Individuals who come in contact with contaminated soils may be exposed via three pathways: ingestion, dermal absorption and inhalation of dust particles.

Incidental Ingestion of Soil

Individuals can be exposed to chemicals by intentionally eating soil (*e.g.*, pica behavior in small children) or by inadvertent hand-to-mouth transfer of soil during gardening, cleaning, and recreational activities. Chronic daily intake from ingesting contaminated soil is calculated using the following equation:

$$I = \frac{CS \times IR \times CF \times FI \times EF \times ED}{AT \times BW}$$

where,
I=intake of chemical from ingestion of contaminated soil/dust (mg/kg/day);
CS=concentration of chemical in soil (mg/kg);

IR=ingestion rate (mg soil/day);
CF=conversion factor (10^{-6} kg/mg);
FI=fraction ingested from contaminated area (unit less);
EF=exposure frequency (days/year);
ED=exposure duration (years);
AT=averaging time, or the period over which exposure is averaged (days);
BW=body weight (kg).

Dermal Contact with Soil

Because dermal contact with chemicals in soil will not result in an adverse effect unless chemicals are absorbed through the skin, the following equation is used to estimate the amount of chemical absorbed through the skin:

$$AD = \frac{CS \times CF \times SA \times AF \times ABS \times EF \times ED}{BW \times AT}$$

where,
AD=absorbed dose of chemical (mg/kg-day);
CS=concentration of chemical in surface soils (mg/kg);
CF=conversion factor (10^{-6} kg/mg);
SA=skin surface area available for contact (cm²);
AF=soil-to-skin adherence factor (mg/cm²-event);
ABS=absorption factor (unitless);
EF=exposure frequency (events/year);
ED=exposure duration (years);
BW=body weight (kg);
AT=averaging time (days).

Modeling Exposures via Inhalation

Exposure from inhalation of chemicals sorbed to resuspended dust or chemical vapors is calculated as a function of the concentration of chemicals in air, respiration rate, and body weight as follows:

$$I = \frac{CA \times IR \times ET \times EF \times ED}{BW \times AT}$$

where,
I=intake of chemical from inhaling contaminated particulates (mg/kg-day);
CA=model-predicted concentration of COCs in air (mg/m^3);
IR=respiration rate (m^3/hour);
ET=exposure time (hours/day);
EF=exposure frequency (days/year);
ED=exposure duration (years);
BW=body weight (kg); and
AT=averaging time (days).

Ingestion of Groundwater

Even if residents do not currently use groundwater for domestic or agricultural purposes, there are often no constraints on the use of such resources as a potable supply in the future. For groundwater to be considered as a potential water source in the future, the water must be both potable and have sufficient production capacity to support domestic or agricultural uses. It should also be assumed that local ground or surface water will not be used for irrigation if annual rainfall is sufficient to support agricultural production without irrigation. Again, this pathway is typically limited to residential receptors, since workers are assumed to consume water from a municipal source.

Daily chemical intake from ingestion of contaminated water can be calculated using the following equation:

$$I = \frac{CSW \times IR \times EF \times ED}{AT \times BW}$$

where,
I=intake of chemical from ingestion of groundwater (mg/kg-day);
CSW=concentration of COCs in groundwater (mg/L);
IR=water ingestion rate (L/d);
EF=exposure frequency (days/year);
ED=exposure duration (years);
AT=averaging time (days); and
BW=body weight (kg).

Modeling Food Chain Intakes

Assessing the magnitude of human exposure to chemicals through the food chain depends largely on being able to predict the bioaccumulation of chemicals in the terrestrial and aquatic food chains. For most chemicals, estimating human exposure to chemicals present in the food chain is accomplished through the use of various predictive equations that estimate exposure from ingestion of contaminated produce (fruits and vegetables), meat, milk/dairy products, and fish. These equations are very conservative and tend to overestimate exposure. Since estimating the concentration of chemicals in various food items is highly variable and EPA does not provide guidance for quantitatively assessing plant or animal uptake of inorganic compounds, EPA recommends that on-site fruits, vegetables, and fish be sampled to determine the concentration of COCs in food items of the human diet. If site-specific data cannot be obtained, intakes from food chain-related pathways may be discussed qualitatively.

Summary

- Since organic chemicals tend to end up in the media in which they are most soluble, a few basic physicochemical properties can be used to predict the behavior and fate of chemicals released into the environment.
- Highly lipophilic compounds tend to sequester in soil, sediment, and biota, and the food chain is the primary pathway of human exposure. Highly soluble compounds are found in water, and ingestion of contaminated water is usually the primary pathway of human exposure. Volatile compounds tend to partition into air, and inhalation is the major source of human exposure.
- Generally, four distinct receptor groups could be exposed to COCs at a given site: residential, commercial/industrial, recreational visitors, and trespassers, each of which is defined by a distinct set of exposure pathways and assumptions.
- To estimate the magnitude of the uncertainty associated with estimating the extent of human exposure to site-related chemicals, both MLE and RME exposure scenarios should be quantified.
- The exposure point concentration used to estimate RME exposures is the upper-bound confidence limit on the arithmetic or geometric mean (or the highest detection if UCLs exceed actual

data), while mean data are used to quantify exposure under MLE conditions.

RISK CHARACTERIZATION

Risk characterization involves estimating the magnitude of any adverse health effects that could occur as a result of the calculated level of exposure to the chemicals under study. It thus combines the results of the toxicity and exposure assessments to provide numerical estimates of health risk. Risk characterization also considers the nature and weight of evidence supporting these risk estimates, as well as the magnitude of uncertainty surrounding such estimates. Risks are typically calculated for both current and future land use scenarios and both the RME and MLE exposure scenarios. This allows for a comparison of the range of risks between average (likely) and upper bound (unlikely but possible), allowing for better risk management decisions.

Carcinogenic Health Effects

Cancer slope factors and the estimated daily intake of a chemical, averaged over a lifetime of exposure, are used to estimate the incremental risk that an individual exposed to that carcinogenic compound may develop cancer using the equation:

$$Risk = Intake \times CSF$$

CSFs are chemical-specific values based on carcinogenic dose-response data. Cancer risks from exposure to multiple carcinogens and multiple pathways are assumed to be additive. To obtain an estimate of total risk from all carcinogens at a site, cancer risks are summed across all exposure pathways for potential carcinogens of concern. Risks are deemed acceptable by EPA if they are within the 10^{-6} to 10^{-4} range (one excess case of cancer in one million exposed individuals to one excess cancer in ten thousand exposed individuals.

Since EPA has not derived CSFs for the dermal route of exposure, oral slope factors are often used to quantify risks associated with exposure to carcinogens via this route. It is inappropriate, however, to use oral CSFs to quantify risks for chemicals that might cause skin cancer through direct action at the point of application or contact. This class of skin carcinogens should be evaluated differently than those that cause cancer through a systemic rather than local mechanism (*e.g.*, arsenic).

Most RfDs and CSFs are based on **administered** doses, while estimates of exposure via the dermal route are expressed as **absorbed** doses (see Equation [4]). Thus, to more accurately quantify risks associated with dermal exposures, toxicity values based on orally administered doses can be transformed to absorbed-dose criteria using the following equations for non-carcinogens and carcinogens, respectively:

$$Absorbed\ RfD = Oral\ Administered\ RfD \times Oral\ Absorption\ Fraction$$

$$Absorbed\ CSF = \frac{Oral\ Administered\ CSF}{Oral\ Absorption\ Fraction}$$

Although fractional absorption data are available for few chemicals, values derived from related chemicals can be applied.

Non-carcinogenic Health Effects

Potential health effects of chronic exposure to non-carcinogenic compounds are assessed by calculating a **hazard quotient (HQ)** for each chemical of concern. An HQ is derived by dividing the estimated daily intake by a chemical-specific RfD as shown in this equation:

$$Hazard\ Quotient = \frac{Intake}{RfD}$$

A HQ greater than one indicates that exposure to that chemical may cause adverse health effects in exposed populations. It is important to note, however, that the level of concern associated with exposure to non-carcinogenic compounds does not increase linearly as HQs exceed one. In other words, HQ values do not represent a probability or a percentage. For example, a HQ value of 100 does not indicate that adverse health effects are 10 times more likely to occur than a HQ value of 10. All one can conclude is that HQ values greater than one indicate that non-carcinogenic health impacts are possible and that the more the HQ exceeds unity, the greater the severity of the adverse health effect may be.

Typically, chemical-specific HQs are summed to calculate pathway **hazard index (HI) values**. This approach can result in the situation where HI values exceed one even when no chemical-specific HQs exceed unity (**i.e.,** adverse systemic health effects would be expected to occur only if the receptor were exposed to several chemicals simultaneously). In this situation, chemicals are segregated by similar effect on a target organ, and a separate HI value for each effect/target organ is calculated. If any of the individual HI values exceed one, adverse, non-carcinogenic health effects are possible. In the absence of knowledge about the possible additive, synergistic, or antagonistic effects of simultaneous exposure to multiple compounds, simple additivity is typically assumed if similar target organs or similar mechanisms of toxicity exist. If mechanisms of toxicity and/or target organ effects are different, then it is not appropriate to calculate hazard indices. For example, several elements are known to interact antagonistically and/or synergistically (*e.g.*, selenium ameliorates the toxic effects of cadmium.)

Summary
- Risk characterization involves estimating the magnitude of the potential adverse health effects of the chemicals under study and making summary judgments about the nature of the health threat to the public.
- Cancer risks from exposure to multiple carcinogens and multiple pathways are assumed to be additive. Risks are generally deemed acceptable if they are within the 10^{-6} to the 10^{-4} risk range.
- Potential health effects of chronic exposure to non-carcinogenic compounds are assessed by calculating an HQ for each chemical of concern.
- While an HQ greater than one does indicate that exposure to that chemical may cause adverse health effects in exposed populations, the level of concern associated with exposure to non-carcinogenic compounds does not increase linearly as HQs exceed one.
- If mechanisms of toxicity and/or end-organ effects are similar, then individual HQs for chemicals that affect the same target organ or system are summed to form a HI.

Uncertainty Analysis

Like other forms of mathematical modeling, risk assessment relies on certain assumptions that have varying degrees of accuracy and validity. Variations in the data and models selected are influenced by several factors including data availability, differences in judgment concerning the balance of conservatism and realism, and lead agency requirements. Risk assessments typically use deterministic exposure and toxicity models that produce single-value estimates of risk. These single-value estimates do not consider the potential variability in estimated exposures due to the inherent variability in exposure assessment input factors and major data gaps. As a result, it is important to estimate the magnitude of overall model error. An uncertainty analysis takes into account the variability in measured and estimated parameters, which allows decision makers to better evaluate risk estimates in the context of the assumptions and data used in the assessment.

Qualitative Uncertainty Analyses

An uncertainty analysis can be done qualitatively or quantitatively. Qualitative or semi-quantitative uncertainty analyses are easy to perform, but yield less precise results. An example of a qualitative discussion of uncertainty follows. Major assumptions used in human health risk assessment include:

- Chemical concentrations remain constant over the exposure period.
- Exposures remain constant over time.
- Selected intake rates and population characteristics used (body weight, lifespan, ingestion rates, *etc.*) are representative of the potentially exposed populations.
- All intake of chemicals is from site-related media, and no other sources contribute to the health risk.
- The principal sources of uncertainty associated with this exposure assessment stem from the assumptions used to model the average and upper-bound exposure scenarios. Since receptor-point concentrations cannot be precisely determined, these two scenarios can be used to provide decision makers with a broader perspective on the true range of risks likely to be encountered by exposure to site-related chemicals (while meeting EPA guidelines, if this is relevant).
- The CSF values used in any risk assessment are conservative by virtue of how they are derived. As noted earlier, CSFs represent all upper-bound estimates of the probability that an individual exposed to a potential human carcinogen during a lifetime will develop cancer. This means that the true risks associated with exposure to chemicals are certainly no greater than the estimated risk and are likely to be lower.
- In general, RfD values represent an estimate of the potential toxicity of a chemical with uncertainty spanning one order of magnitude.

- Uncertainties associated with identifying chemicals of concern include those connected with sampling environmental media and those related to the use of small data sets in the statistical evaluation of data.

Some general sources of uncertainty and the effect they may have on estimated risks are outlined in Table 3.

Quantitative Uncertainty Analyses

Uncertainty can be quantitatively characterized by a sensitivity analysis and a probabilistic simulation (*e.g.*, Monte Carlo or Latin Hypercube Simulation). A sensitivity analysis identifies the parameters that contribute most to overall model variability. Specifically, a sensitivity analysis defines the magnitude of effect a given parameter has on model output and ranks each parameter according to its influence on model output (*e.g.*, cancer risks). In a sensitivity analysis, each parameter is varied over its range of plausible values while all other model variables are held constant to observe the effect each parameter has on model outcome. Since all parameters are assumed to have the same distribution, the most influential parameters on model output are the most sensitive. For example, a sensitivity analysis might reveal that chemical-specific permeability coefficients account for 50 percent of the variability in estimated intake by way of dermal exposure. In this way, one can determine which parameters have the greatest impact on model output. Additional research may then be justified to find more realistic or accurate values for these parameters than, for example, EPA-recommended default values. In the case of parameters that have a high degree of influence on model output, it might even be worthwhile to obtain more site-specific data. Performing a sensitivity analysis allows limited resources to be used in a cost-effective manner so that the uncertainty associated with these parameters can be more accurately described or, if possible, reduced, thus reducing overall uncertainty about the magnitude of exposure and risk.

A Monte Carlo analysis calculates a single output value, or model solution, from a set of randomly-selected input variable values. The advantages of using a Monte Carlo analysis include ease of use, applicability to a broad range of model equations, a general lack of restraints on the form of the input distributions used or the functional form of the model equations, and the ability to characterize the output distribution. A model exists as a series of input parameters, each of which can be described as a random variable with an associated probability distri-

Table 3
General Uncertainty Factors

Uncertainty Factor	Effect of Uncertainty	Comment
Use of cancer slope factors.	May overestimate risks.	Slopes are upper 95th percent confidence limits derived from the linearized model. Considered unlikely to underestimate true risk.
Risks/doses within an exposure route assumed to be additive.	May over- or underestimate risks.	Does not account for synergism or antagonism.
Toxicity values derived primarily from animal studies.	May over- or underestimate risks.	Extrapolation from animal to humans may induce error due to differences in pharmacokinetics, target organs, and population variability.
Toxicity values derived primarily from high doses; most exposures are at low doses.	May over- or underestimate risks.	Assumes linearity at low doses. Tends to have conservative exposure assumptions.
Toxicity values.	May over- or underestimate risks.	Not all values represent the same degree of certainty. All are subject to change as new evidence becomes available.
Effect of absorption.	May over- or underestimate risks.	The assumption that absorption is equivalent across species is implicit in the derivation of the critical toxicity values. Absorption may actually vary with species and age.
Effect of including chemicals without toxicity data.	May underestimate risks.	These chemicals are not addressed quantitatively.
Produce pathway.	Lack of site-specific data prohibits accurate plant uptake; models tend to over-predict concentrations in vegetables.	Uptake factors soil and plant specific; percent amount in regular diet is unknown.
Effect of applying critical toxicity values to soil exposures.	May overestimate risks.	Assumes bioavailability of contaminants sorbed onto soils is the same as detected in lab studies. Contaminants detected in studies may be more bioavailable.
Exposures assumed constant over time.	May over- or underestimate risks.	Does not account for environmental fate, transport, or transfer that may alter concentration.

bution. Thus, the first step is to obtain data on the range over which each parameter used in the risk assessment is reasonably likely to vary (for each parameter, both chemical-dependent and chemical independent parameters), so that the proper distribution of values for each parameter can be determined. Most input parameters for estimating chronic daily intake or absorbed dose are available in the technical literature or in EPA guidance documents, and values for these inputs can be expressed as ranges. For some parameters, however, values are not available in the technical literature. For these parameters, estimates are made from existing data or are based on professional judgment and immediate knowledge of the site. Random samples are taken from these distributions and entered into the uncertainty model to yield a single value estimate of exposure or health risk. This process is repeated several times to produce a distribution of model predictions. Hence, as opposed to a single-value estimate that has no characterization of model variability, an uncertainty analysis describes a **range** of reasonable values for each parameter from which a best estimate (**e.g.**, mean or median) and/or an upper-bound estimate (e.g., 95th percentile) can be determined. The frequency distributions generated from an uncertainty analysis can be used to estimate more realistic health risks.

Summary

- Use of single-value estimates, especially worst-case analyses, does not provide useful information to decision makers who must determine if estimated health risks warrant remediation.
- A quantitative uncertainty analysis provides decision makers with an estimate of the range of possible health impacts so that the optimal, risk-based decision can be made.

REFERENCES

Abou-Setta, M.M., R.W. Sorrell, and C.C. Childers, 1986. A computer program in Basic for determining probit and log-probit or logit correlation for toxicology and biology, *Bull. Environ. Contam. Toxicol.*, 36: 242-249.

Andelman, J.B., 1990. *Total Exposure to Volatile Organic Chemicals in Potable Water*, N.M. Ram, R.F. Christman, and K.P. Cantor (Eds.), Lewis Publishers, Chelsea, MI.

Adriano, D.C. 1986. *Trace Elements in the Terrestrial Environment*, Springer-Verlag, New York.

Bidleman, T.F., 1988. Atmospheric processes, *Environ. Sci. Technol.*, 22(4): 361-367.

Briggs, G.C., 1981. Theoretical and experimental relationships between soil adsorption, octanol-water partition coefficients, water solubilities, bioconcentration factors, and the parachor, *J. Agric. Food Chem.*, 29: 1050-1059.

Chiou, C.T., D.W. Schmedding, and M. Manes, 1982. Partitioning of organic compounds in octanol-water systems, *Environ. Sci. Technol.*, 16(1): 4-10.

Chou, S.F.J., and R.A. Griffin, 1986. Solubility and soil mobility of polychlorinated biphenyls, In *PCBs and the Environment*, Vol. I, J.S. Waid, Ed., CRC Press, Inc., Boca Raton, Florida, pp. 101-120.

Cohen Y., and P.A. Ryan, 1985. Multimedia modeling of environmental transport: Trichloroethylene test case, *Environ. Sci. Technol.*, 19: 412-417.

Crump, K.S., H.A. Guess, and K.L. Deal, 1977. Confidence intervals and tests of hypotheses concerning dose response relations inferred from animal carcinogenicity data, *Biometrics*, 33: 437-451.

Dobbs, A.J., and M.R. Cull, 1982. Volatilization of chemicals—Relative loss rates and the estimation of vapor pressures, *Environ. Pollut.*, (Series B), 3: 289-298.

Dowdy, R.H., and W.E. Larson, 1975. The availability of sludge-borne metals to various vegetable crops, *J. Environ. Qual.*, 4: 278.

Dragun, J., and A. Chiasson, 1991. *Elements in North American Soils*, Hazardous Materials Control Research Institute, Maryland.

Eberhardt, L.L., R.O. Gilbert, H.L. Hollister, and J.M. Thomas, 1976. Sampling for contaminants in ecological systems, *Environ. Sci. Technol.*, 10: 917-925.

Finney, D.J, 1952. *Probit Analysis: A Statistical Treatment of the Sigmoid Response Curve*, 2nd Ed., Cambridge University Press, London.

Fries, G.F., 1987. Assessment of potential residues in foods derived from animals exposed to TCDD-contaminated soil, *Chemosphere*, 16(8/9): 2123-2128.

Gansecki, M., 1991. Comments on the "Background Geochemical Characterization Report for 1989 - Rocky Flats Plant". Mike Gansecki, August 1, 1992.

Geyer, H., P. Sheehan, D. Kotzias, D. Freitag, and F. Korte, 1982. Prediction of ecotoxicological behavior of chemicals: Relationship between physicochemical properties and bioaccumulation of organic chemicals in the mussel *Mytilus edulis*, *Chemosphere*, 11(11): 1121-1134.

Geyer, H., Scheunert, I., and Korte, F., 1986. Bioconcentration potential of organic environmental chemicals in humans, *Regul. Toxicol. Pharmacol.*, 6: 313-347.

Geyer, H.J., I. Scheunert, and F. Korte, 1987. Correlation between the bioconcentration potential of organic environmental chemicals in humans and their n-octanol/water partition coefficients, *Chemosphere*, 16(1): 239-252.

Gilbert, R.O. 1987. *Statistical Methods for Environmental Pollution Monitoring.* Van Nostrand Reinhold, NY.

Gilliom, R.J. and D.R. Helsel, 1984. *Estimation of Distributional Parameters for Censored Trace-Level Water Quality Data*, U.S. Geological Survey Open File Report 84-729.

Grossjean, D, 1983. Distribution of atmospheric nitrogenous pollutants at a Los Angeles area smog receptor site, *Environ. Sci. Technol.*, 17(1): 13-19.

Hattemer-Frey, H.A. and C.C. Travis, 1991. Benzo(a)Pyrene: Environmental partitioning and human exposure, *Toxicol. Industr. Health*, 8(3): 217-219.

Hoffman, H.O., and Gardner, R.H., 1983. Evaluation of uncertainties in radiological risk assessment models, In *Radiological Assessment*, Till, J.E., and H.R. Meyer, Eds., NUREG/CR-3332, U.S. Nuclear Regulatory Commission, Washington, D.C.

Hushon, J.M., A.W. Klein, W.J.M. Strachan, and F. Schmidt-Bleeh, 1983. Use of OECD premarket data in environmental exposure analysis for new chemicals, *Chemosphere*, 12(6): 887-910.

Isensee, A.R., and G.E. Jones, 1971. Absorption and translocation of root and foliage applied 2,4-dichlorophenol, 2,7-dichlorodibenzo-p-dioxin, and 2,3,7,8-dichlorodibenzo-p-dioxin, *J. Agric. Food Chem.*, 19(6): 1210-1214.

Jamall, I.S., A.S. Pacasad, Ed., A.R. Liss, 1988. Modulation of peroxidative injury by dietary trace elements. Cadmium cardiotoxicity as a model, In: *Essential and Toxic Trace Elements in Human Health and Disease*, New York, pp. 415-430.

Kenaga, E.E., 1980. Correlations of bioconcentration factors in aquatic and terrestrial organisms with their physical and chemical properties, *Environ. Sci. Technol.*, 14: 553-556.

Kenaga, E.E., and C.A.I. Goring, 1980. Relationship between water solubility, sorption, octanol-water partitioning, and concentration of chemicals in biota, In: *Aquatic Toxicology*, J.E. Eaton, P.R. Parrish, and A.C. Hendricks, Eds., American Society for Testing Materials STP 707, Philadelphia.

Lewis, S.C., J.R. Lynch, and A.I. Nikiforov, 1990. A new approach to deriving community exposure guidelines from "No-Observed-Adverse-Effect Levels," *Regul. Toxicol. Pharmacol.*, 11: 314-330.

Lieberman, H.R., 1983. Estimating LD_{50} using the probit technique: A basic computer program, *Drug Chem. Toxicol.*, 6(1): 111-116.

Lyman, W.J., W.F. Reehl, and D.H. Rosenblatt, 1982. *Handbook of Chemical Property Estimation*, McGraw-Hill, New York.

Menzer, R.E., and J.O. Nelson, 1980. Water and soil pollutants, In: *Toxicology*, J. Doull, C.D. Klaassen, and M.D. Amdur, Eds., MacMillan Press.

Office of Science and Technology Policy (OSTP), 1985. Chemical carcinogens: A review of the science and its associated principles, *Fed. Reg.*, 50: 10372-10422.

Porter, P.S., R.C. Ward, and H.F. Bell, 1988. The detection limit, *Environ. Sci. Technol.*, 22(8): 856-861.

SAS Institute Inc. (SAS), 1985. *SAS User's Guide: Statistics*, Version 5, Cary, North Carolina.

Sette, A., L. Adorini, E. Marubini, and G. Doria, 1986. A microcomputer program for probit analysis of interleukin-2 (IL-2) titration data, *J. Immunolog. Methods*, 86: 265-277.

Shacklette, H.T., and J.G. Boerngen, 1984. *Elemental Concentrations in Soils and other Surficial Materials of the Conterminous United States*, U.S. Geological Survey Professional Paper 1270, U.S. Government Printing Office, Washington, D.C.

Shor, R.W., C.F. Baes, III, and R.D. Sharp, 1982. *Agriculture Production in the United States by County: A Compilation of Information from the 1974 Census of Agriculture for Use in Terrestrial Foodchain Transport and Assessment Models*, ORNL-5763, Oak Ridge National Laboratory, Oak Ridge, TN.

Swenson, M.J., 1980. *Duke's Physiology of Domestic Animals*, 8th Edition, Comstock Publishing, London.

Travis, C.C. and M.L. Land, 1990. Estimating the mean of data sets with nondetectable values, *Environ. Sci. Technol.*, in press.

Travis, C.C., J.W. Dennison, and A.D. Arms, 1987. The extent of multimedia partitioning of organic chemicals, *Chemosphere*, 16(1): 117-125.

U.S. Environmental Protection Agency (EPA). 1986. *Guidelines for Carcinogen Risk Assessment*. Federal Register 51:33992-34003.

_____. 1989. *Risk Assessment Guidance for Superfund Volume I, Human Health Evaluation Manual (Part A)*, Office of Emergency and Remedial Response, Washington, DC.

_____. 1990a. *Exposure Factors Handbook*, EPA/600/8-82/043, Office of Health and Environmental Assessment, Washington, DC.

_____. 1990b. *Guidance for Data Usability in Risk Assessment*, EPA/540/G-90/008,

_____. 1991a. *Risk Assessment Guidance for Superfund. Volume I - Human Health Evaluation Manual, Part B: Development of Risk-Based Preliminary Remediation Goals*, OSWER Directive 9285.7-01B, Office of Solid Waste and Emergency Response, Washington, DC.

_____. 1991b. *Risk Assessment Guidance for Superfund. Volume I - Human Health Evaluation Manual, Part C: Risk Evaluation of Remedial Alternatives*, OSWER Directive 9285.7-01C, Office of Solid Waste and Emergency Response, Washington, DC.

_____. 1991c. *Risk Assessment Guidance for Superfund, Volume 1, Human Health Evaluation Manual, Supplemental Guidance "Standard Default Exposure Factors*, Draft Final, March 25, 1991, OSWER Directive 9285.6-03, Office of Solid Waste and Emergency Response, Washington, DC.

_____. 1992a. *Guidelines for Exposure Assessment. Fed. Reg.*, 57(104): 22888-22938.

_____. 1992b. *Registry of Toxic Effects for Alpha-Chloralose*, National Library of Medicine (Database Online), Office of Research and Development, Environmental Criteria and Assessment Office.

_____. 1992c. *Supplemental Guidance to RAGS: Calculating the Concentration Term*. Publication 9285.7-081, Office of Emergency and Remedial Response, Washington, DC.

Weil, C.S., 1972. Statistics versus safety factors and scientific judgment in the evaluation of safety for man, *Toxicol. Appl. Pharmacol.*, 21: 454-463.

Weil, C.S., and D.D. McCollister, 1963. Relationship between short and long-term feeding studies in designing an effective toxicity test, *Agric. Food Chem.*, 11: 486-491.

Woodfork, K., and R. Burrell, 1985. A basic computer program for calculation of CH_{50} values by probit analysis, *Computers. Biol. Med.*, 15(3): 133-136.

Yang, Y-Y., and C.R. Nelson, 1986. An estimation of daily food usage factors for assessing radionuclide intake in the U.S. population, *Health Phys.*, 50(2): 245-257.

3

ECOLOGICAL RISK ASSESSMENT

Janet E. Kester, Holly A. Hattemer-Frey, and Gary R. Krieger
Dames & Moore

OVERVIEW

While the EPA has provided agency-wide guidance for performing human health risk assessments since 1986, attempts to develop rationale and consensus for a set of definitive guidelines for the protection of the environment have begun only recently. **Ecological risk assessment (ERA)** is defined as a process that evaluates the likelihood that adverse ecological effects may occur due to exposure to one or more **stressors.** The term stressor describes any chemical, physical, or biological entity or condition that has the potential to induce adverse effects on any ecological component, which might be individuals, populations, communities, or ecosystems.

Although the paradigm being developed for ERA borrows much from the relatively well validated human health risk assessment process, the two also differ in significant (and necessary) ways. First, the subject of human health risk assessment is the human individual, but ERAs may focus on any one or any combination of ecological components. Second, they focus on different **endpoints** (defined as characteristics of an ecological component that may be affected by exposure to a stressor). The endpoints of human health assessments are relatively limited and well-defined (e.g., cancer, systemic toxicity, developmental or reproductive effects), but there are no universally appropriate indices of ecosystem "health" that can be applied in all ERAs. As a result, assessment endpoints must be selected for each site on the basis of scientific, social, and political considerations (which may have little in common).

In addition, ecological risk assessors must be aware of the potential effects of not only chemicals, but also of physical and biological agents on ecological components. **Physical stressors** include global phenomena such as ozone depletion, as well as local and regional phenomena such as habitat destruction or alteration by natural events (drought, fire) or human activities (construction, farming), and extremes of natural conditions (*e.g.*, temperature, moisture, water level and flow rate). **Biological stressors** include disease and predation. Although current ERAs focus primarily on chemical stressors, recent studies have shown that the scale of environmental problems is increasing from local to regional to global, and that nonchemical stressors may be more significant than chemical (EPA, 1990).

The principles and methods proposed in this chapter are derived from or consistent with available guidance, including EPA's *Framework for Ecological Risk Assessment, Ecological Assessment at Hazardous Waste Sites, Ecological Assessment of Hazardous Waste Sites: A Field and Laboratory Reference Document, Risk Assessment Guidance for Superfund, Volume II, Environmental Evaluation Manual, Summary of Ecological Risks, Assessment Methods, and Risk Management Decision in Superfund and RCRA,* and the Oak Ridge National Laboratory *User's Manual for Ecological Risk Assessment.* Additional guidance documents will be published in the next few years.

CURRENT APPROACHES FOR ERA

A wide range of toxicological methodologies exists to examine the effects of chemicals on individual organisms, tissues, cells, and (with the advent of molecular biology) molecules, but the impacts of chemicals on the environment remain far more difficult to evaluate. This difficulty arises not only from a lack of methods for measuring environmental impacts, but also of an adequate scientific and conceptual basis for framing the questions to be investigated. It is exacerbated by legal, financial, and regulatory constraints.

The type of approach selected for ERA depends upon: (1) the complexity of the site, (2) the amount of detail required to support the regulatory decision making process, and (3) resources available to conduct the assessment. The state of the science at present combines field and laboratory approaches to determine what adverse effects may be caused by exposure to chemical stressors:

- **Field surveys.** Visual inspection of a site may reveal impacts on ecosystem parameters such as the abundance, diversity, density, etc. of certain species.
- **Calculation of toxicity quotients (TQs).** Analogous to the **hazard quotient (HQ)** (ratio of daily intake to reference dose or risk-specific dose) used in human health risk assessments, the TQ is calculated as the ratio of measured or modeled COC concentrations in specific media to concentrations (in the same media) that are thought to have no adverse effects. Calculation of TQs is usually used to screen out relatively "clean" areas from further consideration in an ERA.
- **Toxicity testing.** Laboratory or *in situ* bioassays are conducted with site media to measure the effects of site-related chemicals on laboratory or resident test organisms.

These approaches represent increasing levels of complexity and cost; each may be useful in different contexts. To adequately and reliably characterize potential chronic adverse effects on exposed populations, guidance suggests that at least two (and preferably all three) should be performed.

EXPOSURE ASSESSMENT

The objective of the exposure assessment is to estimate the magnitude, frequency, duration, and route of exposure to site-related chemicals by ecological receptors. Accomplishing this task involves the following fundamental steps:

- Characterize the site.
- Identify relevant exposure pathways for terrestrial and aquatic receptors.
- Select key ecological receptors.
- Develop a list of **chemicals of potential concern (COCs)** based on results of monitoring data and the potential toxicity of chemicals to key ecologic receptors.
- Define site-specific assessment and measurement endpoints.
- Calculate dose/intake estimates for key species.

- Using appropriate exposure-response models, characterize potential adverse impacts.

Characterization of the Site

The terrestrial and aquatic conditions of the site and surrounding areas that may be influenced by present and past site activities should be described. Predominant land uses should be considered, along with the physical, geological, and hydrological features of the site. The presence of threatened and endangered (T&E) species and species "in need of management" and their critical habitats should also be investigated. Information on local species and habitat types can usually be gathered from appropriate state, county, and federal sources.

Identification of Relevant Exposure Pathways

Complete exposure pathways are depicted in a site conceptual model. To be complete, an exposure pathway must include a source, a mechanism of chemical release, retention and transport media, a point of potential biota contact, and an exposure route(s) at the contact point. Potential routes by which terrestrial receptors could be exposed to site-related chemicals include: (1) inhalation of vapors or chemicals associated with fugitive dust; (2) ingestion of affected soil, vegetation, prey, and/or surface water; and (3) dermal absorption of chemicals present in soil and/or water. Only those pathways that are complete, and are expected to contribute substantially to exposures by key terrestrial or aquatic organisms, need be addressed. For example, because the diet is expected to be a major source of exposure to highly lipophilic chemicals such as DDT, the relatively small doses received via vapor inhalation or dermal contact with soil may not need to be quantitatively evaluated.

Exposure media relevant to aquatic receptors are surface waters and sediments. Exposure routes for fish are limited to respiration (*i.e.*, uptake of COCs over the water/gill interface) and ingestion of contaminated food (prey) and sediment while foraging. Benthic invertebrates and bottom feeding fish tend to take up COCs by respiration, by feeding on algae attached to substrate particles, and by inadvertent ingestion of sediment during feeding. Therefore, some chemicals contained within the sediments may be retained by benthic invertebrates and consumers of these organisms.

Selection of Ecological Receptors of Concern

Environmental receptors are organisms that may be exposed to chemicals from a site. Ecological *receptors of concern* (ROC) are defined as species that play key roles in ecosystem structure and/or function. ROCs are identified by considering the relevant exposure pathways and the potential or known occurrence of species exposed via those pathways.

Since energy and matter flow through ecosystems in food webs, ROCs can be identified through understanding of the site food web. A food web describes the transfer of matter and energy from one trophic level or organism to another. Food webs can be delineated in hierarchies or trophic levels as follows:

- Primary producers - photosynthetic plants;
- Primary consumers - plant eaters (herbivores/granivores);
- Secondary consumers - herbivore eaters (carnivores);
- Tertiary consumers - eat other carnivores (top carnivores); and
- Decomposers - feed on dead or decaying organisms.

Potential ROCs may include:

- ***Species that are vital to the structure and function of the food web (i.e., principal prey species or species that are major food items for principal prey species).*** In general, loss of a few individuals of a species is unlikely to significantly diminish the viability of the population or disrupt the community or ecosystem of which it is a part. As a result, the fundamental unit for ecological risk assessment is generally the population rather than the individual, with the exception of T&E species. However, significant impacts on species that occupy critical positions in the food web structure may ramify throughout the ecosystem, potentially resulting in disruption of higher trophic level populations that depend upon the affected population for survival and/or stability.

- ***Species that exhibit a marked toxicological sensitivity to the COCs.*** Ecosystem function can be impaired if certain component species are particularly vulnerable to chemical exposure. Selection of ROCs is thus designed to ensure that benchmark criteria are protective of the most sensitive organisms actually present at the site.

- ***Species that have unique life histories and/or feeding habits.*** Significant impacts on such species might eliminate unique ecological niches, with unpredictable results on the ecosystem as a whole.
- ***Species for which toxicological data are readily available in the scientific literature.*** While such species may not be "key" in the sense of occupying a critical ecological niche, the availability of data on their responses to COCs reduces uncertainty in the evaluation as a whole.

Selection of key receptor species is designed to minimize the possibility that other species could be more exposed than the key species themselves and to include representation of sensitive organisms present at the site. Although each key species selected may not necessarily meet all of the criteria defined above, selection of key species collectively is intended to meet all criteria.

T&E species are not included on the list of potential ROCs, as they are given separate consideration. Potential impacts to T&E species must be evaluated if these species occur within site boundaries or there is reasonable potential for these species to occur in or use some portion of the site (*i.e.*, critical habitat exists within the site). The ERA should assess impacts to individuals and loss of critical habitat for those T&E species that could occur on-site or have critical habitat on or near the site. This approach is based upon the premise that loss or injury to even one individual (or any other impact that would adversely affect the species' recovery rate) is unacceptable.

Identifying Chemicals of Potential Concern

Identifying COCs takes into account: (1) the concentration of chemicals present in various environmental media; (2) their predicted mobility, persistence, bioaccumulation, and potential transformation in the environment; and (3) their observed toxicological hazards. A three-phase screening process similar to that used for human health risk assessment is recommended to identify COCs in on-site media that may pose adverse impacts to environmental ROCs. However, it should be noted that COCs for the human health and ecological risk assessments may be different. The first phase involves the screening of detected compounds after detection limits, detection frequencies, and blank samples are examined. Compounds not detected in any site samples will be eliminated from further consideration, while if a compound is detected at an order of magnitude or more above the detection limit, it will be retained regardless of its frequency of occurrence. Concentrations detected in blanks will then be compared with concentrations detected in environmental samples. Environmental sample results are considered positive only if the concentration in the site sample is five times the maximum concentration detected in any blank sample (ten times for common laboratory artifacts).

The second phase involves comparing estimated concentrations in samples to media-specific background data. If the mean concentration of a compound in a given medium is statistically higher (at the 95% level of statistical confidence) than mean background levels as determined by a one-tailed t-test, that constituent will be considered a COC. As a result of this process, compounds likely to be site-related and detected at statistically significant concentrations will be selected as COCs.

Thirdly, concentrations of those chemicals identified in the first two steps will be compared to reference toxicity data (*i.e.,* ambient water quality criteria) found in the literature. Those chemicals whose 95 percent **upper confidence limit (UCL)** on the arithmetic mean or maximum concentration exceed the lowest concentration reported to be toxic will be considered a COC.

Finally, the list of COCs may be narrowed for some groups of compounds, such as **polycyclic aromatic hydrocarbons (PAHs),** by selecting indicator chemicals to represent the group. One or two compounds from each group will be chosen based on toxicity and other relevant characteristics. These compounds will then represent the group. As a screening level assessment, the concentration of all compounds within that group will be summed and toxicity values for the indicator chemical will be compared to the sum of concentrations for the group. This method introduces additional conservatism into the assessment by assuming all compounds in the group are as toxic as the indicator chemical. Obtaining toxicity data on various related compounds, if available, is likely to lower risk estimates.

Results classified as non-detects are typically treated as equal to one-half the detection limit and included in the calculation of mean and upper-bound (95 percent UCL) estimates. This method tends to bias estimates of the mean and standard deviation. Non-detect data from WFS sites should therefore be handled, if possible, using standard, EPA-approved statistical techniques for left-censored data sets. **Log-probit analysis** has been shown to be a robust method for estimating the geometric mean and standard deviation of samples with values in the nondetectable range. In a log-probit analysis, all measurements (both detects and nondetects) are assumed to be samples taken from the same lognormal probability distribution. When samples from a lognormal distribution are plotted on a probit scale, they tend to lie on a straight line. In a probit analysis involving both detects and nondetects, the nondetects are treated as unknowns, but their percentile values are accounted for. Thus, if there are 100 samples, 30 of which are nondetects, the first detectable data point would be plotted at the thirty-first percentile. If sufficient detectable values exist, they can be used to establish the line, using linear regression, that characterizes the entire data set. The geometric mean concentration for the data set (both detects and nondetects) is the 50th percentile value. Thus, a probit analysis allows the geometric mean to be extrapolated from a data set even if it is below detection limits (provided there are sufficient detected values to define the probability distribution). Available probit tables and computerized programs facilitate the use of either log-probit or maximum likelihood estimation analyses in the risk assessment process.

Selection of Ecological Endpoints

Ecological endpoints (EEs) are characteristics of an ecological system that may be affected by site-related COCs and as such epitomize the actual environmental values to be protected. Meaningful EEs are those that characterize the relationship between chemical levels (environmental concentrations) and potential adverse effects.

The term EE has frequently been used to define two related but distinct concepts in ERAs. **Assessment endpoints** (AEs) are "formal expressions of the actual environmental values that are to be protected." Thus, AEs are environmental characteristics that, if shown to be adversely effected, indicate the need for remediation at a hazardous waste site. Measurement endpoints (MEs) are "quantitative expressions of an observed or measured effect of a hazard." Thus, MEs are measurable or quantifiable characteristics that are or can be directly related to an AE and provide the means by which risk assessors can determine if EEs have been significantly affected. MEs can be measured in the field or laboratory (*e.g.*, relative abundance measures) or can be summaries of relevant data reported in the scientific literature (*e.g.*, LC_{50} values). Clearly defined AEs are essential to provide a definitive assessment of the potential toxic effects of COCs.

One ecologically relevant endpoint for most sites is the existence of sufficient, viable, self-sustaining populations of lower-trophic level organisms (*e.g.*, prey). The presence of these organisms is important since they represent the prey base for numerous higher trophic level predators, who may be permanent or frequent inhabitants of the site. Quantitative EEs for aquatic receptors are primarily directed toward fish and key benthic organisms, which serve as a food base for higher level species. If aquatic invertebrates and/or periphyton are sufficiently contaminated, the effects associated with exposure to this contamination could be passed on to fish (*i.e.*, a viable, self-sustaining benthic population must exist to support an associated fish population).

EEs for individuals can include changes in typical death, growth, and fecundity rates or changes in tissue concentrations and behavior. Endpoints for populations include alterations in occurrence, abundance, behavior, reproductive performance, and age/size class structure. Ecologic community endpoints can be assessed by evaluating the numbers of species as well as species diversity and relative abundance indices. Ecosystems endpoints, as a whole, include biomass, productivity, or nutrient dynamics.

Fate and Transport Analysis

Since the movement and transfer of COCs within and among various environmental media can substantially influence the nature and extent of exposure, it is necessary to employ various measurement or predictive techniques to evaluate their fate and transport in the environment. To assist in evaluating fate and transport of chemicals and final exposure concentrations, information on the physical and chemical properties, including persistence, solubility, bioaccumulation potentials, and partitioning characteristics of COCs at the site should be collected. These data, which may be used to estimate changes in concentrations over time and to evaluate cross-media transfer of chemicals, are described in detail in the chapter on human health risk assessment.

Estimating Exposure Point Concentrations

Accurate estimates of chemical concentrations at points of potential exposure are a prerequisite for evaluating chemical intake in potentially exposed receptors. Actual concentrations as determined from sampling data should be used whenever possible.

EPA defines the exposure point concentration used to estimate **reasonable maximum exposure (RME)** as the 95% UCL on the arithmetic mean. An important complication in calculating the RME concentration is that a statistical distribution (usually either normal or lognormal) must be assumed. Supplemental guidance for risk assessment cites EPA's experience that most large environmental data sets are lognormally rather than normally distributed. In this case, the UCL can be calculated using a method presented in **Statistical Methods for Environmental Pollution Monitoring.** However, UCL values based on the lognormal model are often an order of magnitude (ten times) or more above the range of observed concentrations, for data sets that exhibit high variability. This is because the UCLs are extrapolations representing the average concentrations associated with extreme long-tailed (highly skewed) distributions that could have generated the observed sample data with 5% probability. Therefore, where the UCL is higher than any measured value, a more reasonable approach may be to consider the maximum concentration detected for any composite area. Because of the uncertainty associated with estimating the extent of ROCs' exposure to site-related chemicals, it may be useful to follow human risk assessment guidance in developing average as well as RME exposure scenarios. The **RME scenario** is defined as the **highest** exposure that is reasonably expected to occur at a site (*i.e.*, well above average). Since this scenario uses a mix of conservative assumptions and data, the uncertainty associated with its results is unquantifiable, but may be substantial. It should therefore be regarded as an upper bound on exposure and risk.

To provide some estimate of what more common exposures and risks might be, an average or **most likely exposure (MLE) scenario** using mean exposure assumptions should also be considered. Mean chemical concentrations are used to estimate exposure point concentrations. The advantage of evaluating both exposure scenarios is that they provide decision makers with a broader perspective on the true range of risks likely to be posed to ROCs exposed to site-related chemicals. The use of a point estimate of risk, especially an upper-bound estimate, does not provide adequate information to risk managers who must decide if the environmental risks are excessive and therefore if the site warrants remediation.

Estimating Chemical Intakes

The objective of this task is to estimate the magnitude, frequency, duration, and route of exposure to site-related chemicals by ecological ROCs. Accomplishing this task involves estimating total chemical intake by potentially exposed receptors for complete and significant pathways of exposure using the previously defined exposure-point concentrations.

Potential exposures for terrestrial ROCs are determined based on the species' life history and feeding habits. Quantification of exposures involves the use of species-specific numerical exposure factors including body weight, ingestion rate, and fraction of prey, water, and soil consumed from the contaminated area. Both RME and MLE parameter values should be obtained for each ROC. Intakes (mg/kg-day) can be estimated using the equation:

$$I = \frac{(CP \times QP \times FI) + (CA \times QA \times FI) + (CS \times QS \times FI) + (CW \times QW \times FI)}{BW}$$

where,
I=the daily intake of a COC (mg/kg-day)
CP=the measured concentration of COC in whole body prey (mg/kg);
QP=the quantity of prey ingested (kg/day);
CA=the predicted concentration of COC in air (mg/m^3);
QA=the quantity of air inhaled (m^3/day);
CS=the measured concentration of COC in surface soils (mg/kg);
QS=the quantity of soil ingested (kg/day);
CW=the measured concentration of COC in on-site surface water (mg/L);
QW=the quantity of water ingested (L/day);
FI=the fraction of material inhaled or ingested from the site;
BW=species-specific body weight (kg).

Species-specific values for QP, QS, QW, and QA as well as the estimated or measured concentration of COCs in air, surface water, prey, and soil are used to quantify intakes by key terrestrial receptors.

It is not usually necessary to calculate doses for aquatic receptors, since toxicological assessments are done by comparing the measured concentration of COCs in surface water with species-specific and hardness-specific toxicity data obtained from the literature versus Ambient Water Quality Criteria (AWQC). AWQC are not appropriate for all sites, since criteria values for most compounds are based on a water hardness of 100 mg $CaCO_3$/L and are derived using sensitive species (*e.g.*, salmonids) that may not occur within site surface water bodies. As a result, using toxicity data that are based on species known or suspected to occur on-site (versus the most sensitive species) to derive site-specific factors can reduce the uncertainty of risk estimates.

Summary

- Potential routes by which terrestrial receptors could be exposed to site-related chemicals include: (1) inhalation of vapors or chemicals associated with fugitive dust; (2) ingestion of soil, vegetation, prey, and/or surface water; and (3) dermal contact with affected media.
- Exposure routes for fish inhabiting on-site surface water bodies are limited to respiration (*i.e.*, uptake of COCs over the water/gill interface) and ingestion of contaminated food (prey) and sediment while foraging. Benthic invertebrates and bottom feeding fish tend to take up COCs by respiration, by feeding on algae attached to substrate particles, and by inadvertent ingestion of sediment during feeding.
- Key species are those that are (1) vital to the function of the food web, (2) exhibit marked toxicology sensitivity to COCs, (3) have unique life histories and/or feeding habits, and/or (4) are relatively well-defined in terms of toxicological responses.
- For T&E species that assess impacts to individuals and loss of critical habitat for those species that could occur on-site or have critical habitat on or near the site should be assessed.
- Selection of COCs involves three phases. The first phase includes the screening of chemicals after detection limits, detection frequencies, and blank samples have been examined, while the second phase involves comparing estimated metals concentra-

tions in various environmental media (*e.g.*, air, soil, and water) to media-specific background data. Thirdly, those chemicals whose 95% UCL or maximum concentration exceed the lowest concentration reported to be toxic considering only chronic effects will be considered a COC.

- Use of log-probit, maximum likelihood estimation, or other analyses to estimate the mean and standard deviation of a data set with values in the nondetectable range can increase the accuracy of risk estimates.
- EEs are characteristics of an ecological system that may be affected by site-related COCs and as such epitomize the actual environmental values to be protected.
- Since chemicals tend to end up in the media in which they are most soluble, a few basic physicochemical properties can be used to predict the behavior and fate of chemicals released into the environment.
- Given the uncertainty associated with estimating doses by ecological receptors, developing a RME and a MLE exposure scenario is recommended.
- The exposure point concentration used to estimate RME exposures is the upper-bound confidence limit on the arithmetic or geometric mean, while mean data are used to quantify exposure under MLE conditions.
- Typically, the geometric mean is much lower than the arithmetic mean and, therefore, will yield lower risk estimates.

TOXICITY ASSESSMENT

Toxicity assessment consists of (1) hazard evaluation, and (2) dose-response assessment. The hazard evaluation involves a comprehensive review of toxicity data for COCs to identify the nature and severity of toxic properties, especially with respect to ROCs or similar species. Dose-response assessment allows prediction of the amount of chemical exposure that may result in adverse ecological effects.

Once the potential toxicity of a chemical has been established, the next step is to determine the amount of chemical exposure that may result in adverse ecological effects. Thus, the toxicity assessment evaluates the increased likelihood of adverse ecological effects as a result of exposure to site-related chemicals. Since the goal of the ERA is to evaluate long-term population impacts, chronic toxicity values for ecological receptors will be used, since they are a better measure of long-term impacts than are acute data. Sources of toxicity values to evaluate potential adverse effects to plants and terrestrial and aquatic organisms could include Fish and Wildlife Service (FWS) Contaminant Hazard Series Synoptic Reviews, the Agency for Toxic Substance and Disease Registry (ATSDR) Toxicological Profiles, EPA Health Effects Assessment and Water Quality Criteria documents, and other current toxicological literature.

Toxicity Criteria for Aquatic Organisms

AWQC for protection of aquatic organisms are generally based on toxicity data from a range of organisms, including highly sensitive species. The national criteria serve as a basis for each state's own water quality standards. Based on a broad range of toxicological studies using a wide range of different water column organisms and life stages, these criteria were intended to ensure protection of 95% of all aquatic organisms from adverse effects. Chronic AWQC can be used to determine if aquatic life (invertebrates and vertebrates) exposed to site-related COCs in surface water are likely to experience chronic adverse effects. Deriving toxicity data that have been normalized to the mean hardness of site waters is necessary to reflect actual site conditions that may influence the health of ecological receptors, since potential toxic effects of some COCs (metals in particular) to aquatic organisms is highly dependent upon water hardness. Hardness should not be the only criterion used in selecting appropriate toxicity data. Obtaining data on representative species (*i.e.,* those species known to occur in on-site streams or ponds) should also be a concern. The following stepwise procedure is therefore recommended for calculating TQs for aquatic ROCs at specific sites:

- Compare measured/predicted COC concentrations in surface water with AWQC as promulgated. If no exceedances are observed (i.e., TQ 1), then no further analysis is needed, as adverse effects are not expected.
- If the concentration of any COC does exceed its AWQC, then site-specific criteria applicable to species that are known or reasonably expected to occur at the sites should be developed.

Although AWQC are intended to protect the large majority of aquatic organisms, environmental damage has occurred even in areas where AWQC are met. In such cases, the primary source of exposure is **sediment**. Sediments are an important component of aquatic ecosystems because of the habitat they provide for **benthic** (sediment-dwelling) organisms, including commercially important organisms such as shrimp, crayfish, lobster, crab, mussels, and clams, as well as species important in the environmental food web, including many species of worms, amphipods, bivalves, and insects.

Because of the tendency of both organic and inorganic molecules to adsorb to particles, sediments are also repositories for a wide variety of potentially toxic chemicals, both naturally occurring and anthropogenic. Chemicals undetected or found in only trace amounts in the water column can accumulate to high levels in sediments. While this "sink" effect serves to isolate chemicals from aquatic organisms, the sediment can also serve as a reservoir, slowly releasing chemicals from sediment to pore water to organisms. Thus, for the large number of commercial and food chain organisms that spend a portion of their life cycles close to the sediment, sediment provides chemical exposure as well as nurture. In fact, direct transfer of chemicals from sediments to organisms is now considered to be a major route of exposure for many species. Further, consumption of benthic organisms by higher trophic level animals, such as fish-eating birds and humans, can be a major pathway of exposure to chemicals otherwise isolated in the sediment sink.

AWQC may not be protective of benthic species (and ultimately of predators that feed on them) because they are principally based on studies with water column organisms that live and feed in the overlying waters as opposed to benthic organisms that live and feed in association with sediments. As a result, attention has shifted from point to non-point sources of pollution, and protection of sediment quality has emerged as a necessary extension of water quality protection. The EPA has proposed an equilibrium partitioning approach for derivation of chemical-specific sediment quality criteria (SQC) for non-ionic organic chemicals. This method calculates an organic carbon-normalized SQC (SQC_{oc}) from WQC (the assumed pore water concentration) using the organic carbon:water partition coefficient K_{oc}:

$$SQC_{oc} = AWQC \times K_{oc}$$

where

SQC_{oc} = organic carbon-normalized SQC
K_{oc} = chemical-specific organic carbon-water partition coefficient

It allows for differences in chemical bioavailability at different sites by multiplying the generic SQC_{oc} by the site-specific fraction of organic carbon (f_{oc}):

$$SQC_{site\text{-}specific} = SQC_{oc} \times f_{oc}$$

Because organic carbon does not control the adsorption of metals or ionic organic chemicals to sediments to the same extent as non-ionic organic chemicals, this approach is only applicable to non-ionic organics for which AWQC have been developed. Efforts to develop other normalization methods for these other chemical groups are currently in progress.

Toxicity Criteria for Terrestrial Organisms

There are currently no toxicity criteria promulgated for application to terrestrial ecological receptors. Application of toxicity data derived from surrogate species and for endpoints (effects) other than chronic LOAELs or NOAELs introduces uncertainty into the risk assessment. The magnitude of this uncertainty depends largely upon (1) the degree of taxonomic difference between the key and test species; (2) the conditions under which the toxicity data were established; and (3) the endpoint of interest (*e.g.*, chronic LOAEL or NOAEL) and the endpoint measured (*e.g.*, death). Uncertainties associated with extrapolation of toxicity information from literature to site conditions can be offset by applying various **uncertainty factors (UFs)** to the laboratory or literature endpoint values to calculate **toxicity reference values (TRVs),** defined as an exposure estimate for a receptor taxon, including sensitive subgroups, that is not likely to cause appreciable deleterious effects from chronic exposure. The approach developed for a large Superfund site, the Rocky Mountain Arsenal, could be adapted for most species, chemicals, and exposure conditions.

Summary

- Toxicity assessment evaluates the nature and extent of health effects that may result from exposure to site-related chemicals.

- Sources of toxicity values to evaluate potential adverse effects to plants and terrestrial and aquatic organisms could include Fish and Wildlife Service (FWS) Contaminant Hazard Series Synoptic Reviews, the Agency for Toxic Substance and Disease Registry (ATSDR) Toxicological Profiles, EPA Health Effects Assessment and Water Quality Criteria documents, and other current toxicological literature.
- AWQC are available for some common chemicals, and SQC for a few nonionic organic chemicals. EPA's published derivation methods can be used to develop criteria for additional chemicals or to modify existing criteria for site conditions.
- TRVs for ecological ROCs can be developed by judicious use of adjustment factors; one broadly applicable method for TRV derivation was developed for terrestrial organisms at the Rocky Mountain Arsenal Superfund site.

ENVIRONMENTAL RISK CHARACTERIZATION

Risk characterization involves estimating the magnitude of potential health risks and making summary judgments about the nature of potential adverse impacts to ecological receptors. This section evaluates potential adverse effects to key terrestrial and aquatic species based upon their exposure to COCs using two integrated approaches. Risks can be assessed by comparing the measured or estimated exposure levels (*i.e.*, dietary intake or exposure concentration) with chronic toxicity values. Field survey data can also be used as a "reality check" for interpreting the predictive (quantitative) results.

The TQ Approach

The potential for adverse effects on aquatic and terrestrial populations can be rapidly evaluated using a **toxicity quotient (TQ) approach.** TQs are defined as the ratio of chronic daily intake (mg/kg-day) for terrestrial receptors, exposure point concentration (mg/L water or mg/kg sediment) for aquatic receptors, or soil concentration (mg/kg soil) for plants to respective toxicity criteria. The TQ approach is routinely used by EPA as the simplest quantitative method available for estimating risks to ecological receptors.

However, there is no regulatory guideline for an acceptable TQ level. Environmental TQs are roughly comparable to HQs used to evaluate potential human health effects. An important difference between human health and ecological risk assessment is that in the former process, the human individual is of paramount interest, while in the latter the population is the fundamental unit of assessment. Thus, although any exceedance of human toxicological criteria such as RfDs may be considered a basis for immediate action or further investigation, interpretation of TQs greater than 1.0 is not as straightforward.

A qualitative TQ ranking system based on the nature of the TRVs may be helpful in this regard. In general, the severity of potential consequences resulting from exceedance of a toxicological benchmark level depends on (1) the perceived "value" (ecological, social, or political) of the receptor, and (2) the nature of the endpoint measured in the "critical" study used for criterion derivation. Because the unit of concern in ERA is usually the population as opposed to the individual, exceedance of TRVs based on NOAELs for sensitive endpoints may be interpreted as less immediately threatening than exceedance of those based on LOAELs or less sensitive endpoints. On the other hand, protection of individuals of T&E and other protected species (*e.g.,* migratory birds) is mandated by law. Therefore, the significance of TQ values greater than one should be evaluated on a species-, chemical- and site-specific basis.

Summary

- Risk characterization involves estimating the magnitude of the potential adverse health effects of the chemicals under study and making summary judgments about the nature of the health threat to the ROCs in particular and the ecosystem in general.
- TQs, defined as the ratio of intake or exposure to relevant toxicity criteria, can be used to screen for potential effects on ecological health.
- As there are no regulatory guidelines for acceptable TQ levels, the significance of TQ values greater than one should be evaluated on a species-, chemical- and site-specific basis.

UNCERTAINTY ANALYSIS

A clear understanding of uncertainty in risk assessment is critical for accurate interpretation of results. Due to the multiplicity of potential receptor species and general lack of knowledge regarding their life cycles, feeding habits and relative toxicological sensitivity, the uncertainty surrounding estimates of ecological risk may be substantially greater than those associated with human health risk assessment. Uncertainty is inherent in the selection of ROCs and COCs, identification of toxicity criteria, measurement or modeling of environmental chemical concentrations, evaluation of measurement endpoints, *etc.* The principal sources of uncertainty involved in assessing potential hazards and the degree to which these sources are expected to over- or under-estimate the true threat to ecological receptors should be thoroughly discussed.

Uncertainty analyses can be qualitative or quantitative. A detailed description of qualitative and quantitative uncertainty analyses that could be done for the ERA is included in the chapter on human health risk assessment.

REFERENCES

Abou-Setta, M.M., R.W. Sorrell, and C.C. Childers, 1986. A computer program in Basic for determining probit and log-probit or logit correlation for toxicology and biology, *Bull. Environ. Contam. Toxicol.*, 36: 242-249.

Barnthouse, L.W., and G.W. Suter, II, Eds., 1986. *User's Manual for Ecological Risk Assessment*, ORNL-6251, Oak Ridge National Laboratory, Oak Ridge, Tennessee.

Carlson, A.R., G.L Phipps, V.R. Mattson, P.A. Koslan, and A.M. Cotter. 1991. The role of acid-volatile sulfide in determining cadmium bioavailability and toxicity in freshwater sediments. *Environ. Toxicol. Chem.* **10:**1309-1319.

Finney, D.J, 1952. *Probit Analysis: A Statistical Treatment of the Sigmoid Response Curve*, 2nd Ed., Cambridge University Press, London.

Gilbert, R.O. 1987. *Statistical Methods for Environmental Pollution Monitoring.* Van Nostrand Reinhold, NY.

Gilliom, R.J. and D.R. Helsel, 1984. *Estimation of Distributional Parameters for Censored Trace-level Water Quality Data*, U.S. Geological Survey Open File Report 84 - 729.

Grossjean, D, 1983. Distribution of atmospheric nitrogenous pollutants at a Los Angeles area smog receptor site, *Environ. Sci. Technol.*, 17(1):13-19.

Lieberman, H.R., 1983. Estimating LD_{50} using the probit technique: A basic computer program, *Drug Chem. Toxicol.*, 6(1): 111-116.

Ludwig, D.F. et al. 1993. Toxicity reference values for ecological risk assessment. (manuscript in preparation).

Newman, M.C., et al., 1992. *Uncensored*, Version 3.0, Savannah River Ecology Laboratory, Aiken, Georgia.

Norton, S.B., D.J. Rodier, J.H. Gentile, W.H. van der Schalie, W.P. Wood, and M.W. Slimak 1992. A framework for ecological risk assessment at the EPA. *Environmental Toxicology and Chemistry* **11**:1663-1672.

Porter, P.S., R.C. Ward, and H.F. Bell, 1988. The detection limit, *Environ. Sci. Technol.*, 22(8):856-861.

SAS Institute Inc. (SAS), 1985. *SAS User's Guide: Statistics*, Version 5, Cary, North Carolina.

Sette, A., L. Adorini, E. Marubini, and G. Doria, 1986. A microcomputer program for probit analysis of interleukin-2 (IL-2) titration data, *J. Immunolog. Methods*, 86: 265-277.

Suter, G.W., III, 1990. Endpoints for regional ecological risk assessments, *Environ. Manage.*, 14(1): 9-23.

Travis, C. C., and H.A. Hattemer-Frey, 1990. Human exposure to dioxin, *Sci. Total Environ.*, 104:97-127.

Travis, C.C., and M.L. Land, 1990. Estimating the mean of data sets with nondetectable values, *Environ. Sci. Technol.*, in press.

U.S. Environmental Protection Agency (EPA), 1989a. *Ecological Assessment at Hazardous Waste Sites*, EPA/600/3-89-013, Washington, D.C.

_____. 1989b. *Ecological Assessment at Hazardous Waste Sites; A Field and Laboratory Reference Document*, EPA/600/3-89-013, Washington, D.C.

_____. 1989c. *Risk Assessment Guidance for Superfund: Volume II. Environmental Evaluation Manual*, EPA/540/1-89-001, Office of Emergency and Remedial Response, Washington, D.C.

_____. 1989d. *Summary of Ecological Risks, Assessment Methods, and Risk Management Decision in Superfund and RCRA*, Office of Policy Analysis, EPA/230/3-89-046, Washington, D.C.

_____. 1989e. *Risk Assessment Guidance for Superfund: Volume I. Human Health Evaluation Manual (Part A)*, EPA/540/1-89-002, Office of Emergency and Remedial Response, Washington, D.C.

_____, 1991. *Handbook for Remediation of Contaminated Sediments.* EPA/625/6-91/028.

_____. 1992a. *Framework for Ecological Risk Assessment*, EPA/630/R-92/001, Risk Assessment Forum, Washington, DC.

_____. 1992b. Guidelines for exposure assessment, *Fed. Reg.*, 57(104): 22888-22938.

Weil, C.S., 1972. Statistics versus safety factors and scientific judgment in the evaluation of safety for man, *Toxicol. Appl. Pharmacol.*, 21: 454-463.

Weil, C.S., and D.D. McCollister, 1963. Relationship between short and long-term feeding studies in designing an effective toxicity test, *Agric. Food Chem.*, 11: 486-491.

Woodfork, K., and R. Burrell, 1985. A basic computer program for calculation of CH_{50} values by probit analysis, *Computers. Biol. Med.*, 15(3): 133-136.

4

ENVIRONMENTAL CHEMISTRY AND ANALYSIS OF REGULATED COMPOUNDS

Joann M. Slavin, Ursula R. Middel, and Ellen R. Kelly
H2M Labs, Inc.

OVERVIEW

From the moment commonly recognized as "the Big Bang" to the present, the interactions of atoms — the fundamental building blocks of matter — have formed our changing universe. Every natural process on our planet, including human activity, has chemical origins and chemical consequences. The systematic, quantifiable study of chemical elements and processes is a science that underlies and affects almost every other kind of scientific study.

Attempts to understand and manipulate chemical processes have persisted over centuries, with an explosion of technological innovation and research occuring in the last 100 years. The majority of these efforts have focused on using the knowledge gained to produce a useful product, and thus achieve a perceived improvement in some community's standard of living. As this century draws to a close, however, it appears that chemical products and by-products resulting from our manufacturing activities now have the potential to damage the planet and the life that currently exists on it. Environmental chemistry, one branch in the broad study of how atoms arrange themselves in various forms of matter, has grown in importance as citizens and public officials become more aware of the impact of chemicals on the environment.

The laws of science will ultimately dictate what man can — or cannot — do with the environment. The science of chemistry is at the core of our understanding of both the problems posed by man-made pollution and the solutions that are possible. This chapter will treat environmental chemistry, not in terms of a detailed presentation of its theoretical concepts, but through a discussion of the way that the science has been applied in the United States in recent decades, as the country has ventured into the new era of national environmental regulation.

In the twentieth century, scientists have become increasingly sophisticated in utilizing chemical knowledge to better understand the natural environment and man's effect on it. Definitions of "pollution" and "toxic and hazardous materials" depend upon analytical chemical measurements. Social concern over substances that meet those definitions has driven law and regulation for more than twenty years. In that same period, each scientific advance in the methodology and instrumentation that measure chemicals in the environment has resulted in more stringent regulation.

ENVIRONMENTAL CHEMISTRY AND ENVIRONMENTAL REGULATION

The complicated operations of environmental testing laboratories today grow directly out of the laws and regulations that have been passed in the United States over the past quarter century. In earlier decades, outside of university or corporate research centers, independent environmental laboratories existed primarily for the purpose of monitoring public drinking water for bacteriological contamination. This is still an important function for many laboratories. But today, large independent environmental testing facilities analyze all media that can carry chemical contamination to plants, animals and humans: potable water, air, soil, hazardous materials and waste, petroleum products, dredge spoils, industrial and municipal wastewater. Federal and state laws require the use of expensive, sensitive and complex analytical instruments, and dictate differing sets of procedures and protocols to document and control each step of the analytical process under various regulatory programs.

With the establishment of the Environmental Protection Agency in 1970, the United States sought to embark on a coordinated national approach to control and minimize pollutants across all environmental media. During its first decade, four major pieces of legislation (the Clean Air Act of 1970, the Clean Water Act of 1972, the Safe Drinking Water Act of 1974, and the Resource Conservation and Recovery Act of 1976), initiated national regulatory programs. Each required analytical measurements to verify compliance.

DRINKING WATER ANALYSIS

The Safe Drinking Water Act(SDWA) was the federal legislation that had the most immediate impact on the expansion of analytical laboratories and developments in instrumentation for environmental testing. State governments, through their health departments and environmental agencies, had primary enforcement responsibility for the new national water quality standards. They also had authority to certify independent testing laboratories as qualified to analyze the samples required under the new programs. There were then more than 200,000 public water suppliers in the United States, almost 60,000 of which supplied water year round to a community of significant size. Before passage of the Safe Drinking Water Act, only 15 percent of community systems routinely monitored their water for microbial contamination, and even fewer had done any significant chemical testing. The new analytical needs were therefore enormous.

When the SDWA was passed and the first regulations under it were promulgated, the standard analytical equipment available to water suppliers could measure conventional contaminants — primarily metals and other inorganic materials — in the milligrams-per-liter (parts per million) range. However, passage of the federal law had been triggered by growing public recognition that man-made organic chemicals might pose an equally serious threat. In the 1970's, hundreds of different **synthetic organic chemicals (SOC's)**, such as pesticides and herbicides, and **volatile synthetic organic chemicals (VOC's)**, including solvents and degreasers, had begun to be identified in samples drawn from surface waters and underground aquifers. In particular, the group of chemicals known as **trihalomethanes** were discovered in public water supplies, hastening passage of the SDWA. In 1979, this group of contaminants became the first proposed for regulation at the parts per billion level, 1000 times smaller than amounts previously regulated. Ironically, some trihalomethanes appeared in public water as by-products of the chlorination process, which provided safety from bacteriological contamination.

The experience of the first twelve years of implementing the SDWA lead to substantial amendments to the act, which were passed June 1986 and are still in the process of being implemented. The amendments were primarily designed to speed the process of establishing **maximum contaminant levels (MCL's)** for a larger range of chemicals. It is these regulatory levels which determine whether or not a potable water supplier can deliver water from a specific surface water or groundwater source without treatment. The regulatory levels themselves have in turn been affected by the availability of instrumentation, such as gas chromatography, described below, which can detect minute quantities of organic compounds. Linking gas chromatography with mass spectrometry (GC/MS) permitted the handling of far higher concentrations and more complex chemical mixtures found at abandoned and active hazardous waste sites. The capabilities of this technology determined the methods and requirements for analysis of samples built into the two federal regulatory programs for hazardous waste management: RCRA and CERCLA.

HAZARDOUS WASTE ANALYSIS

Although there had been limited earlier efforts by the federal government to address solid waste and resource recovery issues, it was not until passage of the Resource Conservation and Recovery Act (RCRA) in 1976 that Congress established the principle that **hazardous waste** must be carefully managed from its inception to its ultimate disposal, i.e. "cradle to grave." In addition, under the regulatory definition contained in RCRA, to be a hazardous waste, a waste must be a "solid waste" — although in fact the law specifies that these wastes can include solids, semi-solids, liquids or contained gases.

The regulations implementing this law defined hazardous wastes based on their characteristics, and also provided a list of specific hazardous wastes. There are four defining characteristcs by which a waste can be determined to be hazardous: (1) **ignitability**, demonstrated by substances with a low flash point or by friction-sensitive substances that can create fires under certain circumstances; (2) **corrosivity**, manifested by acids and wastes capable of corroding metal; (3) **reactivity**, exhibited by unstable wastes that create explosions or toxic fumes, gases or vapors when mixed with water or acids; and (4) **toxicity**, shown by wastes which are harmful or fatal if ingested or absorbed. A critical test, under the provisions of RCRA regulation to determine toxicity, was the Toxic Characteristics Leaching Procedure (TCLP).

The category of **listed wastes,** those which had already been identified as hazardous by EPA in the RCRA regulations, included wastes from specific industries and wastes from specific manufacturing processes, as well as specific commercial chemical products such as creosote.

Whether or not a waste falls under the stringent provisions of RCRA regulation is an issue with great financial, legal and liability consequences. Analytical verification has therefore become of vital concern to property owners and potentially affected waste generators.

The system of permits established under RCRA require that wastes be analyzed and identified prior to transportation, treatment, storage and disposal. In addition, operators of **treatment, storage and disposal facilities (TSDF's)**, must provide for ongoing environmental sampling and monitoring during operation of their facilities, and even for decades after closure.

This provision of the RCRA regulations is designed to prevent the problems posed by abandoned hazardous waste sites, which led to the passage of the Comprehensive Environmental Response, Compensation and Liability Act (CERCLA), commonly known as the **Superfund** Law, in 1980. The complex mixtures of hazardous wastes found in sites where many different industries may have contributed wastes over decades presented a major new challenge to environmental chemists. The combination of **gas chromatography and mass spectrometry (GC/MS)**, discussed below, has proven to be essential in characterizing sites with this type of organic chemical contamination as part of CERCLA's **Remedial Investigation (RI)** process.

Amendments to both RCRA and CERCLA passed in the 1980's significantly extended the reach and the analytical requirements of both laws. The 1984 RCRA amendments, sometimes known as the Hazardous and Solid Waste Amendments (HSWA), added a whole new regulatory program for underground storage tanks and extended regulatory requirements to small quantity generators of hazardous waste, down to a level of 220 pounds per month. It also prohibited the burning of hazardous waste mixed with fuel oil, which had been a common disposal technique, and extended federal regulatory authority, under Subtitle D of RCRA to solid waste (i.e primarily non-hazardous) facilities such as landfills, land application facilities and waste piles. The Superfund Amendments and Reauthorization Act (SARA) of 1985, added "Community Right To Know" provisions to the law and required industries to file annual reports of their toxic chemical releases.

WATER POLLUTION

The Clean Water Act of 1972 (CWA) set a national goal of "fishable, swimmable" surface waters, and measured the quality of specific water bodies according to their "designated uses." As a result of the act, more than $200 billion dollars has been spent in the construction and operation of public and private wastewater treatment facilities. Conventional pollutants associated with raw municipal wastewater have been reduced significantly by the **publicly owned treatment works (POTW's)** that have been constructed or upgraded across the United States. Direct industrial discharges have also been controlled under the **National Pollution Discharge Elimination System (NPDES)** program of permits, as well as by requirements to pre-treat industrial wastewater prior to its being discharged to a POTW system. State and national permitting and pretreatment programs mandate regular sampling and monitoring programs to verify compliance.

Although water quality improvements have been made under the CWA, population increases and new industry have put continuing strains on water quality. Analytical instrumentation and research associated with drinking-water and hazardous waste regulatory programs has revealed toxic chemical pollution that is not addressed by conventional wastewater-treatment technologies. Non-point source pollution, including stormwater drainage systems, as well as agricultural and construction site runoff, has become a major focus of regulatory attention in the 1990's. In more than 1,000 communities across the country, **combined sewer overflows (CSO's)** in which storm drainage systems are linked to municipal wastewater systems, require control strategies.

AIR QUALITY

The Clean Air Act of 1970 was the first modern federal regulatory program designed to set measurable standards for air quality across the country. It had a relatively small initial impact on the overall growth of environmental testing facilities and capabilities during its first two decades of existence. Its greatest success was in reducing particulate emissions from industrial processes and incinerators (point sources), which resulted in improved air quality in many urban areas. But the controls placed on exhaust emissions from automobiles (non-point sources) as a result of the act were all but offset by the tremendous increase in the total number of vehicles. After twenty years of federal air regulation, the air quality problems that remained were far more difficult to regulate.

The 1990 Clean Air Act Amendments, called "the broadest, most far reaching environmental legislation in the history of the United States," addressed the non-point source problem, as well as the issue of "air toxics" (specific chemical pollutants), in a far more comprehensive way than the original act. In its twenty years of existence, the regulatory program resulting from the original act was able to "list" only eight pollutants as hazardous under the statute definition. Since the 1990 legislation specifically aims to regulate a lengthy list of chemicals, analytical programs to define and monitor progress under the 1990 amendments are expected to provide an increasingly important segment of work for environmental laboratories in the coming decade.

LABORATORY OPERATIONS FOR THE ANALYSIS OF REGULATED SUBSTANCES

Through the analyses conducted as part of each of these major regulatory programs, we have discovered how chemical compounds persist and sometimes change in the environment as a result of physical, biological and chemical processes. With the basic knowledge of the properties of atoms and the nature of chemical bonds, specific techniques and instruments continue to be developed to document and quantify the presence of different types of chemicals in the environment. For the purposes of the following discussion of instrumentation and its uses in today's environmental protection programs, it is important to describe some of the most basic chemical terms and concepts. The reader wishing a detailed presentation of these concepts is advised to consult book length chemical references, such as the sources listed at the conclusion of this chapter.

Chemists have identified 105 **elements** — types of atoms —which occur in nature or have been created in the laboratory. Each element has unique physical and chemical properties. Elements can sometimes be physically combined to form mixtures in which each element retains its own characteristics. When elements are chemically combined, however, a new substance or **chemical compound** with its own characteristics is formed. Compounds are made up of two or more different chemically combined atoms forming a **molecule**, the smallest particle of a substance that retains the unique and distinctive characteristics of the substance.

The binding links which enable molecules to form are **chemical bonds**, and the strength of these intermolecular forces determines whether a particular substance exists as a solid, liquid or gas. Ionic bonds are the simplest kind of chemical bonds. An **ion** is a charged particle formed when an atom loses or gains an electron (a component of the atom carrying a negative charge). When an atom loses an electron, a positive ion (cation) is formed; negative ions (anions) are created as electrons are added to atoms. The **ionic bond** is the link formed by the electrostatic attraction between positive and negative charges. Other types of chemical bonds, including various types of **covalent bonds**, in which two atoms share one or more electron pairs, have been identified and explained as our knowledge of atomic structure has grown.

One major set of instrumentation discussed below applies to elemental, i.e., metals, analysis, with its defined set of analytes. Other instruments address organic analysis, where the number of potential analytes is staggeringly larger. In all cases, instruments apply some sort of energy (thermal, electrical or radiation) to an analyte in order to generate a signal, which must then be appropriately detected, recorded and interpreted. A major challenge, both to the developers of instrumentation and to the environmental chemists who employ it, is to separate the signal of the analyte being sought from the "noise" generated by other chemical activity in the sample matrix. Enhancing or filtering the signal in order to isolate the desired data is essential.

METHODS

For each of the major regulatory programs described above, **methods** were developed to analyze the various media for the organic and inorganic parameters. Each method approved by USEPA specifies the procedures for instrument calibration, sample preparation, the actual analytical operation, and the quality control measures required to verify the integrity of the sample and the accuracy and precision of the analysis.

EPA methods are differentiated according to the media of the sample to be analyzed, and have significant variations under each set of regulations. For example, for organic analyses, the 500 series methods apply to finished drinking waters and raw source water; the 600 series methods are used for municipal and industrial wastewater; and the 8000 series methods pertain to solid waste, although groundwater and wastewater can also be included. The corresponding specific method in each series (e.g., volatile organics are analyzed by methods 524.2, 624 and 8240), uses similar in instrumentation and technology, but has differences in quality control and calibration requirements.

The **EPA Contract Laboratory Protocol (CLP)** was developed for the Superfund Program. CLP specifies a set of methods which are based on the existing methodology for organic (volatile, base-neutral-acid extractable and pesticide/PCB) and inorganic (metals and cyanide) parameters, but which are modified to incorporate certain quality control, calibration and deliverability requirements. The data package includes a full reporting of quality control procedures and data, making it particularly useful if litigation is a possibility. CLP, therefore has become a commonly requested methodology, and has the effect of separating larger labs, which have the equipment, certifications and trained personnel capable of producing data according to this protocol, from the thousands of smaller environmental laboratories which do not.

Because EPA methods, as now written, are not interchangeable, it is very difficult for an analytical laboratory to accomodate all quality control criteria for all of the methods. Thus, after two decades of experience with these still-evolving regulatory programs, EPA's current intent is to create a unified method to minimize the differences in requirements.

INSTRUMENTATION FOR ORGANIC ANALYSIS

The possibilities for creating organic substances by combining carbon atoms with other atoms in chains, rings or combinations are practically endless. A glance at the multitude of available organics listed in handbooks or catalogs makes immediately apparent the difficulty of analyzing for specific substances. Historically, organic analysis relied on reactions which, for the main part, limited detection to functional groups of constituent substances. Early analytical techniques also required substantial levels of the compound to be present. Only with the advent of instrumental analysis was it possible to identify specific analytes at very low concentrations.

In the 1950s, chromatographic methods were developed for use in instrumentation. At the core of the chromatographic separation is a column containing the stationary phase, through which the mobile phase (in which the sample is dissolved) is passed. Chromatographic separation makes use of the chemical and physical differences between compounds, which cause differing affinity to the stationary phase. The affinity of a certain compound to the stationary phase will influence its distribution between mobile and stationary phase, and thus determine how fast that compound will travel through the column. Compounds with different properties will therefore show different **retention times**. Then, to complete an instrument for analytical purposes, it is necessary to add a device that registers the compound after it elutes from the column. These devices are called detectors.

GAS CHROMATOGRAPHY

The first instruments that were developed for environmental analyses of organics were ***gas-liquid chromatographs (GC's)***. In this instrument, the compounds to be analyzed are evaporated and transported through a column with a "carrier gas". For this method, the subject compound must be both sufficiently volatile and stable. That is, the compound should not decompose when injected into the injection port, where the sample is evaporated and introduced into the column inlet.

The Safe Drinking Water Act, which mandated testing of organics starting January, 1977 regulated pesticides, herbicides and solvents. These analytes all met the criteria that made them amenable to GC analysis. They can be grouped in two categories of compounds: ***volatiles*** and ***semi-volatiles.*** Volatiles comprise primarily solvents and light compounds with a high vapor pressure. Compounds are regarded as semi-volatile if they can be vaporized in low quantities at or slightly above 200 degrees C.

DETECTORS

The detector components of instruments are devices that transform the physical or chemical property of the analyte into an electrical signal. In most GC detectors used in environmental analyses, the elution of the compound is indicated by ions or electrons, which are collected. The electrical response is magnified and translated into a **peak** that can be shown on a recorder. In order for a compound to generate ions, the molecules can be exposed to light, as in the **photoionization detector (PID)**, or to flame, as in the *flame ionization detector (FID)*. Another type of detector, **the Hall electrolytic conductivity detector**, reduces any chlorine-containing compound to HCl, using the electrolytic capacity of the resulting HCl to show the presence of the compound. In the **electron capture detector (ECD)** any compound electronegative enough to capture electrons will be indicated by a change in the otherwise constant flow of electrons.

SAMPLE PREPARATION

Detectors used for drinking water analysis have to be able to indicate the presence of a compound at quantities below the regulated level of the compound in drinking water. Even though the sensitivity of the detectors used allows detection of nanograms or even picograms, it is not sufficient for seeing ppb levels that are of interest for drinking water analysis. EPA therefore generated sample preparation protocols that permit a **concentration** of the contaminants before analysis. In the case of semi-volatile compounds, like pesticides for example, it is possible to partition the compound into an organic solvent through a liquid/liquid extraction, and then follow with a concentration step of evaporating the solvent. Due to the concentration, the compound is ultimately dissolved in just a few milliliters of solvent extract instead of in a liter of water. A few microliters of this final extract are then injected into the GC for analysis.

The EPA method for concentration of volatile compounds, including solvents, is called **purge and trap procedure.** The method makes use of the high vapor pressure of the analytes. They are partitioned (purged) into the headspace of the container, called the purge vessel, by a continuous flow of inert gas, like nitrogen, that is bubbled through the water sample. The released volatile substances are then collected on a "trap" filled with solid sorbent material. Upon rapid heating, they can then be desorbed and transferred into the injection port of the GC by the carrier gas.

IMPROVEMENTS IN GC METHODS AND TECHNOLOGY

Since more than one organic compound can elute from the chromatographic column at the same time, a positive identification cannot really be made from GC analysis. But efforts can be made to narrow down the likelihood that the peak found really represents the targeted analyte. If GC analysis alone — without other confirmation — is used, the likelihood of positive identification can be increased by taking the following steps:

1. Use separations and "cleanup" procedures during sample preparation to separate the class of organics of the compound from other interferences as much as possible.
2. If available, use a selective detector, which only registers a specific type of compound, rather than a very general detector, such as the FID.
3. Choose instrument parameters and chromatographic columns to provide optimal resolution for the compound.
4. Confirm positives on a suitably different "confirmatory column."

Technological advances in GC instrumentation have kept pace with increasing demand for higher sensitivity and selectivity. And, as discussed above, changes in regulatory requirements were in turn sometimes triggered by the improved technology.

The largest advance in technological sophistication came about with the introduction of **capillary columns** which replaced **packed columns.** Packed columns contained the exchange medium, the stationary liquid (high molecular weight) phase, as a coating of small inert particles. Problems occurred due to "bleeding" (loss) of liquid phase, which exposed active sites on the support material that in turn caused analyte absorption and poor resolution of compounds. Also, because of large inner volumes and uneven packing, the molecules spread out in broad peaks. However, in capillary columns, the liquid phase is "bonded" to the inner surface of an inert very narrow (typically 0.23mm or 0.53mm) fused silica column. Thus, inertness, even distribution and smaller volume are guaranteed. Much sharper peaks are obtained and, with appropriate column lengths, much better resolution of peaks can be achieved.

Instrument construction has been improved to provide the lower inner volumes of injectors and detectors needed for capillary columns, and better stability to improve reproducibility of responses and retention times. Very early on in environmental analysis, recorders were replaced by integrating computers into the instruments to measure the areas of compounds. Sophisticated computer systems are now used by the analyst that not only set up instrument parameters, do calibrations, identification and quantification, but also produce final reports for the client.

GAS CHROMATOGRAPHY/MASS SPECTROMETRY

Whereas GC analysis was adequate for the analysis of organics in drinking water, the combination of **gas chromotography coupled with mass spectrometry (GC/MS)** was required for the testing of solid waste for the many analytes under RCRA and CERCLA. Even though most of the analyte groups that were included in the target compound list (TCL) of the Superfund program could also be analyzed by GC, that analysis alone would not be practical for two reasons. First, multiple GC runs with various detectors would be needed. Second, GC analysis does not provide sufficient specificity for identifying compounds in a complex medium, for example, in a waste sample that might contain a large number of potentially interfering organics.

In GC/MS analysis, the GC detector, which is nonspecific or at best semi-specific, is replaced by a mass spectrometer that permits **positive identification** of the compound separated by GC, by comparison of the spectrum with the standard spectrum. In the mass spectrometer, the molecules entering the "source" are bombarded by electrons and broken into charged masses [ions (m/e)]. A particular compound always decomposes into the same fragments (ions), which are always present in the same ratios. By separating the masses in a **quadropole** and measuring the ion current for the individual masses, the instrument obtains a spectrum representing the relative abundance of the ions. As long as the instrument response is controlled ("tuned"), the same spectrum is always produced for a particular compound.

The increased complexity of the media analyzed under CERCLA and RCRA requires much greater efficiency of sample preparation. Even the most sophisticated instrument can not handle the injection of an extract for analysis of ppb levels of certain analytes, if those analytes are in the presence of other organics at percentage levels, i.e. at concentrations millions of times stronger. This is analogous to the search for the famous "needle in the haystack". It is the task of the sample preparation analyst to remove most of the "hay," — interferences — before analysis. Elaborate protocols have been developed to achieve this task.

As in the GC analysis, the compounds in GC/MS analysis are grouped into analytical scans for volatile and semi-volatile organics. The volatile compounds do not require cleanup, as they are separated in the concentration step from all other higher boiling substances. The purge and trap method is used, as described for GC analysis. Cleanup procedures are, however, required for semi-volatiles that are extracted with a solvent from liquid or solid matrices.

A drawback of GC/MS analysis is the high cost of operation. In addition to both the GC and MS instruments, a powerful data system is needed to handle the large volume of data collected. Not only is the instrument price high, but also there is the need for experienced operators, which adds to the higher costs associated with GC/MS analysis.

HIGH PRESSURE LIQUID CHROMATOGRAPHY

A relative newcomer in environmental analysis is the **high pressure liquid chromatograph (HPLC)**. It was developed around the same period as gas chromatographs (GC), and has been widely used in the pharmaceutical industry. However it has only recently found application in the environmental field, due to the development of new pesticides that could not easily be analyzed by GC or GC/MS.

The "first generation" pesticides were primarily chlorinated pesticides that are very difficult to break down. Their persistence in the environment has caused devastating effects on wildlife, and ultimately, harmful effects on human beings. The task for developers of new pesticide technology has been to utilize more polar substances that readily decomposed into harmless metabolites in the environment. However, their very polarity makes these new pesticides more difficult or impossible to extract out of water. Therefore, GC analysis for semi-volatiles was no longer an appropriate approach.

In HPLC, as in GC, the compounds are chromatographically separated on a column. But in this case, the carrier is not gas but pressurized liquid. As before, the affinity of the compounds to the stationary phase of the column determines the equilibrium between phases, and thereby the speed with which the compound is carried through the column by the liquid phase.

Generally, the liquid phase consists of a mix of two or more solvents or ionic solutions. Whereas GC analysis uses temperature programs to manipulate compound retention times, in HPLC, the carrier can be changed during the analysis, i.e. the composition of the liquid phase is "programmed" in order to influence the elution pattern of the compounds and thus optimize their separation.

In several of the new HPLC protocols for pesticide analysis, the water sample can be directly injected into the instrument without previous concentration. Detection limits obtained then depend solely on the sensitivity of the detector. One of the detectors, the **fluorescent detector** has ultra-high sensitivity and permits low level detection without any prior concentration step. One drawback, however, is the need to derivatize the compound into a molecule that exhibits fluorescent properties before it enters the detector. Very elaborate "post column derivitization" systems in which the compounds are transformed by chemical reactions are required for this purpose. There are other methods which do require sample preparation before analysis, where the compound is first enriched before injection into the instrument.

HPLC, as explained above, offers the advantage of providing a means of analysis for polar substances. Also, compounds that cannot be volatilzed because they would break down if exposed to higher temperatures, can be analyzed by HPLC methods. In fact an examination of the properties of organic compounds shows that many more fall into the category of being analyzable by HPLC than by GC. This means that more HPLC applications can be expected as EPA expands the list of targeted analytes that have to be investigated in environmental samples.

INSTRUMENTATION FOR INORGANIC CHEMICAL ANALYSIS

Analysis of inorganic elements is an integral part of any full service environmental testing laboratory. Regulation of inorganic constituents in drinking water dates back to the establishment of U.S. Public Health Service Standards in the mid-1960's. Since 1975, EPA's national primary drinking water standards have covered **inorganic chemicals (IOC's),** including metals and other substances ranging from aluminum to zinc. Moreover, many states require potable water suppliers to screen annually and/or semi-annually for heavy metals and primary and secondary inorganics. In addition, inorganic contamination at active or abandoned hazardous waste sites presents some of the most complex and difficult investigative and remedial design challenges under RCRA and CERCLA. When testing these sites, regulatory agencies have required analyses for the **Target Analyte List (TAL)**, which includes twenty three metals and cyanide.

In the new era of environmental regulation, the first advance in inorganics analysis, was the shift from performing separate manual wet chemical analysis for each metal to utilizing atomic absorption and graphite furnace technology. These technologies continue to provide the bulk of metals analysis for both drinking water and CERCLA requirements, and **Graphite Furnace Atomic Absorption (GFAA)** is still the instrumentation which achieves the lowest detection limits in all types of environmental media. Thus, analyses for lead, arsenic, thallium, selenium and mercury are routinely conducted using GFAA.

The **Atomic Absorption spectrophotometer (flame AA)** typically consists of a spectrophotometer set to a specific wavelength of light, corresponding to a particular metal. A light source is split into two beams, one of which acts as a reference. The other beam passes through a torch (usually air/acetylene) to the spectrophotometer. A sample is aspirated into the torch and is ionized in the flame. The ionized sample will absorb a portion of the beam. The amount of light absorbed corresponds to the metal concentration in the sample.

The **graphite furnace (GFAA)** works on the same principle as the flame AA. In this instrument, the liquid sample is introduced into a small graphite tube which is heated in steps to approximately 2700 degrees C, ionizing the metals. The reference beam passes through the graphite tube and is absorbed in proportion to the metal concentration present. The small internal area of the graphite tube increases the sensitivity of the instrument, thus lowering the detection limits.

Flame AA and graphite furnace instruments have been the workhorses for inorganic analysis for years. They have several drawbacks, however. Flame AA and GFAA are single analysis instruments; that is, only one element at a time may be analyzed, and the instruments must be set up with specific operating parameters for each metal to be analyzed. Also, the sample matrix can cause problems with the instruments by introducing interferences which may bias the results. Strictly standardized sample preparation protocols were developed to minimize these interferences.

In the last decade, instrumentation has emerged to provide the low trace-level detection limits, multi-element analysis and fast sample throughput required in today's high volume environmental testing laboratories. **Inductively coupled plasma emission (ICP)** offers accurate and efficient low-level detection, with typical detection limits of five parts per billion for cadmium, 10ppb for silver and chromium, and 20 ppb for zinc. In this instrument, a sample is nebulized into a fine aerosol spray, which is then introduced through the central channel of the torch to an argon plasma of approximately 11,000 degrees K, which ionizes the sample. These ions emit light at the various wavelengths which are characteristic of the elements present in the analyte. This light, falling on a defraction grating, divides into its constituent spectral lines and passes into a photomultiplier tube (PMT). The PMT in turn converts photons into electrons that charge a capacitor. When the capacitor is discharged, the signal emitted is digitized. The number counts generated are then read by a computer to produce the final analytical result.

The next major technological advance in ICP instrumentation was the ability to analyze for different elements simultaneously instead of sequentially. **Simultaneous ICP** sprectrophotometers offer fast, precise, low-level detection for all matrices, including groundwater, wastewater, surface water and leachate. Coupled with an **ultrasonic nebulizer (USN),** which uses ultrasound waves instead of air to break the sample into an aerosol spray, the simultaneous ICP can improve detection limits by a factor of ten to twenty, depending on the element.

The same driving factors that moved organic analysis to link gas chromatographs to mass spectrometers have also moved inorganic analysis toward **ICP/MS**. Here, ICP instrumentation is linked to a quadropole detector, similar to the mass spectrometry instrument used for organic analysis. Identification of elements is determined by mass rather than by light emissions. This linked technology has once again increased the sensitivity of the instrumentation and lowered detection limits further than those available through GFAA. Routine use of such instruments, which can achieve detection in the parts per trillion range, is around the corner.

QUALITY ASSURANCE/QUALITY CONTROL IN THE ANALYTICAL LABORATORY

The values produced in the analytical laboratory are more than just numbers on a piece of paper. Decisions on the nature and extent of site remediation, treatment systems and disposal needs all depend on the values determined by the laboratory. Since the ramifications of erroneous data are severe, an extensive quality control system must be incorporated in the day-to-day production environment of the lab. **Quality assurance**—the ability to prove to clients that the systems used to generate data are under control and that they fully incorporate the necessary quality control measures— is determined through internal documentation, inter-laboratory comparison testing, lab audits, proficiency testing, and data validation.

A full **quality assurance program** should encompass the analysis of proficiency samples regularly to obtain and maintain certification from regulatory agencies, and meeting daily quality control requirements, as well as performing lab audits and blind sample analyses.

The validity of all data generated should be assessed for precision, accuracy, comparability, completeness and representativeness. As part of quality control procedures, ten percent of samples should be run in duplicate, so that **precision** can be measured by calculating any percent difference in the values obtained for the duplicates. **Accuracy** can be evaluated by comparing determined results to true or known values of quality control or check samples. **Comparability** will be assured by the use of standard methodology, while **completeness** is defined as having all the support and audit data to document the reported results. Finally, **representativeness** is assured by collecting and analyzing samples that are indicative of actual conditions.

Environmental laboratories today include sophisticated instrumentation and computer systems which require dependable, qualified lab personnel as the necessary first component of quality assurance and control. All standard preparation, instrument calibration and data reduction is done by these analysts who must follow well-defined consistent procedures in accordance with regulatory requirements and standard scientific practice, documenting each step of the process. In large national and regional laboratories, most lab employees have a minimum of a Bachelor of Science degree and are required to have previous analytical experience prior to working on specific types of samples.

STANDARD OPERATING PROCEDURES

A written guidebook which details the specific procedures used throughout the lab is critical for training new analysts, as well as for providing guidance to all analysts in the proper techniques and procedures to be used. This compendium of **Standard Operating Procedures (SOP)** must be constantly revised to include new methodology or procedural changes. Combined with professional development of staff through internal and external training, the SOP is an important component of the **Quality Assurance/Quality Control (QA/QC) program.**

"IT DIDN'T HAPPEN IF IT'S NOT DOCUMENTED"

Since a large percentage of the work done in today's large laboratories potentially could go to litigation, all aspects of the sample must be documented. This starts with the purchase of the bottles used to collect the samples and ends in the laboratory with the record of the individual who mailed the data package. More and more clients require a "full data package" now than ever before, including all quality control and raw data to substantiate the results reported.

Maintaining the integrity of the sample is critical from both an analytical and a legal standpoint. One key document in the written history of a sample is the **chain of custody form.**

This form is signed and dated whenever a change in the custody of the sample takes place, whether in the field, from field to laboratory, or in the laboratory. Chain of custody forms can be established when sample collection begins in the field or earlier, when the sample bottles are prepared. In the latter case, the chain of custody form is transported to the site with the prepared sample bottles and cooler, sealed with custody tape.

FIELD MEASURES OF QA/QC

Field QA/QC samples are used to provide information on the collection of program samples. This information can be used in the interpretation of analytical results. Different kinds of quality control samples are taken in the field to ensure that the samples analyzed comply with required procedures. A **trip blank** is used to detect sources of volatile organic contamination that could stem from bottle preparation, the quality of the reagent water, or the transportation cycle of the samples. The water used for trip blanks originates in the laboratory. It must be organic chemical-free and verified as such in the lab prior to use. The vials are usually collected for aqueous volatile organics only and accompany the sample bottles during the transportation cycle and storage on-site. **Field blanks** (rinseate blanks) are usually collected for the parameters of interest, at a project-specific frequency. Distilled/ deionized organic-free water is sent to the field with the appropriate bottles for the contaminants of interest. The field blank water is passed through the sampling device to check for potential cross contamination via field equipment. Since the field blank accompanies the blank and sample bottles in transport, it provides an additional check on sample handling and storage in the field.

A *matrix spike/matrix spike duplicate* refers to a sample which is collected for each matrix type at a specified frequency. Precision (reproducibility) and accuracy data is derived from the matrix spike/ matrix spike duplicate *(MS/MSD)* data.

Field personnel may collect a sample in duplicate, but not notify lab personnel. This is called a *blind duplicate* sample, and is collected at a frequency required on a project-specific basis. This provides a check on both field collection technique (collecting representative samples) and also on the laboratory for consistency in data. The exact number of these and other QC samples to be collected varies according to the requirements of specific projects.

All samples are packed to prevent breakage and maintain a consistent temperature of 4 degrees C during transit to the lab. Upon arrival at the laboratory, the samples are checked for proper filling of containers, correct containers and preservative use, adequate cooling and date, time and sampler name. Any deficiencies are noted and the sampler notified. Depending on the severity of the deviation, the sample may or may not be rejected.

DOCUMENTATION CONTROL

Once inside the laboratory, chain of custody documentation continues, including custody of the sample bottles as well as all extracts and digestates. Documentation of all transfers of custody is essential to assure that no unauthorized person has access to the samples. Laboratory facilities must also be secure, with non-employee access prohibited. The samples are maintained in locked areas to prevent tampering.

A well-run laboratory maintains bound, pre-paginated log books. Information recorded in laboratory log books includes temperature records, instrument maintenance, sample preparation, sample analyses and analytical standards.

In a large lab capable of handling many methods and the resulting documentation, the number of log books could be well over 100. Completed log books are archived for easy retrieval.

The Document Control Section of the laboratory is responsible for maintaining control of both the log books, and the data packages generated as the samples are analyzed.

INTERNAL QUALITY CONTROL

The wide variety of analyses performed and matrices dealt with in a large environmental laboratory require use of many different methods to address everything from potable water to highly concentrated hazardous waste. Inherent in each method are initial startup and quality control requirements that must be performed. For each method, **method detection limits (MDL's)** are statistically determined on the basis of replicate analyses performed throughout the entire analytical process, including during sample preparation. **Instrument detection limits (IDL's)** are low level standards applied only to the analytical procedure to determine what level of the analyte can be detected on that particular instrument. **Practical quantitation limits (PQL's)** are concentration levels that can be reliably reported for a sample, taking into consideration the matrix and the interferences that were observed for a "real" sample.

Quality control requirements are specific for the methods selected. The following are general QC requirements:

Method blanks. Method blanks are reagent water or a purified soil matrix that is carried through the entire analytical procedure. No targeted analytes or trace elements should be present in a method blank. Certain methods, however, have allowances for commonly found laboratory contaminants, with an upper limit above which corrective action must be taken prior to sample analysis.

Storage blanks (for volatile organics). Storage blanks may be analyzed to determine if storage in the laboratory has caused contamination.

Spiked sample. A known concentration standard is spiked into a sample to determine the amount of analyte recovered in a matrix. This is calculated and reported as a percent recovery. Some methods call for sampling to be performed in duplicate providing accuracy information as well as precision (reproducibility).

Surrogate standard. A surrogate standard is an analyte with chemical properties that are similar to the analyte of interest. It is assumed that these compounds will respond in a manner similar to the targeted analytes. A surrogate standard is added at the beginning of the method and its recovery is monitored. The recoveries are compared to the method or in-house limits. If the recoveries are not within acceptable limits, corrective action is required. The cause of low recoveries must be investigated and samples with low recoveries must be replicated. The low recoveries may be due to characteristics of the sample matrix which either inhibit or interfere with surrogate compounds.

Internal Standard Area. All GC/MS and some GC methods use the internal standard method of analysis. This method uses the addition of either deuterated analytes or analytes similar to the targeted compounds, and the responses of the targeted compounds are compared to those of a specific internal standard. If they do not fall within a pre-specified range, reanalysis is required to determine if the cause is matrix related.

Duplicate Analysis. Duplicate analysis of a sample determines the reproducibility or precision of the method. The relative range is calculated and recorded in tabular form. Acceptable limits are determined by in-house data. If a sample is not within the allowable range, the cause of the unacceptable results must be investigated.

CALIBRATION

Standards are reference materials purchased in either a neat or solution form to allow for identification and quantification of the compounds to be analyzed. All standards must be certified or traceable to certified standards to ensure the accuracy of the results. An initial calibration is performed over a range of concentration levels that are either pre-determined in the method, or bracket the range of interest of the samples. Calibration curves, derived from the multi-point calibrations, are then used to calculate response factors (if linear) which are used to quantify the samples.

Quality Control Check Standard. A QC check standard is analyzed to verify the accuracy of the calibration standard. This standard is from an independent source or a different lot number than the standard used for calibration. This data is plotted on an accuracy chart and in-house acceptance criteria are determined. If the data do not fall within the acceptance limits, corrective action measures are taken to determine the cause. Analysis is halted until the problem is rectified.

Monitoring of internal quality control measures must be performed in "real time" in order to ensure useable data. Due to tight holding-time contraints for many tests, all QC measurements must be constantly monitored and corrective action measures implemented immediately, so that any necessary re-extraction and re-analysis will be within the holding period.

HOLDING TIMES

A holding time is the time period within which the preparation and/or analysis must be performed. Certain methods begin the holding time from the *validated time of sample receipt (VTSR)* in the laboratory. Other methods initiate holding time from sample collection (e.g. drinking water methods in the Federal Register). Holding-time compliance is very important so that the laboratory is able to assure the client that no analytes were "lost" due to excessive time in the laboratory.

DELIVERABLES

The results of the analyses are provided in many different formats, ranging from a sample report only to a "full" documentation data package. The Contract Laboratory Protocol (CLP), as stated in the EPA Statement of Work (SOW), has a level of quality-assurance detail associated with it. The deliverable requirements include a quality control summary, sample and standard information, and raw qc data, as well as data on a diskette.

Various state agencies have used the EPA CLP as a standard, and made modifications to accomodate individual state requirements. CLP data packages are generally considered more legally defensible and are often requested when a potential for litigation exists.

DATA VALIDATION: COMPLIANCE V. USEABILITY

Some projects require the data produced by the labortatory to undergo validation by an independent third party. Data is validated or verified as to correctness and completeness by review of the quality control calibration and raw data. It is critical for the validator to become familiar with the analytical requirements of the project plan. The main purpose of the validation is to review the data for compliance with the analytical methods chosen and with the specific requirements of the project plan. Any non-conformances are documented and an assessment of the data is made. A decision about its useability is based on the compliance information from the validator, and whether it conforms to the purpose of the data. Results may be non-compliant with respect to the protocol, but still be useable, depending on the needs of the data user.

CONCLUSION

Experienced environmental chemists and other laboratory personnel represent a rich resource for further refining technology and its practical application in the laboratory. Many are already working with EPA to develop consistent protocols and methods which will guide future analytical programs. Environmental chemists and regulatory personnel both recognize that while the recent decades of experience have enhanced our knowledge, they have also revealed large gaps in the kind of data we need to thoroughly understand the action of chemicals in the environment. In order to make more accurate assessments of the effects of our regulatory activity, it will be necessary to accumulate trend data which will allow us to measure physical, chemical and biological changes in various media over extended periods of time. In the past, this data has been very limited, but has now begun to be more systematically obtained.

The development of laboratory instrumentation over the last twenty five years has been one of the forces shaping regulatory standards for substances deemed hazardous or polluting. In the 1990's, improved field instrumentation is already changing approaches to site characterization and cleanup projects. Like the twin strands of the famous DNA double helix, science and regulation will inescapably continue their closely linked relationship.

[The authors would like to thank Vincent Stancampiano for contributions to this chapter.]

REFERENCES

Giddings, J. Calvin, *Chemistry, Man and Environmental Change*, San Francisco, Canfield Press, 1983.

Rice, Rip G., *Safe Drinking Water: The Impact of Chemicals on a Limited Resource*, Lewis Publishers, 1985.

USEPA 530-SW-86-037, *Solving the Hazardous Waste Problem: EPA's RCRA Program*, 1986.

Continue on page 125.

5

AIR QUALITY

James W. Little
Dames & Moore

OVERVIEW

The quality of the air in which we live is something that affects us continuously. Substances in the air that are not "normal" constituents of the atmosphere (oxygen, nitrogen, etc.) are considered to be "pollutants."

The occurrence of air pollutants is certainly not unusual, nor something that originated with the industrial age. Air pollutants, in fact, can result from natural sources such as volcanoes that have nothing to do with the activities of humans. Most concerns about air quality, however, have focused on air pollutant emissions resulting from human activities.

The occurrence of air pollutants is of greatest concern because of potential adverse effects on human health. Potential adverse human health effects attributable to air pollution include acute effects, such as respiratory difficulties, and long-term effects, such as cancer formation. Other adverse effects of air pollution can also occur, either physically damaging effects or displeasing aesthetic effects. Potential adverse effects not directly related to human health include vegetation injury, damage to materials, and visibility degradation.

The "life cycle" of air pollutants starts with the sources that cause pollutants to enter the atmosphere. Once a pollutant becomes airborne, it is subject to a variety of fates. It may react with other substances in the atmosphere to form new substances, or be removed at the earth's surface by uptake in rainfall or by "dry" deposition. Although such transformation and deposition mechanisms may occur close to the emission source, a pollutant released to the atmosphere may also be transported long distances before being transformed or otherwise removed. The dispersion and removal of air pollutants are determined by their physical and chemical characteristics and by the complex motions of the atmosphere.

This chapter provides information on the nature of air quality and on the technological and regulatory measures that are taken to help prevent adverse air quality conditions. The topics covered are:

- types of air pollutants;
- types of air pollution sources;
- estimation of emission rates;
- atmospheric dispersion, transformation, and depletion mechanisms;
- emission control methods;
- ambient air quality evaluation methods;
- stratospheric ozone depletion;
- indoor air quality; and
- the regulatory approach to air quality protection.

TYPES OF AIR POLLUTANTS

At the most basic level, air pollutants can be classified as either in gaseous or particulate form. Common inorganic gaseous air pollutants are carbon monoxide, sulfur dioxide, nitrogen dioxide, and ozone. Hundreds of gaseous organic air pollutant compounds can be present. The particulate matter found in the atmosphere can be made up of many different substances including mineral, metallic, and organic compounds.

Because of various physical-chemical processes in the atmosphere, many substances originally emitted as gaseous compounds may eventually be converted to or adhere to particles. For example, sulfur dioxide emitted from fuel burning sources participates in the formation of aerosol (particulate) sulfates.

Another important consideration in the classification of air pollutants is the distinction between primary and secondary air pollutants. **Primary pollutants** are those that are directly emitted to the atmosphere. A common example is the carbon monoxide emitted from automobile exhausts. **Secondary pollutants,** on the other hand, are formed in the atmosphere as the result of various transformation mechanisms involving primary pollutants or other secondary pollutants. For example, one of the pollutants of most concern in both urban and rural environments is ozone. Ozone is a secondary pollutant produced by the reaction of volatile organic compounds and oxides of nitrogen in the presence of sunlight (a photochemical reaction). The substances that react to form ozone and other secondary pollutants are referred to as precursors.

Another term used in the categorization of pollutants is the term **regulated pollutants.** Many hundreds of pollutants can occur in the earth's atmosphere, but not all potential pollutants are specifically addressed in air pollution laws and regulations. The pollutants that have been singled out for regulatory control are sometimes referred to as regulated pollutants. In the United States, this term primarily applies to pollutants identified in the federal Clean Air Act (discussed in more detail later) and in the regulations adopted to carry out the Clean Air Act.

TYPES OF AIR POLLUTION SOURCES

Air pollution emission sources can be divided into two categories: mobile sources and stationary sources.

Mobile Emission Sources

Mobile source emissions refer to the emissions generated by the combustion of fuels in various types of engine-driven vehicles (automobiles, trucks, buses, motorcycles, aircraft, water vessels, etc.). The principal pollutants emitted from operation of vehicles are carbon monoxide, oxides of nitrogen (primarily nitric oxide and nitrogen dioxide), various species of hydrocarbons, particulate matter, and, depending on the fuel, sulfur compounds. Land-based vehicle emissions are of greatest concern in urban areas where the volume of traffic is highest and where vehicles often operate under conditions (such as stop-and-start driving) that result in higher emission rates. Vehicle emissions typically are the predominant cause of urban smog.

Non-Industrial Stationary Emission Sources

Stationary sources are often thought of in terms of sources at industrial facilities. (The term "industrial" in this context includes manufacturing, mining, power generation, commercial operations, and institutional sources.) Although industrial sources are given most attention in this chapter, recognition should also be made of the many types of non-industrial sources. Non-industrial stationary source emissions consist of emissions related to human activities and emissions from "natural" sources. Some of the emission sources related to human activities are fireplaces, wood-burning stoves, architectural coatings (for example, painting of structures), intentional burning of woodlands for forest management or land clearing purposes, and agricultural tilling. Examples of natural emission sources are wildfires (started by natural means), volcanoes and other geothermal sources, oil seeps, biological decay, and windblown fugitive dust from barren areas.

Industrial Stationary Emission Sources

Stationary industrial sources are often classified as either point sources or fugitive sources. Point sources are those that emit air pollutants through a confined vent or stack. The stack of a fuel combustion source is an example. Fugitive emissions, on the other hand, are those emissions that enter the atmosphere from an unconfined area. Fugitive emissions are often thought of as dust or particle emissions, but the term fugitive can equally well apply to many types of gaseous emissions. Examples of fugitive emissions are the dust particles stirred up by wind blowing over exposed storage piles and the vapors that escape from leaking pumps and valves.

Stationary source emissions at industrial facilities can result from fuel combustion and from various process operations. The following list of industrial activities and categories that produce air pollutant emissions illustrates the wide spectrum of industrial sources. This list is adapted from the U.S. Environmental Protection Agency (EPA) **Compilation of Emission Factors,** a commonly used publication that is referred to again in the section on estimation of emission rates. Examples of specific sources that comprise the category appear in parentheses.

- external combustion sources (coal combustion, fuel oil combustion, natural gas combustion, wood waste combustion)

- solid waste disposal (municipal waste incinerators, open burning, sludge incinerators)
- stationary internal combustion devices (combustion turbines, compressor engines, diesel-fired generator engines)
- evaporation loss sources (dry cleaning; surface coating including can coating, fabric coating, automobile coating, and large appliance coating; magnetic tape manufacturing; distribution and marketing of petroleum liquids such as gasoline; solvent degreasing; printing)
- chemical process industry (chlor-alkali, paint and varnish, pharmaceutical, phosphoric acid, soap and detergents, sulfuric acid, synthetic organic chemicals, synthetic fibers)
- food and agricultural industry (coffee roasting, cotton ginning, grain elevators, meat smokehouses, phosphate fertilizers, bread baking, cattle feedlots)
- metallurgical industry (iron and steel production; primary aluminum production; primary copper, lead, and zinc smelting; secondary aluminum, copper, lead, magnesium, and zinc processing; gray iron foundries)
- mineral products industry (asphalt roofing, bricks and clay products, portland cement, concrete batching, glass and glass fiber manufacturing, aggregate quarrying, metallic minerals mining and processing, coal mining)
- petroleum industry (petroleum refining, natural gas processing)
- wood products industry (chemical wood pulping, wood building products such as plywood and particle board)
- storage of organic liquids (gasoline, fuel oil, process organic chemicals)

ESTIMATION OF EMISSION RATES FOR STATIONARY SOURCES

Determining the types and amounts of pollutants emitted from a stationary emission source is important for several reasons. Emission rates need to be known to assess the effectiveness of pollution control equipment and process emission controls, to determine compliance with regulations and permit limits, and to evaluate the impact of emissions on ambient air quality. Different types of emission estimation methods exist, ranging from the precise to the approximate. The principal methods are as follows:

Stack Tests. A direct and precise means of determining emissions is through measurement of pollutants in the air flow from a stack. Standard methods are prescribed for stack testing depending on the type of source and types of pollutants. To properly interpret stack test results, source operating conditions during the test must be known in detail. Stack tests represent a "snapshot" of emissions (typically a 3-hour measurement period) and may not provide a complete depiction of emissions for conditions other than those occurring during the test.

Continuous Emission Monitoring Systems. Continuous emission monitoring systems (CEMS) are instrumentation systems that continuously draw a sample of air from a stack exhaust gas stream and analyze the sample for one or more pollutants. The advantage of CEMS is that emissions can be determined over a long period of time and under varying operating conditions. Disadvantages include high initial costs and the cost of continuing maintenance.

Emission Factors. Emission factors have been published for dozens of common emission sources. These factors typically represent the consolidation of stack test or other measurement data for a large number of sources. One of the most commonly used emission factors references is the EPA **Compilation of Emission Factors** mentioned previously. This document — often referred to by its publication number, AP-42 — contains emission factors for all of the industrial categories listed in the above discussion on types of emission sources. Due to the increasing concern over hazardous and toxic air pollutants, emission factors are also available for sources of these pollutants. Care must be taken when applying a general emission factor to a specific source to confirm that the characteristics of the specific source match the source characteristics for which the emission factor is applicable.

Mass Balance. The mass balance estimation method requires information on the amount of material going into a process and the amount of material coming out of the process. The difference between input and output can be taken as the quantity of emissions to the atmosphere (if the entire difference is in the form of air emissions). The mass balance method can be especially useful for estimating total emissions over a long period of time where daily material use and production data are tracked in detail.

Materials Data. Emissions can sometimes be calculated on the basis of fuel or materials composition. For example, if a coating is used in a process and the volatile organic content of the coating is known (perhaps from the manufacturer's data sheet for the coating), volatile organic emissions could be calculated by assuming that all of the volatiles in the coating are emitted. Similarly, fuel sulfur content can be used to estimate sulfur dioxide emissions from uncontrolled fuel burning sources by assuming that all of the sulfur in the fuel is converted to sulfur dioxide and emitted.

Design and Vendor Data. The design details of a particular piece of equipment may enable calculations of emission rates or at least supplement calculations based on other methods. Other data from the manufacturer or vendor of the equipment can also be helpful.

ATMOSPHERIC DISPERSION, TRANSFORMATION, AND DEPLETION

Pollutants released from both stationary and mobile sources are subject to atmospheric dispersion, transformation, and depletion mechanisms. A basic understanding of these concepts is important in understanding how emissions ultimately affect ambient air quality.

Dispersion

The dispersion of pollutants in the atmosphere is determined by mean wind flow conditions and by atmospheric turbulence. Turbulence results from such factors as the friction of the earth's surface, physical obstacles to wind flow, and the vertical temperature profile of the lower atmosphere. Important aspects of air pollutant dispersion are the following:

Stability Class. Stability class refers to the degree of turbulence in the atmosphere. For air quality purposes, stability usually refers to the lower layers of the atmosphere where pollutants are emitted. The idea of discrete stability classes (as in the following list) is a simplification of the complex nature of the atmosphere, but has proved useful in predictive studies. (1) A "stable" atmosphere is marked by air that is cooler at the ground than aloft, by low wind speeds, and consequently, by a low degree of turbulence. A pollutant plume released into a stable lower layer of the atmosphere can remain relatively intact for long distances. (2) An "unstable" atmosphere, on the other hand, is marked by a high degree of turbulence. A visible plume released into an unstable atmosphere may exhibit a characteristic looping appearance produced by turbulent eddies. (3) An intermediate turbulence class between stable and unstable conditions is the "neutral" stability class. A visible plume released into a neutral stability condition may display a cone-like appearance as the edges of the plume spread out in a V-shape.

Wind Changes with Height. In most atmospheric conditions, wind speeds tend to increase with height above ground level. In addition, wind direction typically changes with height. Wind shears created by changes in speed and direction affect the transport and diffusion of pollutant plumes.

Inversions and Stagnation Episodes. The term inversion refers to a layer in the atmosphere where temperature increases with height rather than decreases, as is usually the case. This inversion layer serves as a stable "lid" keeping pollutants emitted below the layer from dispersing further upward. Pollutant levels can build up near the earth's surface as a result of this trapping action. A prolonged period of dispersion is sometimes referred to as a stagnation episode. High pollution levels, especially "smog" conditions associated with emissions in urban areas, can occur during stagnation episodes.

Emission Source Dispersion Characteristics and Plume Rise. The same emission rate from two different sources can produce different ground-level concentrations depending on the dispersion characteristics of the two sources. For example, pollutants emitted from a source with a tall stack tend to produce lower ground-level concentrations than the same amount of pollutants emitted from a source with a short stack. Sources with the same stack height can produce different impacts depending on the plume rise above the stack. Plume rise is determined primarily by the vertical exit velocity of the exhaust gases and by the temperature of the exhaust gases. The combination of the physical stack height and plume rise above the stack is referred to as the effective stack height.

Air Quality / 133

Transformation

As previously mentioned, precursor substances can convert to secondary pollutants such as ozone. This is an example of a chemical transformation in the atmosphere. Physical and chemical transformations affect the ultimate impact of the air pollutants that are originally emitted into the atmosphere.

Photochemical reactions in urban atmospheres involving volatile organic compounds and oxides of nitrogen are an especially complex form of chemical transformation. Such reactions are dependent on the types of compounds present as well as on the ratio of compound concentrations. For example, the ratio of organic compound concentrations to oxides of nitrogen concentrations affects ozone concentration levels. Meteorological conditions also play an important role. Peak ozone concentrations in the United States generally occur in warm season months during periods of high temperature, high sunlight intensity (few clouds), and low wind speeds.

Depletion

Pollutants emitted into the atmosphere or formed in the atmosphere are eventually depleted. In addition to chemical transformation which acts as a depletion mechanism for precursors, two other common depletion mechanisms are dry deposition and washout. **Dry deposition** refers to the removal of both particles and gases as they come into contact with the earth's surface. **Washout** refers to the uptake of particles and gases by water droplets and snow, and their removal from the atmosphere when rain and snow fall to the ground. Acid rain (or, as more generally named, acid deposition) is a form of pollution depletion in which pollutants are transferred from the atmosphere to soil or water.

EMISSION CONTROL METHODS

Stationary Emission Sources

Numerous methods exist for the reduction of air pollutant emissions from stationary sources. These methods are summarized here but discussed at greater length in a later section.

Strategies for reducing air emissions have often concentrated on removing pollutants from a stream of air after these pollutants are formed. Such control methods are referred to as add-on controls, flue gas controls, or "end of pipe" controls. More attention is now being directed toward preventing the formation of air pollutants in the first place — an approach that can be cost effective as well as technically effective.

Among the more important control methods for stationary sources at industrial facilities are the following:

Add-On Pollution Control Methods. Add-on control methods entail the use of devices that remove or destroy pollutants after they are generated but before they are discharged to the atmosphere. Examples of such devices are baghouse filters, electrostatic precipitators, wet scrubbers, carbon adsorption beds, and incinerators. Although often highly effective in removing air pollutants, one drawback to this control approach, in addition to its expense, is that frequently a solid or liquid residue is created that must be disposed of in an environmentally protective manner.

An important economic consideration in planning an air pollution control program based on add-on pollution control devices is the recognition that control devices typically become progressively more expensive to install and operate as the required control efficiency increases. For example, the added cost of going from 90 percent control efficiency to 99 percent control efficiency could exceed the cost of achieving the initial 90 percent efficiency. As regulatory pollution control requirements become more stringent, the increasing cost of greater removal efficiencies provides greater impetus for the use of pollution prevention techniques instead of pollution removal techniques.

Cleaner Fuels. For stationary combustion emission sources, switching to cleaner fuels can be an effective pollution control method when technically and economically feasible. Natural gas, for example, is a cleaner burning fuel than most liquid and solid fuels.

Material Substitution. Some pollution emissions can be reduced by switching to alternative materials of production. For example, reformulated cleaning agents and coatings are increasingly available with a much lower volatile organic compound content than was available in previous formulations.

Process Modifications. The most effective process modification in reducing air emissions is the elimination of an air pollutant emitting process. Where complete elimination is not feasible, curtailment or improved efficiency may be possible.

Mobile Emission Sources

Although stationary emission sources may dominate air quality conditions in heavily industrialized areas or in remote areas around a single industrial source, urban air quality is greatly influenced by emissions from vehicles. Mobile source emissions result from fuel combustion in engines and from evaporative fuel losses. Methods for reducing mobile source emissions consist of technological methods and transportation control methods.

Technological control methods include the following: improved engine design to reduce the amount of fuel burned and the amount of pollutants produced; use of pollutant removal devices such as catalytic converters and fuel vapor collectors; use of reformulated fuels such as fuels with a lower volatiles content; use of alternative fuels such as liquified natural gas; development of electric-powered vehicles; inspection and maintenance programs to identify and repair vehicles with excess emissions; and development of mass transit systems.

Transportation control methods include: use of high occupancy vehicle lanes on freeways; car pooling; staggered work hours; improved traffic signal systems; and reversible lanes. In some locations, more extreme measures have been implemented such as restricting operation of vehicles by license tag number during episodes of high pollution levels.

AMBIENT AIR QUALITY EVALUATION METHODS

The most widely used ambient air quality evaluation methods are monitoring and modeling.

Ambient Air Quality Monitoring

Monitoring involves the use of measuring devices to determine the concentration of a specific pollutant, at a specific location, and at a specific time. The chief advantage of monitoring is that an exact level of pollution can be determined, subject only to the accuracy and level of detection limitations of the measurement method. The chief disadvantages are that the expense of monitoring prohibits taking measurements at more than a few locations; the contribution of specific emission sources to the total measured concentration can be difficult, if not impossible, to determine; and, of course, the impact of proposed sources not yet in operation cannot be estimated with monitoring methods.

Monitoring Methods

The basic types of monitoring methods are as follows:

Continuous Direct Monitors. Continuous monitoring systems collect, analyze, and record samples of a few seconds duration on a continuous basis. The continuous ambient monitors in most widespread use are monitors for sulfur dioxide, oxides of nitrogen, ozone, and carbon monoxide. These monitors typically require housing in environmentally-controlled shelters and must be checked frequently to maintain accuracy requirements.

Integrated Samplers. Integrated samplers collect a single sample over some predetermined period of time. The sample is then analyzed to provide the average concentration over the entire period of collection. Peak concentrations during the period cannot be determined. A common example of integrated sampling is the collection of particulate matter samples over a 24-hour period by forcing a known volume of air through a filter. The filter is weighed before and after sample collection to obtain the weight of the material collected. Another example of integrated sampling is the use of adsorption tubes or evacuated cylinders to collect gaseous organic compounds for analysis by laboratory techniques such as gas chromatography.

Remote Sensing. As exemplified by optical sensing, remote sensing involves assessment of ambient concentrations by transmitting a beam of energy across a predetermined distance and detecting and analyzing the return signal. The concentration thus measured represents an average concentration over the path length of the beam. An example of this method is *fourier transform infrared (FTIR) sensing* which can be used to detect many organic compounds.

Requirements for a Monitoring Program

The following considerations apply when ambient monitoring is needed:

Site Selection. A monitoring site (or sites) should be selected representing the approximate location where greatest impacts from sources of interest are expected. Site selection should also take into account ease of access, power availability, and adequate exposure to wind flow from all directions.

Duration of Monitoring. The minimum duration of a monitoring program depends on the purpose of monitoring. The duration may be a few days for special purpose monitoring, or year-round for monitoring stations operated by state and local agencies to determine compliance with ambient air quality standards. Seasonal considerations may also apply. Ozone concentrations, for example, are at their peak during warm season months. Therefore, detection of highest ozone concentrations requires monitoring that extends from as early as March to as late as November in the United States.

Percentage of Valid Data Collection. The conclusions that can be reached from monitoring data are determined in part by the amount of valid data obtained during the monitoring period. If data are needed to represent conditions throughout an entire year, an acceptable valid data collection percentage is required for each quarter of monitoring.

Quality Assurance. Adequate quality assurance is an important part of a monitoring program. Quality assurance involves use of approved equipment and test materials, performance of calibrations on a regular basis, frequent maintenance, and detailed documentation and recordkeeping.

Air Quality Modeling

Air quality modeling refers to the use of mathematical representations of pollutant dispersion and transformation for estimating ambient pollutant concentrations. The chief advantages of modeling are that modeling can be used to estimate concentrations at many hundreds of locations at very low cost (through use of computer programs), to evaluate the contribution of specific emission sources to total concentrations, and to predict the impact of proposed new sources not yet in operation. The chief disadvantages of modeling lie in the fact that a mathematical model can not exactly replicate the complexities of atmospheric dispersion and transformation. Models, therefore, have inherent accuracy limitations. These limitations tend to be greatest when using models to estimate concentrations at considerable distances from emission points, and when using models to estimate pollutant levels resulting from physical-chemical transformation and depletion mechanisms. Modeling is often performed in connection with permit applications to demonstrate compliance with ambient air quality limits.

Types of Data Required for Modeling

The input data required for a modeling analysis include the following:

Meteorological Data. Meteorological input data are required to characterize the dispersion characteristics of the atmosphere into which a plume is released. Simple screening models make use of hypothetical meteorological data. More detailed models use data based on actual weather observations. Data from National Weather Service surface observation stations and upper air sounding stations are typically used with detailed models. For certain types of modeling, specific on-site meteorological monitoring may have to be performed to obtain required data. The meteorological parameters needed for use with standard models consist of wind speed and direction, near-surface temperature, atmospheric stability index (to characterize turbulence conditions), and mixing height (indicating the vertical extent of the atmospheric layer within which dispersion takes place).

Emission Source Data. The types of emission source data needed for modeling include emission rates for each source modeled; location coordinates for each source; and (for sources emitting through a stack or vent) stack height, stack diameter, exit gas temperature, and exit gas volumetric flow or velocity.

Receptors. The term receptors refers to the locations at which concentrations are predicted by modeling calculations. A typical modeling evaluation starts out with a coarse receptor grid (wide spacing between receptors) and then focuses on potential areas of high concentration using a more closely spaced receptor grid.

Available Models

Air quality models can generally be categorized as screening models and refined models. Screening models are simple models, typically using hypothetical meteorological data, that are often used to determine the need for more detailed modeling and to help plan a detailed modeling evaluation. Refined models are more complicated models that require use of detailed input data. Included within the category of refined models are advanced special purpose models such as those applicable to complex terrain, coastal dispersion, simulation of photochemical reactions in urban areas, and long-range transport.

Among the most commonly used models in the United States are those approved for use by EPA, including those designated as "guideline" models. Examples include the SCREEN2 model for initial screening use and the **Industrial Source Complex Short Term (ISCST2) model** for detailed evaluations.

Good Engineering Practice Stack Height Concept

Wind flow across a tall structure or terrain obstacle can generate a turbulence pattern referred to as downwash or wake effects. If an emission source stack is located close enough to the structure or obstacle, the plume from the stack can be affected by wake effects depending on the height of the stack. The term **good engineering practice (GEP) stack height** refers to the minimum emission source stack height needed to avoid the downwash effects of nearby structures or terrain obstacles. If the actual stack height is less than GEP height, then consideration of downwash effects may be necessary in a modeling evaluation. Many of the EPA models have downwash calculation methods. Downwash effects are important because they can produce much higher predicted concentrations than would be the case in the absence of downwash.

STRATOSPHERIC OZONE DEPLETION

At the lowest level of the atmosphere (the troposphere) where humans and other organisms live, ozone is considered an air pollutant. In the stratospheric level of the atmosphere, however, ozone provides a valuable protective function. Ozone in the stratosphere (often referred to as the ozone layer) blocks out ultraviolet radiation from the sun that can cause harmful biological effects, such as increased risk of skin cancer in humans.

Upper atmosphere studies in recent years have indicated an apparent depletion of ozone concentrations during certain seasons of the year above different parts of the earth. This recurring depletion of stratospheric ozone concentrations has been linked with emissions of several types of halogenated compounds — compounds containing chlorine, fluorine, or bromine.

The compounds of greatest concern are **chlorofluorocarbons (CFC's)** commonly used as coolants in air conditioning and refrigeration systems and as industrial cleaning agents. Ironically, one of the reasons that such compounds have been in use is that they are relatively inert and do not cause harmful health effects or participate readily in photochemical reactions that cause ozone formation in the troposphere. Because of this low reactivity property, chlorofluorocarbons can remain undepleted in the atmosphere for long periods of time and can eventually be transported from the surface of the earth where they are emitted up to the stratosphere. Once in the stratosphere, these compounds can be dissociated by intense solar radiation. The dissociated components, especially chlorine, can then enter into a chain of chemical reactions that result in ozone depletion.

Several types of chlorofluorocarbons have ozone-depletion potential. Among those with the greatest potential are CFC-11, CFC-12, CFC-113, CFC-114, and CFC-115. In addition to chlorofluorocarbons, other compounds with ozone-depletion potential are certain halons (bromine-containing compounds), carbon tetrachloride and methyl chloroform (both of which are common solvents), and several types of **hydrochlorofluorocarbons (HCFC's)**.

The United States and many other industrial nations are taking steps to reduce further depletion of stratospheric ozone. These steps include a rapid elimination in the production of compounds with the highest ozone-depletion potential. Such compounds will still be in use for several years, but will be subject to stringent requirements such as capture and recycling requirements when servicing air conditioners. Eventually, the compounds with highest ozone-depletion potential will be replaced by compounds posing less risk to the ozone layer.

INDOOR AIR QUALITY

The emphasis of this chapter has been on air quality conditions in the outside air, referred to as ambient air quality. The quality of air indoors where many people spend most of their lives is also of obvious importance.

Because of the many factors that affect indoor air quality, characterizing and predicting indoor air pollutant levels in any given building can be very difficult. These factors include pollutant levels in the outdoor air that is circulated indoors, the types of construction materials and furnishings in use and their age, whether smoking is allowed and how many smokers are present, ventilation rates, and the types of filters (if any) that are used in the ventilation system. As an example, certain types of furnishings (especially when first installed) can release low levels of organic compounds.

A relatively recent term — the "sick building syndrome" — has been coined to describe indoor air quality conditions that cause irritation or even illness among sensitive individuals. Building designers and companies that provide building construction materials and furnishings have undertaken studies to pinpoint the causes of sick building syndrome and to help reduce its occurrence.

REGULATORY APPROACH TO AIR QUALITY PROTECTION

The following discussion summarizes policies in the United States as an example of the regulatory approach to control of air quality. In the U.S., the federal Clean Air Act is the basic air pollution law applicable throughout the nation. To implement the Clean Air Act, the U.S. Environmental Protection Agency adopts, revises, and rescinds regulations as needed. States and local governments also adopt laws, ordinances, and regulations that in some cases may be more stringent than federal requirements.

Overview

Legislation

The Clean Air Act was substantially revised by the Clean Air Act Amendments of 1990 that became effective on November 15, 1990. Including the 1990 amendments, the major parts (referred to as titles) of the Clean Air Act are as follows:

Title I. A comprehensive section that (1) requires development and review of national ambient air quality standards, (2) establishes requirements for areas not meeting ambient air quality standards (non-attainment areas); (3) details procedures for preventing significant deterioration of air quality in areas that are meeting standards; (4) mandates the development of emission and performance standards for many types of new sources; (5) establishes a program for controlling releases of "hazardous air pollutants"; and (6) outlines federal enforcement policies.

Title II. Pertains to motor vehicles and vehicle fuels.

Title III. Contains general provisions related to such areas as administrative proceedings and citizen suits.

Title IV. Contains acid deposition control requirements that affect emissions of sulfur dioxide and nitrogen oxides from utilities and independent power producers.

Title V. Establishes a national operating permit program which requires annual permit fees and periodic permit renewal.

Title VI. Mandates procedures for controlling the manufacture and release of stratospheric ozone depleting chemicals.

Key provisions of the federal Clean Air Act (incorporating the Clean Air Act Amendments of 1990) are as follows:

National Ambient Air Quality Standards. Ambient air quality concentration limits necessary to protect human health and welfare. National ambient standards have been set for sulfur dioxide, nitrogen dioxide, carbon monoxide, ozone, particulate matter, and lead. Nonattainment areas (areas not in attainment with national ambient air quality standards) must take steps to achieve attainment.

Prevention of Significant Deterioration (PSD). Applicable to attainment areas, a policy for minimizing the incremental increase in pollutant levels above baseline conditions. Applicants for proposed "major" sources in PSD areas must demonstrate that emissions will be controlled with the best available control technology.

New Source Performance Standards. Emission limiting standards that must be achieved by designated types of emission sources.

Hazardous Air Pollutants. A separate list of pollutants in addition to those regulated by ambient air quality standards. A total of 189 hazardous substances and categories of substances are listed in the Clean Air Act Amendments of 1990. Designated emission sources must apply maximum achievable control technology to comply with national emission standards for hazardous air pollutants.

Operating Permit Program. A program requiring permits for the operation of many emission sources. When it becomes effective, this program will require compliance reports, periodic permit renewal, and annual fees based on the quantity of emissions.

Mobile Source Standards. Requirements for vehicle emission control equipment and for motor fuel characteristics.

Acid Deposition Control Requirements. Procedures for the reduction of sulfur dioxide and nitrogen oxides from power plants. These requirements include the concept of an "allowance" program for sulfur dioxide emissions.

Stratospheric Ozone Protection. A program to implement a phase-out in the manufacture and use of chemicals that cause depletion of ozone in the stratosphere.

Types of Air Quality Regulations

Air quality regulations are basically divided into two types: ambient based regulations and emissions based regulations. Ambient air quality regulations establish pollutant concentration limits in the ambient atmosphere where people, vegetation, building materials, etc. come into contact with air pollutants. Ambient based regulations typically do not specify restrictions on the amount of pollutants that can be emitted by an individual source. Examples of ambient based regulations are national ambient air quality standards and allowable increases in ambient concentrations specified by PSD "increments."

Emission based regulations, on the other hand, establish emission standards and other control technology standards that must be met by specific types of sources. These regulations do not typically contain limits on the ambient concentrations that may result if emission based standards are met. Examples of emission based regulations are the federal new source performance standards and the national emission standards for hazardous air pollutants.

Types of Pollutants from a Regulatory Perspective

As previously discussed, air pollutants can be broadly classified as ***primary pollutants*** and ***secondary pollutants.*** The distinction between the two types of pollutants is important to the development of regulatory strategies for reducing ambient pollutant concentrations. Ambient concentrations of a primary pollutant such as sulfur dioxide can be controlled by reducing emissions from direct emission sources such as combustion sources that burn sulfur-containing fuels. Controlling the concentrations of a secondary pollutant such as ozone is more complicated. Control strategies for ozone must be based on information or assumptions about the types of precursor emissions that contribute to ozone formation, the effect on photochemical reactions of different mixtures of precursors, and the precursor emission sources that can be controlled with the greatest effect and at the least cost. Control of precursor emissions can involve a large segment of the public as exemplified by automobile inspection and maintenance programs in certain areas with high ozone concentrations.

Another term used to describe certain types of pollutants is the term criteria pollutants. ***Criteria pollutants*** are those for which a national ambient air quality standard has been established.

Still another air pollutant term often used is the term **hazardous air pollutant.** Related to federal laws and regulations, this term currently refers to the 189 substances and substance categories listed in the Clean Air Act. Note, however, that the act contains procedures for adding and deleting hazardous air pollutants so that the original list of 189 may expand or contract.

Finally, another commonly used term is **toxic air pollutants** or **air toxics.** This term is much more vaguely applied than are the other terms. Typically, toxic air pollutants are those pollutants referenced in state and local regulations and policies to encompass substances that may be harmful and that are not regulated by an ambient air quality standard. Toxic air pollutants include the federally defined hazardous air pollutants, but can include many other substances as well.

Attainment and Nonattainment Areas

An attainment area for a given pollutant is a geographic area that complies with the national ambient air quality standards for that pollutant. Conversely, a nonattainment area for a given pollutant is one that does not comply.

Since attainment and nonattainment designations are pollutant-specific, the same geographic area can be both an attainment area and a nonattainment area. For example, an area might be an attainment area for sulfur dioxide while at the same time a nonattainment area for ozone. Understanding the pollutant-specific nature of attainment-nonattainment designations is important for two reasons. First, depending on its location, a new or modified emission source may be subject to both attainment area regulations and nonattainment area regulations. Second, if a new or modified source is located in a nonattainment area, it will not be subject to nonattainment regulations if it does not emit the nonattainment pollutant (or the precursors to the nonattainment pollutant in the case of a secondary nonattainment pollutant such as ozone).

Attainment areas are sometimes referred to as **PSD areas.** This terminology is in reference to the fact that prevention of significant deterioration (PSD) regulations apply only to pollutants for which an area is in attainment. Restating one of the points above, a new or modified emission source may be subject to both PSD and nonattainment regulations depending on its location and on the pollutants (and pollutant quantities) it emits.

State and Local Programs

In addition to the federal Clean Air Act and the federal regulations implementing the act, state and local laws, regulations, and policies also affect air quality control. For convenience, all programs below the federal level will be referred to here as state programs, but local requirements should not be overlooked wherever they apply.

With the approval of the EPA, states can set up programs to enforce federal requirements, and in fact are usually encouraged to do so. When implementing a federal requirement, state regulations and procedures must be at least as stringent as the federal requirements they implement, if not more stringent. In addition, states can adopt regulations and policies that address subjects not even covered by federal programs. For example, state programs for control of toxic air pollutants often go well beyond anything required at the federal level.

In the context of state air pollution control programs, an important term is **state implementation plan, or SIP.** The SIP consists of the assortment of written rules, regulations, policies, etc. that a state draws up to meet binding federal requirements. The SIP must be approved by EPA initially and whenever it is revised either for general purposes or for purposes of accommodating changes at specific emission sources affected by the SIP.

Prevention of Significant Deterioration Requirements

The **PSD concept** appears in Title I of the Clean Air Act and is implemented through federal and state regulations. The essence of the concept is that, in areas where ambient air quality standards are being achieved, air quality concentrations should not be allowed to increase above certain defined levels. These allowable increases are referred to as PSD increments. Such increments have been established for three pollutants: sulfur dioxide, particulate matter, and nitrogen dioxide. As will be discussed, however, other pollutants are also subject to PSD regulations.

PSD permitting requirements are examined here as a case study of the complexities that can arise in securing a construction permit for a new project. The detailed permitting process that has arisen around the PSD concept is often the driving force in the environmental permitting schedule for a new or modified emission source, and can also have substantial impacts on project costs. Allowing for PSD permitting compliance is therefore an important element of industrial project planning. All applicable PSD requirements must be satisfied before a construction permit can be issued.

PSD Area Categories

Three PSD geographic area categories exist. PSD Class I areas are pristine areas such as large national parks where only minor increases in pollutant concentrations are permitted. Much greater increases are allowed in PSD Class II areas to accommodate industrial growth, and still larger increases in PSD Class III areas. Thus far, no PSD Class III areas have been designated. Most locations in the United States fall within PSD Class II areas. PSD Class I areas were created by definition in the Clean Air Act and exist in most regions of the U.S., although not in every state.

Emission Sources Subject to PSD Review

A proposed new emission source or source modification is subject to PSD regulations if it is a "major" source or "major" modification. A new source is major if it is one of 28 listed sources and will emit more than 100 tons per year (ton/yr) of a regulated pollutant, or if it will emit more than 250 ton/yr of a regulated pollutant regardless of its source type. A planned modification is major if it will occur at an existing major source and will cause emission increases of regulated pollutants above "significant" emission rate levels defined in the PSD regulations.

Pollutants Subject to PSD Review

The pollutant for which an emission source is major is obviously subject to review under the PSD regulations. Other regulated pollutants that will be emitted above defined significant emission rates are also subject to review. Significant emission rates have been defined in PSD regulations for several pollutants.

Potential PSD Permitting Requirements

Permitting requirements potentially applicable to PSD-type projects include the following. These requirements would normally apply only to pollutants that will be emitted in "significant" amounts.

Preconstruction Monitoring. If necessary to characterize air quality and dispersion conditions in the project site area, the permitting agency may require the applicant to carry out an ambient air quality monitoring program prior to submittal of an application. However, exemptions from monitoring may be granted at the discretion of the permitting agency. One basis for an exemption is that representative monitoring data already exist for the site area.

Air Quality Modeling. Air quality dispersion modeling is typically required as part of preparing a permit application subject to PSD regulations. The purpose of modeling is to (1) show compliance with PSD increments, (2) show compliance with ambient air quality standards, (3) show compliance with state toxic air pollutant ambient concentration requirements (if any), and (4) provide the basis for a preconstruction monitoring exemption request.

Best Available Control Technology Evaluation. For each pollutant subject to PSD review, the applicant must show that proposed emission controls represent best available control technology (BACT). In addition, non-regulated toxic air pollutants must sometimes be assessed in the BACT evaluation. The selection of BACT from the various control method alternatives that may be applicable typically takes into account economics (the initial costs and annual operating costs of alternative methods), energy use, and environmental considerations including non-air quality impacts (such as water quality and waste disposal impacts).

Additional Evaluations. PSD regulations also require that an assessment be made concerning the effect of project emissions on vegetation, soils, and visibility. In addition, the potential air quality effects of any growth associated with the project must be addressed.

PSD Class I Areas. The PSD Class I areas are federal lands under the administration of an agency serving as the federal land manager. Satisfying the air quality concerns of federal land managers has become an increasingly important (and increasingly difficult) part of the PSD permitting process. Projects located more than 100 kilometers (km) from a PSD Class I area may be exempt from a PSD Class I review. However, projects located at greater distances may be reviewed.

Emission Offsets

The term *emission offsets* refers to the concept of achieving emission reductions at an existing emission source to compensate for increased emissions from a new source. In a nonattainment area, emission offsets may be a prerequisite for obtaining a permit.

The need for emission offsets most often appears in the context of reducing emissions of ozone precursors, such as **volatile organic compounds (VOCs)**. The following discussion will therefore focus on ozone nonattainment areas. However, the emission offset concept can apply to other pollutants as well. For example, sulfur dioxide emission offsets have been used to satisfy concerns about impacts on PSD Class I areas.

Offset Pollutants Related to Ozone

In most areas of the U.S., strategies aimed at controlling ozone levels have concentrated on reducing emissions of VOCs. Oxides of nitrogen (NO_x) were also recognized as contributing to ozone formation, but were not addressed as intently as volatile organic compound emissions. Therefore, emission offsets related to ozone almost always meant VOC offsets.

The role of NO_x emissions in ozone formation has become officially established, however, as a result of the Clean Air Act Amendments of 1990. As a result, NO_x emission offsets are expected to become of increasing concern with respect to proposed new emission sources in ozone nonattainment areas.

Offset Ratio

The term *offset ratio* refers to the relative quantity of emission reductions needed to compensate for increased emissions. In ozone nonattainment areas, VOC emission offsets must be greater than one for one. That is, the offsetting reduction in emissions must be greater than the proposed increase in emissions.

The Clean Air Act lists required nonattainment area VOC emission offsets depending on the degree to which the ozone ambient standard is exceeded. VOC emission offset ratios range from 1.1 to 1 in "marginal" nonattainment areas (areas in which the ozone standard is barely exceeded) to 1.5 to 1 in "extreme" nonattainment areas. Specific NO_x offset requirements and ratios are not established in the Clean Air Act.

Offset Credibility Requirements

To be credited as an acceptable emission reduction, offsets must meet four tests: they must be surplus, enforceable, permanent, and quantifiable.

Surplus. The term surplus means that the proposed emission reduction has not been credited or required for any other purpose.

Enforceable. Offsets must be established through a mechanism that can be enforced. A permit revision, for example, is an enforceable action.

Permanent. Emission reductions must remain in effect. Implementing an emission reduction for a limited time and then restoring emissions to the pre-reduction level would not be considered permanent.

Quantifiable. Although seemingly a straightforward requirement, quantifying an emission reduction may not be a simple procedure. This is partly because offsets are to be based on actual rather than allowable emissions. Operating records do not always allow a simple determination of actual emissions.

General Air Quality Permitting Process

The air quality permitting process can be divided into the application preparation phase and the application review phase.

Application Preparation Phase

A recommended early step during the application preparation phase is a pre-application meeting (or meetings) with the permitting agency. At this time, potential requirements related to preconstruction monitoring, modeling, control technology, and emission offsets can be discussed.

One of the next steps is to assess the need for preconstruction monitoring. If an exemption from monitoring does not appear possible, then setting up a monitoring program should begin immediately — starting with development of a monitoring plan and gaining acceptance for the plan from the permitting agency.

Most projects subject to PSD regulations will require a modeling analysis. Development of a modeling plan (or "protocol" as it is often called) is recommended for many projects, and in some states is required. The protocol should be reviewed and approved by the permitting agency before the applicant undertakes extensive modeling. The protocol lists the models that will be applied and the pollutants that will be modeled, identifies the meteorological input data that will be used, and either lists the sources that will be modeled or explains how sources will be selected for modeling.

Application Review Phase

When an application is submitted, the permitting agency will first perform a completeness review. If an application is considered incomplete, additional information is requested from the applicant.

When an application is considered complete, the permitting agency will review the application in detail and will eventually develop a preliminary determination. The preliminary determination describes the agency's findings and typically contains what amounts to a draft permit with proposed permit conditions.

Completion of the preliminary determination signals the beginning of public review (which also includes review by other cognizant government agencies). PSD regulations require that a public notice be issued announcing a 30-day public review period. If sufficient concern is expressed by the public during this period, the permitting agency may decide to hold a public hearing. If held, a public hearing must be preceded by a 30-day notification period.

The public review period is also a time when the permit applicant can assess the acceptability of proposed permit conditions. Negotiations with the permitting agency are often held to reach an approach acceptable to both parties when a proposed permit item is in question.

Following public review and any negotiations between the applicant and the permitting agency, the permit is either issued or denied. If the permit is issued, project construction can begin.

Individuals or entities not satisfied by the permitting agency's permit decisions can file an appeal. Such appeals are usually heard first by an administrative officer. If the outcome of this administrative appeal is still unsatisfactory to the party who filed the appeal, this party can then file suit in the judicial system if sufficient grounds for a suit exist.

Schedule for Construction Permits

The time required for preparation of a permit application and for application review and permit issuance is highly variable depending on the type of project, the location of the project, preconstruction monitoring requirements, etc. For major projects subject to PSD and/or nonattainment area permitting requirements, obtaining permits can take many months. The key points to remember are that construction cannot begin until the construction permit is obtained and that adequate time for permitting should be allowed when planning a project.

Clean Air Act Operating Permit Provisions

The permitting discussions up to this point have been directed primarily at construction permit requirements. Permits are also required for facility operation. The Clean Air Act Amendments of 1990 added a new national operating permit program to be administered by individual states. By the mid-1990's, this new operating permit program should be in effect nationwide.

Owners of sources affected by operating permit regulations will have to submit an application and compliance plan to obtain a permit. After a permit has been issued, monitoring and reporting requirements will apply. In addition, annual fees will be imposed. The Clean Air Act recommends a minimum fee of $25 per ton (1989 dollars) of each regulated pollutant emitted, with fees not imposed on individual pollutant emissions above 4,000 tons per year. States can require greater fees, but can not adopt lower fees without adequate justification. Fees are to be adjusted annually in accordance with changes in the Consumer Price Index.

The specific types of sources subject to national operating permit requirements will be defined in implementing regulations. All sources emitting more than 100 ton/yr of a single regulated pollutant will require a permit. Sources subject to regulations such as new source performance standards and national emission standards for hazardous air pollutants will also require an operating permit under the new program. Minor sources may also eventually require a national operating permit. When the operating permit program is fully implemented, thousands of emission sources will be affected.

Hazardous Air Pollutant Control Requirements

The Clean Air Act Amendments of 1990 require EPA to identify sources of hazardous air pollutants that should be regulated and to develop stringent emission standards for these sources. The control methods that will be required are referred to as **maximum achievable control technology (MACT)** and will be adopted as **national emission standards for hazardous air pollutants (NESHAP).** MACT standards will be developed for both existing and new sources. These standards will be implemented over a period of several years. The MACT/NESHAP program constitutes one of the most far-reaching provisions of the Clean Air Act.

EPA has designated more than 170 categories of emission sources for which national emission standards for hazardous air pollutants will be developed. Examples of emission source categories that will be subject to MACT standards are the following: synthetic organic chemical manufacturing industry (SOCMI), styrene-butadiene rubber and latex production, chromium electroplating, polystyrene production, wood furniture surface coating, chlorine production, iron foundries, petroleum refineries, pharmaceuticals production, pulp and paper production, and semiconductor manufacturing. Virtually all of the major industrial manufacturing and production sectors in the U.S. will have some type of emission source subject to MACT standards.

Clean Air Act Enforcement Provisions

Tougher enforcement provisions are now in the Clean Air Act as a result of the 1990 amendments. Several types of violations can be treated as criminal violations, and Clean Air Act criminal violations are considered felonies. A sentence of up to 15 years in prison can be imposed for the most extreme violations. In addition, for the first time EPA has the authority to assess an administrative penalty without initiating a judicial proceeding. Commencement of a civil judicial action also remains an option to EPA. Fines for violations can be as much as $25,000 per day for each violation.

REFERENCES

Air Pollution Engineering Manual, Buonicore, A.J. and Davis, W.T. (eds), Van Nostrand Reinhold, New York, 1992.

Building Air Quality, Government Institutes, Rockville, MD, 1992.

Clean Air Act, 42 U.S.C. 7401 *et seq.* (including the Clean Air Act Amendments of 1990, P.L. 101-549).

Clean Air Handbook, Brownell, F.W. and Zeugin, L.B., Government Institutes, Rockville, MD, 1991.

Compilation of Air Pollutant Emission Factors, Volume I: Stationary Point and Area Sources (through Supplement E, October 1992), U.S. Environmental Protection Agency, Office of Air Quality Planning and Standards, Publication No. AP-42. [Available through the Government Printing Office.]

Fundamentals of Environmental Science and Technology, Knowles, P.C. (ed), Government Institutes, Rockville, MD, 1992.

Guideline on Air Quality Models (through Supplement B, 1993), U.S. Environmental Protection Agency, Office of Air Quality Planning and Standards, Publication No. 450/2-78-027R).

Journal of The Air & Waste Management Association (formerly, *Journal of the Air Pollution Control Association*), published monthly by the

Air & Waste Management Association, Three Gateway Center, Four West, Pittsburgh, PA 15222.

Rethinking the Ozone Problem in Urban and Regional Air Pollution, National Research Council, National Academy Press, Washington, D.C., 1991.

6

AIR POLLUTION CONTROL TECHNOLOGIES

Perry W. Fisher, Ph.D., Kaushik Deb
Dames & Moore

OVERVIEW

Central to any environmental effort striving to reduce air pollutant emissions from industrial activity, air pollution control (APC) technologies play a very pivotal role. In 1991, projected APC system orders for North America alone totalled $3,860 million, for Europe $4,085 million, and for the rest of the world, $4,077 million.[1] Coupled with the passage into law of the most aggressive piece of U.S. environmental legislation to date—the Clean Air Act Amendments (CAAA) on November 15, 1990—this multi-billion dollar industry is on the brink of unprecedented growth. It is with evolving APC technologies that the goals of the CAAA regulations and rules will be achieved.

AIR POLLUTANT CHARACTERIZATION

Most industrial activities—from utilities, pulp and paper manufacture, steelmaking, automobile manufacture to small-scale businesses like chemical resin production, laundries and bakeries—produce airborne emissions of particulates or gases. Typically, particulates are classified as either SPM (suspended particulate matter), TSP (total suspended particulates) or simply PM (particulate matter). For human health purposes, the fraction of particulates which has been shown to contribute to respiratory diseases is termed PM_{10} (i.e., PM with sizes less than 10 microns). From a control standpoint, particulates can e fully characterized by their following attributes:

(i) Particle size distribution (typically determined by Bahco analysis).
(ii) Particle concentration or loading in the exhaust airstream (expressed as mg/m^3).
(iii) Some physical properties for specific control applications, e.g., particle density, resistivity, etc.

Gaseous pollutants are similarly characterized by chemical species identification, e.g., inorganic gases such as sulfur dioxide (SO_2), nitrogen oxides (NO_x), and carbon monoxide (CO) or organic gases such as chloroform ($CHCl_3$) and formaldehyde (HCHO). The rate of release or concentration in the exhaust airstream (in parts per million or comparable units) along with the type of gaseous pollutant greatly predetermines the applicable control technology.

Apart from the difference in types of air emissions, i.e., particulates and gases, it is important to note that under actual plant conditions both mixed-phase and phase transference of particulates and gases can occur. Thus, from a given source which needs to be controlled, air pollutant emissions often occur as both particulate matter and gases. In this regard, the phenomenon widely referred to as gas-to-particle conversion (GPC) occurs fairly commonly.

Under such circumstances, it is imperative that the proposed control should address only the phase of the pollutant of concern dictated by the requirements of the emissions reduction program. This is necessary because most control systems are designed to be pollutant-specific and fairly phase-specific. For instance, a system for particulate control will be designed to attain a specified efficiency for particulate matter reduction; in addition, it may or may not achieve some minimal reductions in gaseous emissions through a variety of physical and chemical processes. To convince a regulatory body that the same control is effecting an emissions reduction for gaseous pollutants will usually require appropriate source sampling data.

Besides phase, air pollutants are also characterized as:

(i) Criteria Pollutants—PM, PM_{10}, SO_2, CO, NO_x volatile organic compounds (VOCs), and lead (Pb); and

Air Pollution Control Technologies / 157

  (ii) Hazardous Air Pollutants (HAPs)includes, but is not limited to, many individual species of VOCs and PM.

Many criteria pollutants are fossil-fuel combustion derived, while HAPs can virtually be emitted from any segment of industrial activity.

IN-PROCESS AIR POLLUTION CONTROL

Generally, APC technology refers to end-of-pipe or add-on controls to reduce air emissions of specific pollutants. However, with the evolutin of technology and a better understanding of pollutant formation mechanisms (e.g., combustion elements such as burner and furnace configuration and design, flame temperature, catalyst properties, chemical kinetics, etc.) it is presently possible for many facilities to retard the formation of air pollutants in place by means of oxidation and other chemical reactions, rather than add-on controls. The benefits of such an approach include:

- Elimination of the capital and O & M (operation and maintenance) costs of additional pieces of control equipment;
- Elimination of process perturbations due to transient shutdowns (for maintenance or otherwise) of add-on controls; and
- If the piece of control equipment has no recovery potential (solvents, cement dust, etc.), it is a production overhead in the plant managers psyche.

However, the advocacy of in-process APC is severely hampered by the fact that the option is heavily dependent on commercially available and proven technology, and is also not readily amenable to retrofit situations. Further, the technology can be very process-specific and even plant-specific. Large corporations spend millions of research dollars fine-tuning their processes; inadvertently or deliberately this may result in abatement of specific pollutants. In many cases, details about these processes are considered to be proprietary information and protected under a confidentiality agreement with the state regulatory agency.

Various approaches by which an industrial facility may reduce its air emissions with in-process controls are described below:

  A. **Combustion Reactions:** Most of the common fossil fuels fired for steam and/or power generation include grades of coal (bituminous, sub-bituminous, anthracite, lignite and others), natural gas and #2, #5, and #6 fuel oils. Other fuels may include bark, wood waste, sludge, pulp liquor and refuse derived fuel (RDF). This excludes incinerators which burn a wide variety of wastes and are mostly controlled through fuel preparation and post combustion emission control. In-process APC for combustion includes the following approaches depending upon specific target pollutants:

(i) Fuel Substitution—Based upon established and widely used air pollutant emission factors for different fuels[2], the descending order of so-called clean fuels for most criteria pollutants arenatural gas, #2 fuel oil, blends of other oils, and grades of coal. Thus, the most rational in-process control, if feasible, involves fuel substitution, use of dirtier topping fuels with a cleaner primary fuel, and other permutations so as to optimize fuel requirements, costs and air emissions.

(ii) Combustion Modification—Historically, NO_x and CO have been very successfully controlled by combustion modification techniques. Although both pollutants are essentially combustion byproducts, it should be noted that NO_x and CO reduction schemes generally work against each other. Levels of CO which ar formed as an intermediate product of combustion increase with most combustion-modification-based NO_x control strategies.[3] Tables 1 and 2 present prevalent combustion modification approaches and strategies for these pollutants for two different boiler and fuel combinations. Various combustion modification techniques used widely for reducing combustion-derived NO_x emissions are reduced NO_2 and O_2 levels, low excess air (LEA), reduced peak temperature, reduced exposure time, optimum burner design, and staged combustion.

Also, continuous monitoring of the flue gases for either excess oxygen or CO, fuel analyses, fuel preparation, pressure and temperature measurements, flame appearance and even periodic stack tests will determine whether any combustion modification scheme is operating properly. For control of individual or total VOCs, temperature modulation (to prevent thermal degradation of the organics) and flares are often used for control purposes.

Table 1

Combustion Modification Schemes for NO_x and CO Control

Boiler: Pulverized Coal/Cyclone

Control Approach	Applicable Controls	NO_x Control Feasability	Commerical Availability
Decrease Primary Flame Zone Oxygen	• Low Excess Air (LEA)	Increase in CO	Available
	• Staged Combustion – Overfire Air injection (OFA)	N/A for cyclone	N/A
	– Reduced Air Flow to Burners	Increase in CO	Available
	– Burners out of Service (BOOS)	Not effective	N/A
	• Low NO_x Burners (LNB)	N/A for cyclone retrofit	N/A
Decrease Residence Time at High Temperature	• Flue Gas Recirculation (FGR)	Not effective	Available
	• Reburining	Effective in R&D not fully field tested	N/A
	• Burner Redesign	Not effective	Available
	• Load Reduction	Adverse operational impacts	Available

Note: Above unit is a standby and peak demand unit which experiences substantial load variations.

Boiler: Natural Gas/Watertube with Air Preheater

Control Approach	Applicable Controls	NO_x Control Feasability	Commerical Availability
Decrease Primary Flame Zone Oxygen	Low Excess Air (LEA)	Increase in CO	Available
	Staged Combustion	Not effective	Available
	Low NOx Burners (LNB)	Effective	Available
Decrease Residence Time at High Temperature	Reduced Air Preheat (RAP)	Significant loss in boiler efficiency	Available
	Flue Gas Recirculation (FGR)	Effective	Available
	Load Reduction	Not effective	N/A

Note: Above unit has modulated firing rates in response to variable plant steam demands.

Table 2

Post-Combustion Strategies for NO_x and CO Control

Boiler: Pulverized Coal/Cyclone

Control Approach	Applicable Controls	NO_x Control Feasability	Commerical Availability
Post-Flame Region NOx Reduction	Selective Catalytic Reduction (SCR) – Ammonia	Excessive energy requirements for temperature elevation, performance degradation at variable loads.	Available
	Selective Non-Catalytic Reduction (SNCR) – Exxon Thermal DeNO$_x$ – NO$_x$—out by Fuel Tech	Difficult to ensure residence time ~0.5 sec at 1600 to 1900°F	Available

Note: Above unit is a standby and peak demand unit which experiences substantial load variations.

Boiler: Natural Gas/Watertube with Air Preheater

Control Approach	Applicable Controls	NO_x Control Feasability	Commerical Availability
Post-Flame Region NOx Reduction	(SCR) – Ammonia	Excessive energy requirements for temperature elevation, performance degradation at variable loads.	Available
	(SNCR) – Exxon Thermal DeNO$_x$ – NO$_x$—out by Fuel Tech	Difficult to ensure residence time ~0.5 sec at 1600 to 1900°F	Available

Note: Above unit has modulated firing rates in response to variable plant steam demands.

B. **Process Reactions:** *Various* process reactions which are employed for the purpose of in-process control include:

(i) Reactant substitution and/or reformulation;
(ii) Better control of reaction environment, e.g., lower reaction temperatures;
(iii) Use of passive catalysts which interfere with pollutant formation; and
(iv) Solvent recovery and reuse, if possible.

With the understandable reluctance of most facilities to divulge any details pertaining to their processes, dissemination of information with respect to this approach is limited.

ADD-ON AIR POLLUTION CONTROL

A. **Particulate Control:** Historically, particulate control has been one of the primary concerns of regulatory agencies and industries alike, since emissions of particulates are readily perceived in short-range depositions of flyash and soot and in impairment of visibility. Differing ranges of control can be achieved with the following equipment:

(i) Cyclones—wet, dry, axial flow, multicyclones, etc.
(ii) Fabric Filters—shaker type, reverse-air, pulse jet, etc.
(iii) Wet Scrubbers-venturi type, packed bed, etc.
(iv) Electrostatic Precipitators—field number types, hot-side, cold-side, etc.

Upon proper characterization of the particulate matter emitted by a specific process, the appropriate piece of equipment can be selected, sized, installed, and performance tested. It is beyond the scope of this chapter to detail specific types of equipment currently available; however, the fundamental principles of operation for each of the broad classes of particulate control devices are presented below:

(i) Cyclones—PM is removed by centrifugal forces generated by providing a path for the carrier gas to be subjected to a vortex-like spin. Cyclones are very effective in removing coarser fractions of PM. The equipment can be arranged in either parallel or series to both increase efficiency and decrease pressure drop. Figure 1 shows the schematic of a typical dust cyclone.

(ii) **Fabric Filters**—For industrial applications devices are typically designed with non-disposable filter bags. As the dusty airstream flows through the filter media (typically cotton, polypropylene, teflon or fiberglass), PM is collected on the bag surface as a dust cake. Fabric filters are generally classified based on the filter bag cleaning mechanism employed. Figure 2 shows a pulse-jet type fabric filter system.

(iii) **Wet Scrubbers**—A counter-current spray liquid is used to remove particles from an airstream. Device configurations include plate scrubbers, packed beds, orifice scrubbers, venturi scrubbers, and spray towers, individually or in various combinations. Wet scrubbers can achieve high collection efficiencies at the expense of prohibitive pressure drops.

(iv) **Electrostatic Precipitators**—ESPs operate on the principle of imparting an electric charge to particles in the incoming airstream, which are then collected on an oppositely charged plate across a high voltage field. The dust cake is then collected from the plate by striking it with rappers. The dust collection efficiency is a strong function of dust resistivity. Figure 3 shows a schematic of the principle of electrostatic precipitation.

Tables 3 and 4 show typical characteristics, advantages and limitations of the above types of equipment. Sizing considerations, predictive equations for collection efficiency, failure rate, and economic considerations are important criteria which are discussed extensively in the literature.[8]

B. **Gaseous Pollutant Control:** Unlike particulate control a discussion of gaseous pollutant control does not lend itself readily to a description of equipment types, ranges of efficiency and pressure drop due to the dependence of gaseous pollutant control on the specific chemistry involved. Rather, a discussion of individual processes manifest in the various types of control euipment employed for these pollutants is more ap-

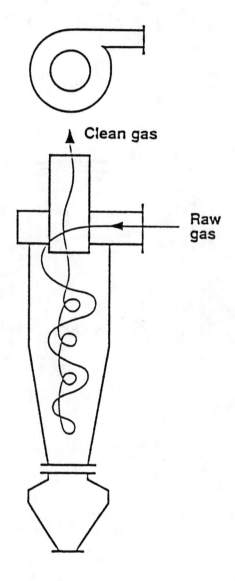

Figure 1

Dust Cyclone

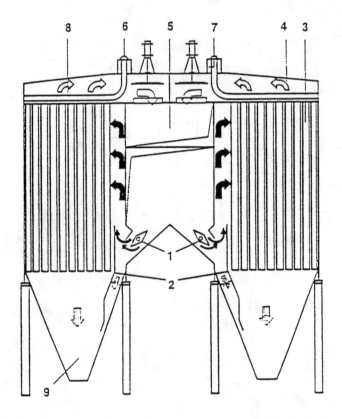

1 Raw gas inlet	4 Clean gas space	7 Compressed air reservoir
2 Baffle plate	5 clean gas duct	8 Nozzle lance
3 Filter bags	6 Diaphragm valve	9 Dust hopper

Figure 2

Pulse-jet Fabric Filter System

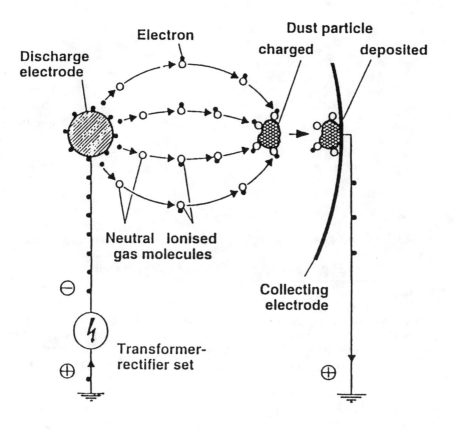

Figure 3

Principle of Electrostatic Precipitation

Table 3

Typical Characteristics of Particular Control Equipment

Particle Collection Device	Inlet Concentration (mg/m³)	Typical Overall Collection Efficiency (weight %)	Pressure Loss (in. H_2O)	Maximum Gas Flow Rate (10^3 cfm)	Relative Space Required
Cyclone	2000	85	0.5–3	17.7	Medium
Multiple cyclone	2000	95	2–6	70.6	Small
Fabric filter	200	99	2–6	70.6	Large
Electrostatic precipitator	200	99	02.–1	706.2	Large
Wet scrubbers					
— Gravity Spray	2000	70	1	35.3	Medium
— Centrifugal	2000	90	2–6	35.3	Medium
— Impingement	2000	95	2–8	35.3	Medium
— Packed bed	600	90	1–10	17.7	Medium
— Submerged orifice	200	90	2–6	17.7	Medium
— Venturi	200	99	10–30	35.3	Small

Source: ASHRAE, 1983.

Table 4

Advantages and Disadvantages of Particulate Control Equipment[9]

Advantages and Disadvantages of Cyclone Collectors

Advantages:
1. Low cost of construction.
2. Relatively simple equipment with few maintenance problems.
3. Relatively low operating pressure drops (for degree of particulate removal obtained) in the range of approximately 2- to 6-inch water column.
4. Temperature and pressure limitations imposed only by the materials of construction used.
5. Dry collection and disposal.
6. Relatively small space requirements.

Disadvantages:
1. Relatively low overall particulate collection efficiencies, especially on particulates below 10 mm in size.
2. Inability to handle tacky materials.

(continued on next page)

Table 4 *(continued)*

Advantages and Disadvantages of Fabric-Filter Systems

Advantages:
1. Extremely high collection efficiency on both coarse and fine (submicrometer) particulates.
2. Relatively insensitive to gas-stream fluctuation; efficiency and pressure drop relatively unaffected by large changes in inlet dust loadings for continuously cleaned filters.
3. Filter outlet air capable of being recirculated within the plant in many cases (for energy conservation).
4. Collected material recovered dry for subsequent processing or disposal.
5. No problems with liquid-waste disposal, water pollution, or liquid freezing.
6. Corrosion and rusting of components usually not problems.
7. No hazard of high voltage, simplifying maintenance and repair and permitting collection of flammable dusts.
8. Use of selected fibrous or granular filter aids (precoating), permitting the high-efficiency collection of submicrometer smokes and gaseous contaminants.
9. Filter collectors available in large number of configurations, resulting in a range of dimensions and inlet and outlet flange locations to suit installation requirements.
10. Relatively simple operation.

Disadvantages:
1. Temperatures much in excess of 288fC (550fF) requiring special refractory mineral or metallic fabrics that are still in the developmental stage and can be very expensive.
2. Certain dusts possibly requiring fabric treatments to reduce dust seeping or, in other cases, assist in the removal of the collected dust.
3. Concentrations of some dusts in the collector (~50 g/m^3) forming a possible fire or explosion hazard if a spark or flame is admitted by accident; possibility of fabrics burning if readily oxidizable dust is being collected.
4. Relatively high maintenance requirements (bag replacement, etc.).
5. Fabric life possibly shortened at elevated temperatures and in the presence of acid or alkaline particulate or gas constituents.
6. Hygroscopic materials, condensation of moisture, or tarry adhesive components possibly causing crusty caking or plugging of the fabric or requiring special additives.
7. Replacement of fabric possibly requiring respiratory protection for maintenance personnel.
8. Medium pressure-drop requirements, typically in the range 4- to 10-inch water column.

(continued on next page)

Table 4 *(continued)*

Advantages and Disadvantages of Wet Scrubbers

Advantages:
1. No secondary dust sources.
2. Relatively small space requirements.
3. Ability to collect gases as well as particulates (especially "sticky" ones).
4. Ability to handle high-temperature, high-humidity gas streams.
5. Capital cost low (if wastewater treatment system, not required).
6. For some processes, gas stream already at high pressures (so pressure-drop considerations may not be significant).
7. Ability to achieve high collection efficiencies on fine particulates, (however, at the expense of pressure drop).

Disadvantages:
1. Possible creation of water-disposal problem.
2. Product collected wet.
3. Corrosion problems more severe than with dry systems.
4. Steam plume opacity and/or droplet entrainment possibly objectionable.
5. Pressure-drop and horsepower requirements possibly high.
6. Solids buildup at the wet-dry interface possibly a problem.
7. Relatively high maintenance costs.

Advantages and Disadvantages of Electrostatic Precipators

Advantages:
1. Extremely high particulate (coarse and fine) collection efficiencies attainable (at a relatively low expenditure of energy).
2. Dry collection and disposal.
3. Low pressure drop (typically less than 0.5–inch water column).
4. Designed for continuous operation with minimum maintenance requirements.
5. Relatively low operating costs.
6. Capable of operation under high pressure (to 150 lbf/in^2) or vacuum conditions.
7. Capable of operation at high temperatures [to $704fC$ ($1300fF$)].
8. Relatively large gas flow rates capable of effective handling.

Disadvantages:
1. High capital cost.
2. Very sensitive to fluctuations in gas-stream conditions (in particular, flows, temperatures, particulate and gas composition, and particulate loadings).
3. Certain particulates difficult to collect owing to extremely high- or low-resistivity characteristics.
4. Relatively large space requirements required for installation.
5. Explosion hazard when treating combustible gases and/or collecting combustible particulates.

(continued on next page)

Table 4 *(continued)*

6. Special precautions required to safeguard personnel from the high voltage.
7. Ozone produced by the negatively charged discharge electrode during gas ionization.
8. Relatively sophisticated maintenance personnel required.

---

propriate. In this regard, the four general processes used for gaseous pollutant control are:

(i) Adsorption;
(ii) Absorption;
(iii) Catalytic Oxidation; and
(iv) Thermal Oxidation.

Each of the above processes is briefly described below:

(i) Adsorption—This is a physico-chemical phenomenon in which the gas is concentrated on the surface of a solid or liquid. Subsequently, the captured gas can be desorbed with hot air or steam either for recovery or for thermal destruction. Usually, activated carbon is the adsorbing medium, which can be regenerated upon desorption. Adsorbers are widely used to preconcentrate a low gas concentration prior to incineration unless the gas concentration is very high in the inlet airstream. Adsorption also is employed to reduce odors from gases which have potential odor problems. The only major limitation for an adsorption system is the requirement for minimization of PM and/or condensation of liquids (e.g., water vapor) that could mask the adsorption surface and drastically reduce its efficiency. Figure 4 shows the schematic for an adsorption system.

(ii) Absorption—Absorption differs from adsorption in that it is not a physico-chemical surface phenomenon, but an approach in which the absorbed gas is ultimately distributed throughout the absorbent (liquid). The process depends only on physical solubility and may include chemical reactions in the liquid phase (chemisorption). Common absorbing media used are water, caustic, sodium carbonate, and nonvolatile hydrocarbon oils, depending on the type of gas to be absorbed. Usually, gas-liquid contactor designs which are employed are plate columns or packed beds.

(iii) Catalytic Oxidation—Predominantly used for the destruction of VOCs and CO, these systems operate in a temperature regime of 400 to 1100F in the presence of a catalyst. Without the catalyst the system would require much higher temperatures to operate. Typically, the catalysts used are a combination of noble metals deposited on a ceramic base in a variety of configurations (e.g., honeycomb-shaped) to enhance good surface contact. Catalytic systems are usually classified based on bed types such as fixed-bed (monolith or packed-bed) and fluid-bed. These systems generally have very high destruction efficiencies for most VOCs, resulting in the formation of CO_2, water, and varying amounts of HCl (from halogenated hydrocarbons). The presence of contaminants such as heavy metals, phosphorus, sulfur, chlorine, and most halogens in the incoming airstream act as poison to the system and

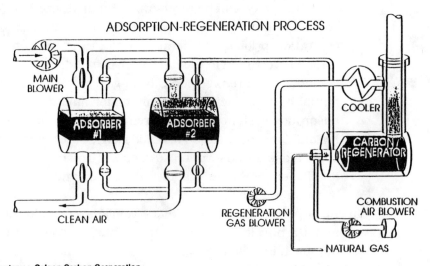

Courtesy: Calgon Carbon Corporation

Figure 4

Carbon Adsorption System

can foul up the catalyst. Recently, however, a new generation of specialty catalysts (termed HDC and made by Allied Signal) containing a unique formulation has been developed which helps breakdown most hydrocarbons and halo hydrocarbons to CO_2 + H_2O + HCl in the operating temperature range of 700 to 1000F. Figure 5 shows a typical catalytic incineration system.

(iv) Thermal Oxidation—Without the use of catalysts most thermal systems operate at around 1500°F or higher. Since the operating temperatures are roughly 500 to 1000°F higher than catalytic systems, thermal units have a much higher auxiliary fuel requirement for preheating the waste gas stream. However, if the waste gas stream has a high calorific value (and thus participates in an exothermic reaction reducing heating requirements), the auxiliary fuel requirement is appreciably reduced. Generally, there is a trade-off between higher capital costs and catalyst replacement costs for catalytic systems and higher operating costs for thermal oxidation systems. Both systems, however, afford very high destruction efficiencies for VOCs and CO in optimal operation modes. Figure 6 shows a typical thermal incineration system.

Table 5 lists the advantages and limitations of the technologies discussed above. As with particulate control devices, design considerations, predictive equations for efficiency, and economic considerations are extensively discussed in the literature.[10]

Courtesy: Anguil Environmental Systems Inc.

Figure 5

Catalytic Incineration System

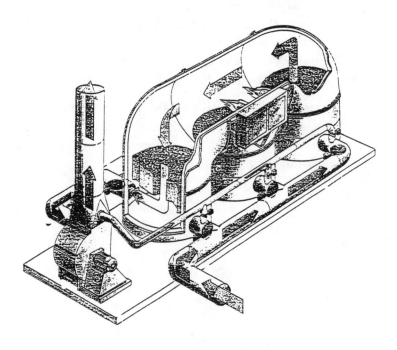

Courtesy: Salem Industries Inc.   **Figure 6**

Thermal Incineration System

Table 5
Advantages and Disadvantages of Gaseous Pollutant Control Processes[11]

Advantages and Disadvantages of Adsorption Systems

Advantages:
1. Possibility of product recovery.
2. Excellent control and response to process change.
3. No chemical-disposal problem when pollutant (product) recovered and returned to process.
4. Capability of systems for fully automatic, unattended operation.
5. Capability to remove gaseous or vapor contaminants from process streams to extremely low levels.

Disadvantages:
1. Product recovery possibly requiring an exotic, expensive distillation (or extraction) scheme.
2. Adsorbent progressively deteriorating in capacity as the number of cycles increase.
3. Adsorbent regeneration requiring a steam or vacuum source.
4. Relatively high capital cost.
5. Prefiltering of gas stream possibly required to remove any particulate capable of plugging the adsorbent bed.
6. Cooling of gas stream possibly required to get to the usual range of operation [less than 49°C (120°F)].
7. Relatively high steam requirements to desorb high-molecular-weight hydrocarbons.

Advantages and Disadvantages of Absorption Systems (Packed and Plate Columns)

Advantages:
1. Relatively low pressure drop.
2. Standardization in fiberglass-reinforced plastic (RFP) construction permitting operation in highly corrosive atmospheres.
3. Capable of achieving relatively high mass-transfer efficiencies.
4. Increasing the height and/or type of packing or number of plates capable of improving mass transfer without purchasing a new piece of equipment.
5. Relatively low capital cost.
6. Relatively small space requirements.
7. Ability to collect particulates as well as gases.

Disadvantages:
1. Possibility of creating water (or liquid) disposal problem.
2. Product collected wet.
3. Particulates deposition possibly causing plugging of the bed or plates.
4. When FRP construction is used, sensitive to temperature.
5. Relatively high maintenance costs.

(continued on next page)

Table 5 *(continuted)*

Advantages and Disadvantages of Combustion Systems

Advantages:
1. Simplicity of operation.
2. Capability of steam generation or heat recovery in other forms.
3. Capability for virtually complete destruction of organic contaminants.

Disadvantages:
1. Relatively high operating costs (particularly associated with fuel requirements).
2. Potential for flashback and subsequent explosion hazard.
3. Catalyst poisoning (in the case of catalytic incineration).
4. Incomplete combustion possibly creating potentially worse pollution problems.

SUMMARY

Air pollutant emissions are primarily controlled either through in-process control or with add-on equipment. In-process control for combustion-derived pollutants is more widely used, and the various approaches on which it is based are either commercially available or readily accessible through information sharing. For process-related pollutants which can be very process- or even plant-specific, much less is known due to prevalent confidentiality clauses utilized by most facilities in submitting this information to regulatory agencies. With respect to add-on controls, a wide variety of such approaches exist for both particulate and gaseous pollutant emissions, most of which, if appropriately applied, can significantly reduce air contaminant emissions. Most add-on control equipment is widely commercially available.

ENDNOTES

1. R. W. McIlvaine, The 1991 Global Air Pollution Control Industry, *J. Air Waste Mng. Assoc.*, v. 41, no. 3, March 1991, pp. 272–275.
2. *A Compilation of Air Pollutant Emission Factors: Stationary Sources*, USEPA AP-42, 1991.
3. *Sourcebook of NO_x Control Technology Data*, USEPA, PB91-217364, July 1991.
4. R. A. Wadden and P.A. Scheff, Engineering Design for the Control of Workplace Hazards, McGraw-Hill, 1987.

5. H. E. Hesketh, Air Pollution Control—Traditional and Hazardous Pollutants, Technomic Publishing, 1991.
6. W. M. Vatavuk, Estimating Costs of Air Pollution Control, Lewis Publishers, 1990.
7. Handbook of Control Technologies for Hazardous Air Pollutants, EPA/625/6-91/014, June 1991.
8. See notes 4-7 above.
9. *Perry's Chemical Engineers' Handbook*, 6th Edition, McGraw-Hill, 1984.
10. See notes 4-7 above
11. See note 9.

ADDITIONAL REFERENCES:

Brownell, William F. and Lee B. Zeugin, *Clean Air Handbook*, Government Institutes, Rockville, MD, 1991.

Control Technology for Hazardous Air Pollutants, Government Institutes, Rockville, MD, 1992.

7

SOLID AND HAZARDOUS WASTE TREATMENT AND DISPOSAL TECHNOLOGIES

Kenneth W. Ayers, P.E.
Willis Coroon

OVERVIEW

Throughout history, man disposed of his wastes in one of two manners: either by returning the materials to the earth, or following the discovery of fire, by burning. These two methods were completely acceptable to the agrarian society that existed in the United States prior to the Industrial Revolution. As centers of population developed and agriculture was replaced by industrialization, the quantities of wastes to be disposed also increased at staggering rates. However, the methods of handling these wastes remained the same until the latter part of this century; that is, they were either burned, buried, or simply dumped on the surface of the ground.

Starting in 1976 with the passage of the Resource Conservation and Recovery Act (RCRA), continuing with the passage of the Comprehensive Environmental Response, Compensation, and Liability Act (CERCLA or Superfund) in 1980 and the Hazardous and Solid Waste Amendments in 1984, and culminating with the Superfund Amendments and Reauthorization Act (SARA) in 1986, the management and disposal of solid and hazardous wastes have come under increasingly tighter federal and state regulation. These regulations have not only limited the types and quantities of wastes that can be interred in the ground but have also, in many instances, prescribed the type of technology that must be employed to treat the wastes. The outcome of this heightened regulation of disposal practices has been a tremendous impetus to develop new technologies and to adapt existing ones in an attempt to effectively and economically treat solid and hazardous wastes. As science and industry respond to the challenges posed by the present waste management problems, the number and types of technologies available seem to increase on a daily basis. This chapter attempts to provide an overview of both proven and emerging technologies available for the disposal of solid and hazardous wastes.

OVERVIEW OF WASTE TREATMENT TECHNOLOGIES

When considering the appropriate technology to employ to treat a specific waste, the first step is to determine whether the waste is hazardous or if it is simply a solid waste. Since this determination is driven by the requirements of the Resource Conservation and Recovery Act (RCRA), it is necessary to examine the RCRA definitions of hazardous and solid wastes. RCRA defines **solid wastes** as:

> "...any garbage, refuse, sludge from a waste treatment plant, water supply treatment plant, or air pollution control facility and other discarded material, including solid, liquid, semisolid, or contained gaseous material resulting from industrial, commercial, mining, and agricultural operations, and from community activities, but does not include solid or dissolved materials in domestic sewage, or solid or dissolved materials in irrigation return flows or industrial discharges which are point sources subject to permits... or source, special nuclear, or byproduct material as defined by the Atomic Energy Act of 1954, as amended."

Again referring to RCRA, **hazardous wastes** may be considered as:

> "...a solid waste, or combination of solid wastes, which because of its quantity, concentration, or physical, chemical, or infectious characteristics may:

(A) cause or significantly contribute to an increase in mortality or an increase in serious irreversible, or incapacitating reversible, illness; or

(B) pose a substantial present or potential hazard to human health or the environment when improperly treated, stored, transported, or disposed of or otherwise managed."

From these rather all encompassing definitions, it is clear that hazardous wastes are a subset of the broader solid waste category. Within this basic definition, hazardous wastes are normally further classified as **volatile organic compounds (VOCs), semi-volatile organic compounds (SVOCs),** metals, radioactively contaminated materials, or a mixture of any or all of these types. Tables 1-3 present representative examples of these classifications.

The technologies to manage hazardous and solid wastes fall into four general categories: thermal treatment, biological treatment, physical/chemical treatment, and methods for containment/disposal. The effectiveness of the application of each of these technology groups to a specific waste varies depending on the type of waste, the concentration and mixture of individual components of the waste, the physical phase (solid or liquid) of the material, the media (if any) in which the waste is contained, the desired level of treatment, and the final method of disposal of any remaining residue. Another consideration in selecting a treatment technology is where the wastes are to be treated. Wastes may be treated in place (in situ), within the confines of the site, or at a facility off-site (ex situ). The final major consideration is whether the waste is being treated as a virtually homogeneous stream emanating directly from a manufacturing process or is being removed from an earlier disposal site for additional treatment to comply with the CERCLA or RCRA requirements. In Figure 1, the Technology Decision Flow Chart proposes one method of selecting the appropriate technology for a given application.

Much of the current concern over the treatment of hazardous wastes centers on the remediation of waste disposal sites required under both CERCLA and RCRA. Therefore, the technologies applicable to them will be the focus of this chapter. Following this discussion, those technologies that are amenable to the treatment and disposal of solid wastes will be explored. The presentation is structured in this manner since most technologies that may be applied to hazardous waste are

Table 1
Representative Volatile Organic Compounds (VOCs)

1,1,1-Trichloroethane	Cis-1,2-Dichloroehtylene
1,1,2,2-Tetrachloroethane	Cis-1,3-Dichloropropene
1,1,2-Trichloroethane	Dibromochloromethane
1,1-Dichloroethane	Dibromochloropropane (Dbcp)
1,1-Dichloropropylene	Dibromomethane
1,2,3-Trichloropropane	Dichlorodifluoromethane
1,2-Dichloroethane	Dichloroethane
1,2-Dichloropropane	Dichloroethylene
1,2-Transdichloroethene	Dichloromethane
1,3-Dichloropropane	Dichloropropene
1,3-Trichloropropane	Ehtyl Ether
1,4-Dichloro-2-butene	Ethyl Methacrylate
2-Butanone (Mek)	Ethylbenzene
2-Chloroethyl Vinyl Ether	Iodomethane
2-Chloropropane	Isopropanol
2-Hexanone	M-psa
3-Hexanone	M-xylene
4-Methyl-2pentanone	Methane
Acetone	Methylene
Acrolein	Methylene Chloride
Acrylonitrile	O-xylene
Benzene	P-xylene
Bromodichloromethane	Polyvinyl Chloride
Bromodichloroethane	Styrene
Bromoform	Tetrachloroethene
Bromomethane	Tetrachloroethylene
Carbon Disulfide	Toluene
Carbon Tretrachloride	Trans-1,2-Dichloroethene
Chlorobenzene	Trans-1,3-Dichoropropene
Chloroethane	Trichloroethene
Chloroform	Trichlorofluoromethane
Chloromethane	Vinyl Acetate
Cis-1,2-Dichloroehtane	Vinyl Chloride

Table II

Representative Semi-volatile Organic Compounds (SVOCs)

1,2,3-Trichlorobenzene
1,2,4,5-Tetrachlorobenzene
1,2,4-Trichlorobenzene
1,2-Dichlorobenzene
1,2-Diphenylhydrazine
1,3-Dichlorobenzene
1,4-Dichlorobenzene
1-Chloroaniline
1-Naphthylamine
2,2-Dichlorobenzidine
2,3,4,5-Tetrachlorophenol
2,4,5-Trichlorophenol
2,4,6-Trichlorophenol
2,4-Dichlorophenol
2,4-Dichlorotoluene
2,4-Dimethylphenol
2,4-Dinitrophenol
2,4-Dinitrotoluene
2,6-Dichlorophenol
2,6-Dinitorotoluene
2-Chloronaphthalene
2-Chlorophenol
2-Naphthylamine
2-Nitoaniline
2-Nitrophenol
2-Picoline
3-Methylchlolanthrene
3-Methylphenol
3-Nitroaniline
4,4-Ddd
4,4-Dde
4,4-Ddt
4,6-Dinitro-o-cresol

4-Aminobiphenol
4-Bromophenyl Phenyl Ehter
4-Chloro-3-methylphenol
4-Nitroaniline
4-Nitrophenol
7,12-Dimethylbenz(A)Anthracene
A,A-dimethyl-b-phenylethlamine
Acenanthrene
Acenaphthene
Acenaphthylene
Acetoaphthylene
Aldrin
Alpha-bhc
Amiben
Aniline
Anthracene
Benzidine
Benzo(A)Anthracene
Benzo(A)Pyrene
Benzo(B)Fluorathene
Benzo(Ghi)Perylene
Benzo(K)Fluoranthene
Benzo(J)Fluoranthene
Benzo(K)Pyrene
Benzoic Acid
Benzothiazole
Benzyl Alcohol
Bis(2-Chloroehtoxy)Methane
Bis(2-Chloroethyl)Ether
Bis(Ehtylhexyl)Phthalate
Butyl Benzyl Phthalate
Chlordane
Chrysene

Table II *(continued)*

Creosote	Isophorone
Delta-bhc	Kepone
Dhd	Malathion
Di-n-octyl Phthalate	Methoxychlor
Dibenzo(A,H)Anthrazene	Methyl Ethyl Benzene
Dibenzofuran	Methylmethanesulfonate
Dibutyl Phthalate	N-methylpyrrolidene
Dimethyl Phthalate	N-nitroso-di-n-butylamine
Dinitrophenol	N-nitrosodimethylamine
Dinoseb	N-nitrosopiperidine
Diphenylamine	Naphthalene
Dnb	Nitrobenzene
Endosulfin I	Oxazolidone
Endosulfin Ii	Parathion
Endosulfin Sulfate	Pcb
Endrin	Pentachlorobenzene
Endrin Aldehyde	Pentachloronitrobenzene
Eptc	Pentachlorophenol
Ethyl Methanesulfonate	Phenacetin
Ethylamylketone (Eak)	Phenanthrene
Thylene	Phenol
Dibromide	Phenothiazine
Fluoranethene	Pronamid
Fluorene	P-dimethylaminoazobenzene
Heptachlor	Resorcinal
Heptachlorepoxide	Shell Sol 140
Hexachlorobenzene	Tdx
Hexachlorobutadiene	Tertbutylmethylehter
Hexachlorocyclohexan	Tretrahydrofuran
Hexachlorocyclopentadiene	Tnb 1,3,5-Trinitrobenzene
Hexachloroehtane	Toxaphene
Hexadecanoic Acid	Vernolate
Indeno(1,2,3-Cd)Pyrene	

Table III

REPRESENTATIVE METALS

Aluminum	Nickel
Antimony	Plutonium
Arsenic	Potassium
Barium	Radium
Berylium	Selenium
Boron	Silicon
Cadmium	Silver
Calcium	Sodium
Cesium	Strontium
Chrome	Technetium
Chromite	Thallium
Chromium	Thorium
Cobalt	Tin
Copper	Titatium
Iron	Tritium
Lead	Uranium
Magnesium	Vanadium
Manganese	Zinc
Mercury	Zirconium
Molybdenum	

184 / Environmental Science and Technology Handbook

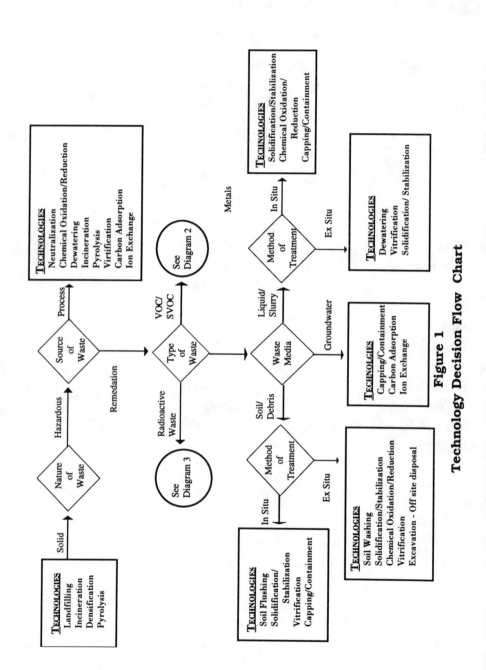

Figure 1
Technology Decision Flow Chart

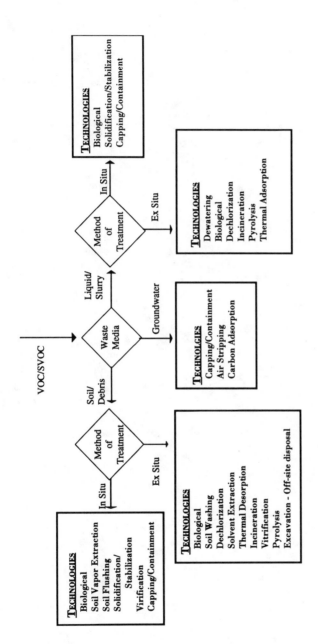

Figure 1 *(continued)*

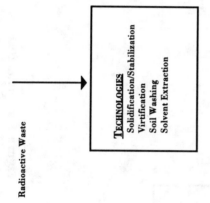

Figure 1 *(continued)*

also potentially available for solid wastes. The major difference is that the technologies currently being applied to hazardous wastes focus on the destruction of specific contaminants within the waste while solid waste management technologies focus primarily on volume reduction.

THERMAL PROCESSES

Outside of the traditional containment methods (landfilling and capping), application of some type of thermal process has, until recently, been the most common form of treatment for hazardous wastes. Thermal treatments have lost some popularity recently due to the threat of emmissions from incomplete combustion. Except for vitrification, thermal technologies are ex situ processes requiring the wastes to be transported to the processing unit.

Incineration

When most people think of thermal treatment, they think of incineration, which may be defined as the burning of substances by controlled flame in an enclosed area. Incineration detoxifies hazardous wastes by destroying organic compounds, reduces the volume of the wastes, and converts liquid wastes to solids by vaporizing any fluids present in the wastes. Its predominant use is for the destruction of VOCs and SVOCs. Incinerators have been extremely capable of destroying organic compounds in waste. Removal efficiencies as high as 99.9999 percent have routinely been achieved (this is often referred to as the "six-9's" treatment level.)

Incinerators can be designed to handle wastes in any physical state and have proven effective in treating solids, liquids, sludges, slurries, and gases. The effectiveness of an incinerator depends on three factors:

- Temperature of the combustion chamber,
- Residence time in the chamber, and
- Amount of mixing of the material with air while in the chamber.

Normal combustion temperatures range between 900° and 1,500°C and in some instances much higher. Many incinerators for hard to burn compounds employ two combustion chambers. The first chamber converts the compounds to gas and initiates the combustion process. In the second chamber, combustion of the gases is completed.

The inert portion of the wastes remains as ash after incineration. For liquids, the amount of ash remaining is generally insignificant. For solid wastes, the volume of ash can be as much as 30 percent of the original volume. If the ash contains metals or radioactive material, it must be further treated prior to disposal. The most frequently employed method of treating the ash remaining from the incineration process is solidification/stabilization.

Several types of incinerators are available for treating wastes. The most common is the rotary kiln. The kiln of a **rotary kiln incinerator** is a cylindrical shell mounted on its side on a slight incline. As the kiln rotates, the wastes pass through and are combusted. Rotary kiln incinerators are capable of accepting wastes in all phases. **Liquid injection incinerators** introduce the material under high pressure through a nozzle which atomizes the wastes. This allows air to mix with the waste and the combustion process to take place. **Fluidized bed incinerators** burn finely ground solids or liquids in a bed of inert material suspended above the floor of the combustion chamber. Any ash remaining after combustion is removed when the bed is changed. A recent development has been the use of **infrared incineration** technology. This system uses electrically powered silicon carbide rods to raise the wastes to combustion temperatures in the primary combustion chamber. Infrared incinerators can accept solids and sludges. Liquids must be mixed with sand or soil prior to introduction into the combustion chamber.

Incinerators may be mobile, transportable, or stationary/fixed. Mobile incinerators are normally relatively small units that are mounted on a flat bed trailer and transported to the job site. Transportable incinerators are larger units that can be disassembled into manageable components and moved from job site to job site by a caravan of trucks. Stationary/fixed incinerators are permanently erected at a site, and the wastes are brought to the site for treatment.

The EPA and some states are now allowing wastes with lower levels of contamination and some BTU content to be mixed with fuel oil and burned in boilers and industrial furnaces (this ruling also includes cement and light aggregate kilns). Boilers and furnaces used in this manner must meet the same permitting and emission requirements of hazardous waste incinerators.

While incineration remains one of the most effective methods of treating organic wastes, its application is declining. This is largely due to the public's fear of hazardous materials escaping from the incinerator's stack and being deposited on the surrounding community.

Thermal Desorption

Thermal desorption is the process of heating a waste in a controlled environment thereby volatilizing any organic constituents. Thermal desorption works especially well for VOCs but can also be employed for SVOCs. Removal efficiencies ranging between 65 and 99 percent have been achieved depending upon the type of waste.

Prior to entering the thermal desorption unit, the wastes are screened to eliminate coarse pieces. If the wastes have a high moisture content, this excess moisture is also removed. The wastes are then passed to a furnace which operates at temperatures ranging between 300° to 600°C. The gaseous organic compounds volatilized by the process are then either collected on a medium such as activated carbon for further treatment or passed through an incinerator connected in-line with the thermal desorption unit.

Thermal desorption is growing in popularity due to its lower energy requirements than conventional incineration. These lower energy requirements and the reduction in downstream pollution control equipment result in the costs of thermal desorption being less than those for incineration. Public acceptance of thermal desorption has been better since combustion does not occur and the final treatment of the compound may take place off site thereby lessening the fear of release of dangerous compounds.

Pyrolysis

Pyrolysis is a chemical change brought about by the action of heat. This differs from incineration, which is the combustive destruction of a material in direct flame in the presence of oxygen. Pyrolysis can be thought of as destructive distillation in the absence of oxygen (or other oxidant). It converts wastes containing organic material (VOCs and SVOCs) to combustible gas, charcoal, organic liquids, and ash/metal residues. The organic liquid fraction produced during the pyrolytic reaction has the potential to form the basis of synthetic crude. Because of the chemical transformations that take place, most of the off-gasses resulting from pyrolysis do not pose a hazard to workers.

The effectiveness of pyrolytic destruction depends upon:

- Residence time within the retort,
- Rate of temperature rise,
- Final temperature, and
- Makeup of the feed material.

Pyrolytic units, which operate at temperatures ranging between 500°C and 800°C have achieved 99.9999 percent destruction/removal efficiencies and volume reductions in excess of 50 percent.

Plasma Torch

Plasma torch processes apply the principles of pyrolysis at very high temperatures (5,000° to 15,000°C). The wastes are fed into the thermal plasma and become disassociated into their basic atomic components. The atoms recombine in the reaction chamber to form carbon monoxide, nitrogen, hydrogen, and small quantities of methane and ethane. The process results in the formation of some acid gases which are removed by scrubbers. Any solids produced are either incorporated into the molten bath at the bottom of the chamber or removed from exhaust gasses by particulate scrubbers or filters. The plasma torch technology is currently applicable only to fine particle wastes, liquids, and pumpable wastes.

Vitrification

The vitrification process can be applied to wastes containing both organic and inorganic components. Unlike other types of thermal technologies, vitrification can be applied both in situ and ex situ. This versatility of application has moved vitrification to the forefront of technologies considered for remediation projects.

The principles behind vitrification are essentially the same as those underlying the production of glass. High temperature electrodes are used to melt the wastes. Organics are transformed by pyrolysis and collected and destroyed in secondary processes. The inorganic components are immobilized in the resulting glass matrix.

Ex situ applications closely resemble typical glass production plants. The wastes are introduced into the furnace along with silica, soda, and lime. The organics are driven off, captured, and treated while the inorganics are incorporated into the glass.

In situ use of vitrification involves insertion of large electrodes into the soil. Graphite is spread on the soil surface between the electrodes to complete the circuit. A negatively pressurized hood is placed over the site to collect any off-gases for later treatment. High voltage (4,160 volt, 3,000 kW electrical source is required) is applied across the electrodes to produce temperatures reaching 3,600°C. The use of in situ vitrification is limited by:

- High groundwater tables,
- Buried metal objects, and
- The need for sufficient quantities of glass-forming material in the soil.

The vitrification process has shown great promise for treating radioactive and mixed wastes. The radioactive wastes are immobilized in the glass matrix and can then be stored until the radioactivity decays to a safe level. In the case of mixed wastes, vitrification drives off the nonradioactive components allowing them to be treated as hazardous wastes while immobilizing the radioactive component.

BIOLOGICAL TREATMENT

As the costs of many treatment technologies continue to increase and the public expresses heightened concerns over the side streams produced by various technologies, the viability of biological treatment for VOCs and SVOCs has dramatically increased. While biological treatment is addressed in detail elsewhere in this volume, the following provides a brief overview of its applicability to hazardous and solid wastes.

Bioreactors

Bioreactors have been the cornerstone of wastewater treatment processes for decades. The wastes are introduced to a biomass of microorganisms wihin the reactor which metabolize the soluble organic components. In most instances, additional nutrients and oxygen (aeration) must also be added.

Reactors may be either fixed film or slurry phase. Fixed film reactors are similar to the traditional trickling filters or **rotating biological contactors (RBCs)** of the wastewater industry. In either case the microorganisms are supported on the medium of the filter. The wastes are passed over the filter (or in the case of RBCs the filter is passed over the waste) allowing the microorganisms to come into contact with the wastes and breakdown the organic material. Slurry phase reactors are tanks into which the wastes, nutrients, and microorganisms are placed. The tank is mixed and may be aerated. In many instances, contaminated groundwater is used to create the waste slurry.

Both fixed film and slurry phase treatments are either batch or continuous mode. The effectiveness of the treatment depends on the:

- Concentration of the wastes,
- Contact (retention) time with the microorganisms,
- Availability of oxygen in the system, and
- Temperature within the reactor.

Solid Phase Bioremediation

Solid phase bioremediation, often referred to as land farming, treats wastes using conventional soil management practices to enhance the microbial degradation of the wastes. The wastes are placed directly on the ground or in shallow tanks, if required by RCRA restrictions. Nutrients and microorganisms are normally added to the wastes which are routinely tilled during the treatment process. This tilling improves aeration and the contact of the organisms with the wastes. While treatment may occur throughout the upper three to five feet of the soil, most occurs within the top foot, called the ***zone of incorporation.***

Soil Heaping

Soil heaping is piling wastes in heaps several feet high on an asphalt or concrete pad. Nutrients, microorganisms, and air are provided through perforated piping placed throughout the pile. The pile is covered with visqueen or a similar material to contain VOCs, to stabilize the microorganism's environment, and to control soil erosion. VOCs can be further controlled by applying a vacuum to the pile and treating the exhaust.

Composting

Composting is another application of bioremediation. In this process the wastes are normally mixed with a structurally firm bulking material such as chopped hay and wood chips. As with the other bioremediation technologies, nutrients, air, and microorganisms must be added. The three major types of composting are ***open windrow, static windrow,*** and ***reactor systems.*** The differences among the three relate to how aeration is accomplished. In the open windrow system, the compost piles are mechanically turned at specified intervals. Air is mechanically forced into the compost piles under the static windrow system. When reactors are used, the compost is mechanically mixed to ensure aeration.

In Situ Bioremediation

One of the advantages of bioremediation is that it can be effectively applied to treat wastes in place. The process usually entails introduction of nutrients, microorganisms, and air to the soil/waste through a series of injection wells or infiltration trenches. If the soil does not have sufficient moisture content, water may also have to be added. In situ bioremediation is often applied in conjunction with groundwater pump and treat systems and soil flushing activities.

CHEMICAL/PHYSICAL PROCESSES

Traditional Physical/Chemical Treatment Processes

Separation

Separation is used to divide the wastes into two or more distinct waste streams based upon size, density, or material type. Normally accomplished by either manual or mechanical means, separation allows for a more efficient operation of the subsequent technologies while reducing the quantities of material to be treated.

Screening is the most common technology employed for separation. Five general categories of screens are available:

- Grizzly screens, sets of parallel bars used for removal of coarse material;
- Revolving screens, a cylindrical frame covered with wire cloth;
- Shaking screens, a rectangular frame with wire cloth;
- Vibrating screens, normally used when higher capacities are required; and,
- Oscillating screens, used at lower speeds than vibrating.

These latter screens are used for separating particles by grain size. Other techniques often used for separation are air separation, flotation, and magnetic separation.

Factors to be considered in selecting a separation technique include:

- Availability and mobility of equipment;
- Type of material to be handled;
- Volume of waste;
- Ease of cleaning, maintenance, and decontamination of the equipment;
- Required feed rate; and,
- Whether the separation is wet or dry.

Dewatering

Dewatering is the process of removing or reducing the moisture content of a material, usually sludges or slurries prior to the application of another technology. By reducing the water content:

- the ease of handling is greatly improved,
- the volume of the material is decreased, and
- the energy costs associated with thermal destruction can be significantly reduced since the liquids do not have to be evaporated.

Dewatering is accomplished by: thickening, centrifuges, vacuum filters, pressure filters, and lagoons.

Thickening. Thickening is accomplished by passing the sludge or slurry through a thickener which is a specially designed gravity settling tank. With the addition of chemical conditioning agents such as lime, ferric chloride, and polymers, the solid concentrations of sludges and slurries may be increased from less than 1 percent to 2 to 15 percent.

Centrifuges. Centrifuges operate on the principle of density separation accomplished by spinning the wastes and allowing the solids to disassociate from the liquids. Three types of centrifuges are commonly available: disc, basket, and scroll. Centrifuges produce sludges with solids concentrations ranging between 9 and 25 percent.

Vacuum Filtration. A vacuum filter normally consists of a porous cylinder that is passed over a trough containing the sludge. A vacuum is applied to the cylinder, and the sludge is picked up as the cylinder rotates. The vacuum draws the water from the sludge which is then scraped from the cylinder. Vacuum filters produce solid concentrations of 20 to 40 percent.

Pressure Filters. Pressure filters are capable of dewatering sludges to solids between 30 and 45 percent. Compressing the filter plates forces the water from the sludge. Types of pressure filters include screw jack, hydraulic cylinder, and pressure leaf.

Lagoons. Lagoons work on the same principle as gravity thickeners and are used when larger volumes of waste require treatment. Their operation may be enhanced by the addition of vacuum-assisted underdrain systems.

Densification

When wastes are placed directly into a landfill or other type of containment with no further treatment, the waste volume is normally reduced through the process of densification. Densification is the mechanical compacting of a waste to reduce the volume of its void space. The most common densification process is the compaction of municipal solid wastes performed by trash collectors. Similar processes can be applied to any type of waste to decrease the final disposal volume.

Neutralization

In many instances, wastes, especially liquids, sludges, and slurries, may be highly acidic or alkaline. When this occurs, the first step in treatment is to bring the waste to a near neutral pH to ease handling and to improve the effectiveness of any follow-on processes. This can be accomplished by:

- mixing several waste streams to achieve a neutral pH;
- adding lime slurries to acidic wastes;
- adding caustic soda or soda ash to acidic wastes;
- adding CO_2 to alkaline wastes; or,
- adding sulfuric acid to alkaline wastes.

Neutralization is normally accomplished by mixing the waste with the appropriate compound in a reactor vessel. In some instances, sludges may be treated in situ by mixing soda ash or lime with the sludge in the pond or lagoon.

Oxidation/Reduction

Metals, and to some extent, SVOCs, can often be removed from a liquid waste by application of the chemical principle of oxidation/reduction. Oxidation/reduction is used to alter organics and inorganics to enhance their later removal. The rate of the reaction depends upon the temperature, the amount of oxidant used, and the concentration of the contaminant.

Oxidation has been shown to effectively degrade many organics. Originally air was used as the primary oxidant. Because of its limited success, stronger oxidizing agents such as ozone, permanganate, chlorine dioxide, hydrogen peroxide, hypochlorous acid, and chlorine are now used. Oxidation can be applied both in situ and ex situ. An additional advantage of in situ oxidation is that it often enhances biological degradation.

Reduction is often employed with metals to transform them to a form that will allow precipitation of the metal with lime. While the most common use of reduction is to remove hexavalent chromium, it can be used effectively for the removal of most metals. Reduction, as with oxidation, can be applied both in situ and ex situ. Removal efficiencies in excess of 98 percent have been attainable in field applications.

Ion Exchange

Ion exchange systems selectively exchange ions from a chosen material in the waste stream for the exchange medium. Ion exchange systems can be either cation exchangers or anion exchangers. Cation exchangers are used to remove positively charged materials (metals) while anion exchangers remove negatively charged materials (typically organics). The principle use has been to remove metals from surface waters and groundwater. The metals are concentrated on the column and are removed by backflushing with water. The ion exchange columns can be recharged by passing dilute acid through the beds. Removal efficiencies of greater than 99 percent have been achieved with effluent qualities of less than 0.1 ppm of metal.

Activated Carbon Adsorption

Activated carbon can be employed to physically adsorb contaminants such as VOCs and SVOCs in liquid and gaseous waste streams. Powdered activated carbon can be added to batch processes or a granular form can be contained in columns through which the wastes are passed. Activated carbon columns can also be employed to remove VOCs following vapor extraction systems.

Soil Washing

Soil washing is the process of excavating contaminated soil, placing the soil in a reaction vessel with water, and mixing or agitating the soil. Soil washing accomplishes two purposes. The first is to physically remove contaminants from the soil through the mechanical action. The contaminant removal may also be enhanced by dissolving some of the materials present. Secondly, the total quantity of material requiring further treatment can be reduced by separating the fines, which are normally more highly contaminated than the larger particles. Soil washing has been shown to be effective for metals and SVOCs. Removal efficiencies ranging from 70 to 99 percent have been reported. It should be noted that the higher removal efficiencies appear to be influenced by volatilization of some compounds. The major drawback with soil washing is that it produces a liquid waste stream that must be treated.

Soil Flushing

Soil flushing is an in situ process used to remove both organic and inorganic contaminants from the soil. The typical process involves flooding the site with the appropriate washing solution which is allowed to percolate into the contaminated area. The flushing solution collects the contaminants by solubilization, formation of emulsions, or chemical reactions. The ***elutriate*** (the flushing solution and any contaminants collected) is then collected by well points or subsurface drains. Following collection of the elutriate, metals may be precipitated from the elutriate and the remaining fraction applied to the land surface as a cost effective treatment alternative.

Flushing solutions include:

- water;
- acidic solutions of sulfuric, hydrochloric, phosphoric, or carbonic acids;
- basic solutions such as sodium hydroxide; or,
- surfactants.

Water is normally used for water soluble compounds. Acidic solutions are generally used to recover metals and basic organic compounds, while basic solutions are also used for metals. The overall effectiveness of soil flushing depends on the amount of contact between the flushing agent and the contaminant, the selection of the flushing agent, and the hydraulic conductivity of the soil.

Soil Vapor Extraction

VOCs and SVOCs can be removed from the soil by application of several vapor extraction techniques. All of the systems involve the injection of air into the soil and the collection the air and volatiles in recovery wells.

Vacuum Extraction

Clean air is injected into the soil through wells to fill the void spaces in the soil. Contaminants with high vapor pressure partition into these air-filled voids in the soil. By applying a vacuum to collection wells, vapor-filled air is extracted from the soil. The extracted air/vapor mixture is then treated with activated carbon or a similar air stripping process to collect the organic material.

The spacing of the injection and extraction wells is critical in a soil vapor extraction system. Initially, only vertical wells were used. However, recent tests have indicated that the use of horizontal wells may greatly enhance the performance and reduce the operational time for vapor extraction systems. Vacuum extraction systems have achieved removal efficiencies as high as 99.8 percent. These systems do not operate effectively in saturated soils or in dense clays with low permeabilities.

Steam Stripping

Steam stripping is an enhancement of the basic vacuum extraction process. By injecting steam along with air into the soil beneath the contaminated zone, the removal of certain types of compounds may be enhanced. VOC removals of up to 99 percent with SVOC removals of 50 percent have been reported. The major drawback with steam stripping is that it requires a geomembrane to be placed over the site to help contain any volatilized contaminants.

Radio Frequency Heating

Some types of contaminants such as jet fuels and gasoline, which volatilize at temperatures between 70° and 250°C, may be more easily removed after the soil is heated. The use of electromagnetic energy in the radio frequency band, introduced into the soil by electrodes, is one method to accomplish this heating. The frequencies used range between 6.78 MHz and 2.45 GHz. Heating takes from two to eight days to accomplish. Radio frequency heating requires four separate systems for complete installation: an energy deposition array, a power generating source, a vapor barrier, and a gas handling system. This process has demonstrated removal efficiencies of between 98 and 99 percent for VOCs and between 94 and 99 percent for SVOCs.

Dechlorinization

Treatment of PCBs and dioxins is a major concern when dealing with hazardous wastes. The dechlorinization process was developed as a response to the need for an alternative to incineration. The basic principle of dechlorinization is the removal of the chlorine atoms from the compounds. The process uses specially synthesized reagents to destroy chlorinated molecules or to form other compounds that are less detrimental to the environment.

While several processes have been developed, the most frequently used processes employ **polyethylene glycols (PEG)** that have been reacted with alkali metals to effect the dechlorinization. The two most common processes are the **Alkali Polyethylene Glycol Process (APEG)** and the **Potassium Polyethylene Glycol Process (KPEG).** Both processes can be used in situ or ex situ. Removal efficiencies of up to 99.5 percent have been demonstrated.

Key factors in the implementation of any dechlorination process are:

- Temperature of the material,
- Contact time between the contaminant and the reagent, and
- Moisture content of the soil.

Water can adversely affect the rate of reaction. Raising the temperature of the wastes from 20°C to 80°C increases the reaction efficiency from 50 percent to 90 percent. The contact time for ex situ treatment ranges between 4 and 8 hours while in situ contact times average 7 days. Both types of reagents (APEG and KPEG) appear to have the same effectiveness. The primary concern with dechlorinization is that the reaction byproducts have not been well characterized to date.

Solidification/Stabilization

The processes of solidification and stabilization are designed to improve the handling and physical characteristics of a waste by:

- Producing a solid,
- Reducing the contaminant solubility, and

- Decreasing exposed surface area.

While the terms solidification and stabilization are often used interchangeably, they do have different connotations. Solidification normally refers to the elimination of free liquid and increasing the bearing strength of the material to form a monolithic solid. Solidification does not necessarily involve chemical bonds between the material and the solidifying agent but does denote the waste being bound within the final matrix. Stabilization is the process of reducing the hazardous properties of a material by converting the contaminants into their least mobile or toxic form.

Several solidification/stabilization systems are readily available for use with metals and to a lesser extent with VOCs and SVOCs. These include:

- Pozzolan-portland cement,
- Lime-fly ash pozzolan systems,
- Thermoplastic encapsulation,
- Sorption, and
- Organic binding.

The pozzolan-portland cement systems entrap the wastes in a concrete matrix. Lime-fly ash pozzolan systems employ the same mechanism; however, the resulting material is not as structurally strong as the pozzolan-portland cement matrix. Thermoplastic microencapsulation blends the wastes with melted asphalt. Sorption is the process of adding a material to the waste to absorb free liquids. Typical sorbants are activated carbon, clays, zeolites, anhydrous sodium silicate, and gypsum.

The most critical element in achieving proper solidification/stabilization is complete and uniform mixing of the wastes and the agent. Complete mixing is easily achieved with ex situ applications. For in situ applications the method of adding the agent and mixing are crucial. Two methods are currently available for adding solidification/stabilization agents during in situ applications. The agent can be injected into the waste in a liquid or slurry form. If the wastes are shallow, the agent can be spread on the surface and allowed to penetrate to the necessary depth. This latter method is not often used.

TREATMENT TRAINS

Examination of the application of various technologies to specific wastes quickly leads one to conclude that a single technology is seldom capable of completely treating a complex waste. To add to the complexity of treating hazardous waste, many of the technologies produce side streams that must in turn be treated. In this event, several technologies are employed sequentially in what is commonly called a "treatment train." Probably the two most common uses of treatment trains are for wastes containing both organics and metals and for wastes containing radioactive material in combination with other contaminants. This latter group of wastes is typically known as "mixed wastes," since they mix both hazardous and radioactive materials.

The following are examples of treatment trains that have been employed to treat hazardous wastes:

- Soil washing followed by bioremediation, incineration, or solidification/stabilization;
- Thermal desorption followed by incineration, solidification/stabilization, or dechlorination;
- Soil vapor extraction in combination with in situ bioremediation or flushing, solidification/stabilization, or soil washing;
- Dechlorination with soil washing;
- Solvent extraction in conjunction with, solidification/stabilization, soil washing, or incineration;
- Bioremediation followed by solidification/stabilization; and,
- In situ flushing in conjunction with in situ bioremediation.

MATERIALS HANDLING

Regardless of the technology selected, including situ technology, some amount of materials handling must be accomplished. It may involve site access or site preparation tasks, construction of treatment systems and support structures, or the actual movement of wastes to the treatment process. Materials handled range from hazardous chemicals, to solid wastes, to soil and debris. Without proper consideration to the material handling requirements, the treatment of solid and hazardous wastes can become more difficult and less efficient than necessary.

In many instances, equipment used for general construction activities can be successfully applied to waste treatment. The following factors should be considered in determining the type of materials handling equipment required:

- Type and quantity of materials,
- Technology being employed, and
- The characteristics of the site (climate, soil type, topography).

CONTAINMENT/DISPOSAL

Until the late 1980s, the most common method of handling wastes had been to place the treated or untreated materials into an engineered landfill. The Land Disposal Restrictions (LDRs) of RCRA, which had taken affect beginning in 1989, have drastically altered the types and quantities of wastes that can be placed in landfills. The LDRs mandate at least some type of treatment of most hazardous wastes prior to placement in a landfill. The recent requirements for solid waste landfills have also refocussed the pattern of use of landfills for nonhazardous wastes.

Even with these restrictions, landfilling is still the final disposal option for many wastes. A subset of this technology is the capping and containment of wastes in place at a contaminated site. While not as prevalent as in the past, containment is still the only practical alternative for waste sites, such as abandoned landfills whose volumes often exceed a million cubic yards of waste.

Simple Soil Cover

The simplest containment remedy that is employed in special circumstances is covering the waste with a layer of soil. This is most often used for widespread, low-level contamination that would otherwise require excessive amounts of soil to be excavated and treated. The general practice is to remove the upper foot of contaminated soil, treat the contaminated material, and replace it with clean fill. This practice has been effectively used to control metal exposures from mine tailings and low-level radium contamination.

RCRA Caps

The principle of containment is to prevent the hazardous components of the waste from migrating from the site. One method of controlling the migration of the contaminants is to limit the infiltration and subsequent leaching of water into and out of the waste material. This can be accomplished by constructing a RCRA cap over the contaminated area.

RCRA caps are composed of multi-layers of soil and synthetic membranes which limit the infiltration of water and allow for the collection and removal of any methane gas which may be generated within the fill. RCRA caps consist of a top layer of soil, normally six inches, which is provided to promote the growth of ground cover to inhibit erosion. The next layer is 18 to 24 inches of select fill used to help support the vegetation and to protect the impermeable layers below it. Directly beneath the select fill is a filter layer. This layer prevents the migration of fine particles into the drainage layer.

The drainage layer is designed to collect any infiltration from the select layer and channel it from beneath the cover. The drainage layer consists of either 12 inches of clean granular material or an acceptable geonet. Beneath the drainage layer, a geomembrane is placed. This geomembrane must be at least 20 mils thick and be below the active frost layer. The next layer is the impermeable clay layer. This layer is a minimum of 24 inches thick with a permeability of less than 10^{-7} cm/sec.

Beneath the clay layer of the cap is the random fill used to bring the existing top of the landfill to final grade and to establish proper slopes and drainage patterns. The final consideration in using a RCRA cap is to provide for the collection of gases produced as the wastes decompose. Gas collection systems can be either passive blanket systems or well points. A passive blanket system normally consists of a layer of clean granular fill that is used to collect the gas. The gas is released from the fill through a series of outlet ventilation pipes. When wells are used, they are placed through the depth of the waste, and the gas can be either actively collected (by drawing a vacuum on the wells) or passively collected.

In instances where the surface of the groundwater table is above the bottom of the waste material, a barrier must be established to prevent the flow of the groundwater from leaching material from the site. This is most commonly accomplished by constructing a slurry wall around the waste. The slurry wall is normally composed of cement or bentonite slurry placed into a trench around the site. In several instances, the containment has been achieved by mixing cement slurry with the soil around the site.

The final consideration when capping a site is how to manage any leachate that may emanate from the site. This is generally done by constructing a collection system around the site at the depth of the bottom of the wastes. The collection system is generally comprised of a trench filled with granular material with perforated collection pipes. The leachate is collected and treated prior to discharge to the local sewer system or surface stream.

RCRA Landfill

Prior to 1993, only landfills for hazardous waste were required to have a double lined bottom with a leachate collection system and groundwater monitoring wells. Beginning in October 1993, nonhazardous solid waste landfills must also meet these requirements. Basically, the liner requirements look like an inverted RCRA cap. That is a bottom layer of at least two feet of clay with a permeability of less than 10^{-7} cm/sec with a geomembrane placed over it. A leachate collection system is required above the geomembrane. This collection system can be either granular material or a geonet. Above the leachate collection system is a filter layer to prevent the leachate collection system from plugging. In some instances, variations from this design may be allowed if the proposed liner system meets the performance standards established for the site. When the disposal site has been filled to capacity, it must be covered with a RCRA cap as described above.

NATURAL ATTENUATION AND EXCAVATION WITH OFF-SITE DISPOSAL

While natural attenuation and excavation with off site disposal are not true technologies, they are viable alternatives that must be considered when selecting a hazardous waste treatment technology. Natural attenuation refers to the decay or breakdown of contaminants without interference from man. Normally natural attenuation is not an option in a highly populated area or for high concentrations of wastes. However, in certain instances, it may be the preferred alternative. Since no effective treatment technology is currently available for treating radioactive wastes, natural attenuation is the only alternative. In situations where the access to the waste or site can be controlled and the concentrations are barely above action levels, natural attenuation may also be employed. This normally applies to contaminated groundwater when wellhead treatment is provided. When natural attenuation is employed for groundwater remediation, additional institutional controls must be put into place to prevent access to the aquifer through unauthorized wells.

While LDRs have greatly limited the use of excavation and off site disposal, they may still be used in certain instances. The best example is again radioactive wastes. Since no effective treatment is available, off-site disposal is the only alternative while awaiting the completion of the decay process.

RADIOACTIVE AND MIXED WASTES

Radioactive wastes require special mention in that there is currently no known technology to neutralize the radioactive component of the wastes. The major types of radioactive wastes are: radium, thorium, uranium, and plutonium, and these materials in combination with other hazardous wastes. The only way to treat radioactive and mixed wastes is to attempt to separate the hazardous components, treat this portion separately, and then safely contain the radioactive portion until it decays to an acceptable level.

The most promising technologies for dealing with radioactive wastes are: soil washing with water, chemical extraction, physical screening, vitrification, solidification, ion exchange, carbon treatment, and containment in a specially design landfill. It is clear to see that the focus for radioactive wastes is on separation or isolation of the radioactive component.

GROUNDWATER TREATMENT

Contamination of groundwater resulting from constituents leaching from either buried wastes or leaking from ponds and lagoons has become a major problem in the United States. Controlling the source of the contamination is the first step in treating the problem. The second is the actual treatment of the contaminated groundwater. There are two approaches to treating groundwater contamination. The first is the application of either pump and treat technologies or in situ remediation to restore the aquifer to its original condition. The other approach is to provide wellhead treatment to ensure potable drinking water and to allow the aquifer to self cleanse. While the latter is highly controversial, the former has not been shown to be successful, and its costs are often extremely prohibitive given the indefinite duration of the pump and treat process.

A detailed discussion on groundwater is presented in another chapter of this volume.

RECYCLING AND REUSE

Recycling and reclamation of materials should be given serious consideration when selecting the technology for managing both hazardous and solid wastes. Recycling not only reduces the volume of material requiring treatment and final disposal but also saves natural resources and reduces the energy requirements necessary to produce products from virgin materials.

Many of the technologies discussed in this chapter lend themselves to some form of recycling or reclamation. Among these are the extraction and washing processes as well as some of the other physical processes. Recycling is explored in detail elsewhere in this text.

SOLID WASTES

Currently, most solid wastes are being managed by landfilling in conjunction with some type of densification. Nearly every city and county in the United States has a landfill that is operated by the local municipality or by a private contractor under the auspices of the local governing body. The Solid Waste Disposal Facility Criteria released by EPA in October 1991 contains operating and design guidelines that may drastically alter this historic local approach. These guidelines require all facilities to meet minimum design and monitoring requirements or cease operation. Many local landfills will not be able to comply with these regulations because of prohibitive costs; thereby forcing them to close. This will result in the regionalization and privatization of many solid waste landfills.

Incineration has been promoted as the best means of volume reduction available for managing solid wastes. However, public concerns over the safety (that is the control of emissions from the unit) have greatly curtailed the once predicted use of incineration. Pyrolysis was also once considered to be the answer to solid waste disposal problems. It, however, has experienced the same problems as incineration.

At the present time, the solution to the increasing problems of solid waste disposal is unclear. Many of the technologies discussed in this chapter may be applicable to solid wastes, but they are not cost effective under today's public definition. The public refuses to allow new facilities, landfills or incinerators, to be sited in their communities (the Not-In-My-Backyard [NIMBY] syndrome) and at the same time exhibits no great inclination to recycle or limit their use of disposable products.

PROCESS WASTE STREAMS

As discussed earlier in this chapter, many of the technologies described are applicable to industrial wastes streams. Since these wastes are normally more homogeneous than the wastes requiring remediation, basic physical/chemical processes can be effectively applied. Additionally, recycling, reclamation, and substitution are increasingly being employed. The recent push for pollution prevention and the fear of future CERCLA or RCRA actions have greatly reduced the quantities of hazardous waste being produced by industries.

SUMMARY

The control and treatment of both solid and hazardous wastes continue to be two of the great problems facing modern man. To answer this challenge, the scientific and engineering communities are striving to develop new technologies and to find innovative applications for existing ones. While positive steps such as recycling, reuse, and product substitution are slowing the generation of wastes, further emphasis is still needed to change the American mindset from its current disposable mentality.

ADDITIONAL REFERENCES

1. *The Superfund Innovative Technology Evaluation Program: Technology Profiles.* EPA/540/5-88/003, U.S. EPA, Washington, D.C. (1988).
2. *Design and Construction Issues at Hazardous Waste Sites: Conference Proceedings.* EPA/540/8-91/012, U.S. EPA, Washington, D.C. (1991).
3. *Survey of Materials-Handling Technologies Used at Hazardous Waste Sites.* EPA/540/2-91/010, U.S. EPA, Washington, D.C. (1991).

4. *Second Forum on Innovative Hazardous Waste Treatment Technologies: Domestic and International.* EPA/540/2-90/010, U.S. EPSA, Washington, D.C. (1990).
5. *Handbook on In Situ Treatment of Hazardous Waste Contaminated Soils.* EPA/540/2-90/002, U.S. EPA, Washington, D.C. (1990).
6. *Basics of Pump-and-Treat Groundwater Remediation Technologies.* EPA/600/8-90/003, U.S. EPA, Washington, D.C. (1990).
7. *Assessment of Technologies for the Remediation of Radioactively Contaminated Superfund Sites.* EPA/540/2-90/001, U.S. EPA, Washington, D.C. (1990).
8. *Hazardous Waste Incineration: Questions and Answers.* EPA/530-sw-88-018, U.S. EPA, Washington, D.C. (1988).
9. *Cleaning Up the Nation's Waste Sites: Markets and Technology Trends.* EPA 542-R-92-102, U.S. EPA, Washington, D.C. (1993).

8

UNDERGROUND AND ABOVEGROUND STORAGE TANK TECHNOLOGY

A.D. Young, Jr.
Consultant

OVERVIEW

Since the early days of storing, transferring and distributing liquid products such as motor fuels or various chemical or specialty products used in manufacturing processes, underground tanks have been routinely used for smaller quantity storage. Greater volume bulk quantities are usually stored aboveground in larger size tanks. In facilities such as service stations, underground tanks provide greater safety against fire potential and enable more efficient use of land area. Current estimates are that over 4,000,000 underground tanks are in active use today. There is no reliable estimate of the number of aboveground tanks now in service.

Historically, both tank types have been of steel construction. Many systems have been in place for years, and serious failures, due primarily to corrosion damage over time, have caused numerous environmental pollution incidents. The problem became so acute with underground systems that Congress, in 1984, amended the Resources Conservation and Recovery Act (RCRA) to regulate the use of underground storage tanks storing petroleum products. The Environmental Protection Agency issued the appropriate regulations effective December 1988. Various states also promulgated local UST operating rules, as authorized by RCRA.

UNDERGROUND STORAGE TANKS

The following sections examine the evolution of the underground storage tank (UST) technology.

Causes and Solutions for UST Tank Failure

A survey conducted by the American Petroleum Institute in 1978 revealed that corrosion damage was the cause of over 90% of underground tank failures. Other causes were primarily due to faulty installation or manufacturing defects.

Corrosion of metal in contact with the ground is a natural phenomenon; the soil normally contains all the elements necessary to produce a corrosive atmosphere. Corrosion occurs when dissimilar metals in the soil are in such close proximity to one another that an electromotive force results between them, and electrical current is generated. The flow of electricity through the soil between the metal surfaces produces a corrosion effect at one metal surface, known as the "anode," with no corrosion at the second surface, called the "cathode." The phenomenon was well understood, and technology was available to combat it. The protection technique employed a concept known as "cathodic protection", in which an electromotive force was induced externally, and was opposite in direction and force to the natural current flow in the soil. This method was widely used in pipelines, buried utility structures, ship hulls, water tank installations, and in certain specialized industrial applications. However, for many years, there was little or no interest in applying the corrosion protection strategies tosmall UST systems. By the end of the 1970s, however, to combat the possibility of tank failure, many tank owners began retrofitting existing tank installations with cathodic protection systems, installing new steel tank designs incorporating this concept, or using designs having noncorrosive materials.

Regulations

The new federal regulations, 40 CFR 280 (effective December 1988), specified design performance standards for new installations, required upgrading actions for existing installations, and established timetables for compliance. The regulations also dictated that all UST systems must have a form of leak detection in place, selected from a group of approved methods listed in the rules. All states were required to enforce the federal rules, but could, with EPA approval, issue their own, more stringent, rules. The rules also directed that other UST practices, such as tank testing, installation practices, closure, and corrosion protection must follow procedures designed to insure better protection against tank failure.

Technical Developments

Driven by market forces, regulation, and rising public interest in improved environmental protection, there have been numerous improvements in UST equipment design and operation practices over the last fifteen years. Tank designs, once almost exclusively based on steel construction meeting Underwriters' Laboratory Standard UL 58, now include over 12 different basic designs of differing materials or construction forms. Together with a wide array of ancillary equipment or processes now available for leak protection, leak detection, secondary containment, monitoring, and testing, there can be as many as 40 different combinations of design choices for a single size storage tank installation. While the regulations specify minimum standards, current technology affords the tank owner a wide range of choices (with a corresponding wide range of price tags) which go beyond compliance with the basic rule.

UST Tank Design

Henry Ford's famous comment about customer preference for the color of his Model T automobile: "any color they want, as long as it's black" might apply equally to early UST designs. For over 60 years, most tanks were carbon steel, 3/16 or 1/4 inch plate, rolled and welded to size, coated with a black asphaltum rust inhibitor. They were distributed by local fabricating companies whose primary customers were oil companies. Little thought was given to corrosion prevention, leak monitoring or spill containment, all considered routine ancillary designs today.

The first major change in tank design technology came about in the 1960s, with the introduction of the fiberglass reinforced plastic (FRP) tank. This design was developed by the fiberglass manufacturing interests through the persuasion of the oil industry. They were looking for a viable storage tank alternative to steel to reduce the costs arising from the growing problem of corrosion damage. The FRP tank was a thin-wall epoxy shell impregnated with fiberglass strands for strength, and encased in a series of support ribs for stiffness and handling support. The FRP tank eliminated concerns over corrosion, since none of the construction materials were affected by that phenomenon. The inherent structural strength of the FRP tank shell, however, was much less than that of a similar steel tank. In order to provide support for the liquid weight and to resist stresses on the shell during product movement or tank "breathing," very careful selection and placement of backfill material was essential. The non-corrosive shell material protected against corrosion effects, while the backfill support provided the structural integrity to insure reliable long-term service from the FRP tank.

In the early 1970s, FRP tanks began to make serious inroads in the market for steel tanks. To counter this threat to their business volume, the Steel Tank Institute, a trade association of the steel tank fabricating companies, embarked on a research program to develop a competitive corrosion-resistant steel vessel. The resulting design, which appeared in the mid-1970s, was an epoxy-coated, cathodically protected UL 58 type tank, labeled the **STI-p3:** "STI" for the institute name, and "p3" for the three levels of corrosion protection that were provided. The shell was coated with a hard epoxy film to insulate the tank surface, and a bag of magnesium anodes was attached to each end of the tank to provide sacrificial anode cathodic protection of the tank shell. At the piping openings, flange bushings were of non-conductive material to isolate the piping and to prevent excessive demand on the cathodic protection electrochemical field.

During this same period, a third new design entered the competition to replace plain steel tanks: the **Buffhide Tank,** developed by Buffalo Tank Company, one of the few major steel companies involved in tank fabrication. This vessel was a UL 58 steel shell coated with 100 mils (approximately 1/8 inch) of a FRP sprayed-on exterior film. The coating provided corrosion protection, while the steel shell offered structural strength.

A design imported from West Germany offering a different technical concept also appeared during this period: the **Haase tank,** a double-walled vessel. The inner and outer shells were of epoxy material (no fiberglass), measured approximately 1/4 inch thick, and were separated by an interstitial space of about 1 1/2 inches, filled with a stone aggregate. The tank was shaped like an onion and contained all internal piping already in place. During installation, a partial vacuum was impressed in the interstitial space for continual monitoring of shell integrity. Due to its shape and construction, the largest size Haase tank was limited to 4000 gallons. It had a 13 foot diameter at that capacity, which made it impossible to ship on flatbed trucks under many highway bridges. While the Haase tank was the first commercially available double-wall design, the limitations on size and the lack of an Underwriters' Laboratory approval listing in earlymodels relegated this tank to a minor role in emerging UST technical development.

The concept of double-walled tanks had been of interest to others in the tank fabrication business at that time, as it became evident that pollution prevention and protection against tank failure were becoming serious social and political issues. More and more tank owners were willing to pay higher prices for tank designs incorporating environmental safeguards. Another early entry utilizing the double-wall feature was offered by a company named Total Containment, Inc. Its design incorporated an inner shell of steel (using a standard single-walled UL 58 tank) on which was mounted an outer shell of high density polyethylene (HDPE). The concept first began as a retrofit of a single-wall tank, where the outer shell was added at the job site in the field before the tank was installed in the hole. It was later offered as a pre-engineered unit available from the manufacturer's facilities, and its newest designs have subsequently become major players in the present-day tank market.

During the 1980s, state-of-the-art tank design progressed to new dimensions, with the proliferation of practical double-wall storage tanks in steel, FRP, and HDPE configurations. The first steel model appeared in 1983, when the manufacturers introduced an inner shell with a 300 degree outer shell, leaving the top section exposed as a single-walled segment. Customer demand, however, soon mandated that both shells be complete 360 degree units, and the partial outer shell model was short-lived. An FRP double-walled model appeared very soon thereafter. As competition increased, a variation of the original steel model was soon introduced, in which the inner shell was thinner gauge metal than the outer, thus lowering the cost.

Double-wall designs brought with them some new complications, as well as new benefits. Concern over possible corrosion in the interstitial space of steel vessels led to a design in which argon gas was used in that space to displace the oxygen and prevent damage. An HPDE model used hydrated lime to combine with condensation and to diminish the corrosive effects of moisture. However, since the FRP double-wall tank, was not subject to corrosion, it was able to utilize its interstitial space for simple continuous leak detection by filling the space with water as a hydrostatic monitoring medium. (Actually, to prevent cold weather damage, a brine solution was normally used.)

In the decade of the 1970s, companies continued to install lower cost unprotected carbon steel tanks, but, as the pressures from the environmental movement mounted, more and more FRP, Buffhide, and STI-p3 tanks found their way into new installations. By the early 1980s, casual use of unprotected steel tanks had almost disappeared, and research and development work on new and different tank concepts accelerated.

As recently as 1992, the latest entry in the new model tank derby, again developed by the Steel Tank Institute, was the **Permatank.** This was a double-walled tank incorporating both steel, as the inner shell, and FRP, as the outer shell. Separated by an interstitial mesh support, a permanent vacuum was impressed on the interspace to monitor shell integrity. It was an interesting combination of two construction materials which had hitherto been strictly competitors.

UST Piping

In any liquid storage tank system, the piping must withstand hydraulic stresses and shock loads; the tanks themselves are rarely exposed to these problems. But, in the early days, no great thought was given to piping materials or layout. Schedule 40 galvanized steel pipe was routinely used, and the pipe layout was usually determined at the site, as the job progressed. Concerns over piping support, crossovers, direction changes, and fittings were casual at best. Corrosion at threaded sections, loose connections, or excessive changes in piping directions were common problems. As a result, frequent incidents of tank system failure and product leakage occurred within piping networks. With the introduction of submerged pumps, pumping pressures and hydraulic forces became greater, and piping networks failed even more often.

By the end of the 1970s, interest in better piping design and installation grew apace with improvements in tank technology. FRP pipe manufacturers offered new and different piping materials: non-corrosive and flexible to a degree unknown with steel pipe. Proper joints, fittings, and connections continued to be a nagging problem, while failures due to corroded fittings virtually disappeared. To help cope with installation defects, FRP pipe manufacturers produced excellent field manuals for contractors on proper pipe installation procedures -- something unknown in the days of galvanized steel pipe.

As contractors became more proficient in handling the new materials, failure from pipe stresses diminished. At the same time, more innovations in pipe design were introduced. Offering specialized fittings for elbows, tees, and risers, pipe manufacturers developed double-walled piping, a valuable innovation for tank system installations. In recent years, the introduction of jacketed double-wall flexible pipe in materials other than FRP wall structures has revolutionized the ability of contractors to lay out pipe networks without rigid straight runs and sharp turns required with less flexible materials. Hydraulic shock problems are greatly diminished with the newer flexible materials.

Design features complementing both tank and piping have also played a role in improved protection against tank failure. The Steel Tank Institute introduced a concept that effectively eliminated a nagging problem: discovery of failure of piping at the entry point into the tank. For years, tank openings for pipe entry were located across the top of the tank at various intervals of 3 to 7 feet apart. When a pipe loosened at tank entry and product leaked into the ground, the entire tank top had to be excavated to find out where the defective pipe and fitting were located. Thiswas expensive, inefficient, and time-consuming. Specification STI-86 changed this concept by having all piping enter the tank at one section on the tank top through a manway welded to the tank. By opening the manway hatch, all the pipes entering the tank could be observed. In addition, if a pipe section was leaking, the lost product could be contained within the manway shell and easily recovered without polluting the surrounding soil.

Overfill and Spill Protection

A continuing problem with UST systems is the potential for overfilling or spilling product during deliveries. Recurring spills at the same location over time can add major quantities of pollutants to the soil as small spills accumulate more and more product in the same place. Combating this possibility requires a number of simple strategies. First, better personal supervision of the transfer process itself eliminates many of the spills resulting from carelessness or inattention. Personnel training and insistence on close monitoring of the system while product flows is a most important spill prevention technique.

To supplement the oversight of those handling the product, there are a number of tank system design improvements effective in reducing the frequency of spills from overfilling the tank. A simple ball check device inserted in the tank at the throat of the vent pipe creates a back pressure when the tank is nearly full. As the ball is floated up into the vent pipe, further delivery flow into the tank is restricted. This ball check valve and vent system can be set at a height in the tank to activate at any desired setting of tank quantity -- 90%, 95%, or another volume. A second flow restriction concept is the installation of a fill pipe containing a flapper gate within the pipe. As the liquid rises in the tank and approaches a preselected level of full quantity, the liquid outside of the fill pipe raises a hinged plate on the outer surface of the pipe. This lowers the flapper gate inside and closes the pipe against further flow from above.

Since small spills cannot be entirely eliminated, some protection must be provided against leaks that occur when disconnecting piping or in the event that the overfill devices fail to insure tight, dry transfers. For this function, fill boxes are made liquid tight. (The ***fill box*** is an inground container surrounding the delivery pipe at the filling connection. Early designs had no bottom, allowing spills to drain into the soil above the tank.) Spills when disconnecting the delivery hose will drain into this containment volume, and be redirected back into the fill pipe through a small pipe drain opening.

Secondary Containment

A further protective design against spill losses, or major leaks due to tank failure, is provided by various methods of secondary containment around the tank system: impervious soils, geotextile linings of the tank hole, or double-walled tank designs. These design concepts are all intended to provide a barrier against uncontrolled loss of spilled or leaked product, and to contain the product for effective recovery.

Where multiple tanks are installed in the same hole, impervious lining of the hole before setting the tanks allows a tank owner to utilize single-wall tanks. This achieves some cost reduction for the installation over the use of multiple double-wall tanks. The lining material must be properly designed for the specific installation and should be manufactured and installed by contractors qualified to perform this work. The material selected must be chemically compatible with the product being stored. The installation must be made with due regard to avoiding debris and sharp objects in the ground that might puncture the fabric. Seams must be carefully sealed to provide long life protection against rupture or separation with age. Upon its completion, the lining design must incorporate a mechanism for monitoring and draining any liquids that may collect in the contained area.

Since a double-wall UST tank is, by definition, its own secondary containment, many tank owners prefer to use this design in a multiple tank installation. This is a more efficient, albeit more costly, means of providing containment than lining the hole. In some older situations, double-wall tanks were installed, but the piping was of single-wall design; pipe runs were secondarily contained using lining material. As improved designs for double-wall pipe have been introduced, these hybrid secondary containment designs have become less common.

Monitoring and Leak Detection

In preparing the UST regulations, EPA was particularly concerned with the ability of tank operators to detect leaks before serious environmental damage occurred. The rules stress the need for effective leak monitoring, and require tank owners to incorporate one or another of a number of approved leak detection methods be in place by the end of 1993. The methods include a combination of a daily inventory program and periodic tank testing, or:

1. physical monitoring equipment installed at the facility,
2. the use of intank gauging equipment with periodic overnight integrity testing, or
3. the use of a program of statistical inventory reconciliation analysis.

Driven by these regulatory fiats, many vendors offer a wide variety of equipment and procedures to help the tank owner meet the requirements. In 1993, the various directories of manufacturers and vendors list more than 65 firms offering leak monitoring systems or components of systems; 15 years earlier, there were fewer than eight, and at that time, no one was offering composite monitoring systems. Currently, based on the intank gauge as the centerpiece of detection control, systems have been introduced which incorporate both interstitial and external liquid and vapor detection capabilities, computer data readout and control functions, and automatic alarms. Various principles have been employed in tank gauging equipment: capacitance probes, magnetostrictive probes, laser equipment, fiber optic designs, and simple float devices. Accuracy, or claims of accuracy in liquid measurement, range from 0.1 inch to 0.001 inch. Probes provide temperature readings and detect the presence of water in petroleum product storage tanks.

To monitor the environment external to the tank system, both within and outside of the secondary containment structures, observation wells have been used extensively. **Observation wells** are 2 inch or 4 inch slotted pipe sections driven to depths below tank bottoms or inserted into the ground water in the vicinity of the facility. They enable physical examination of ground conditions either by manual or automated techniques. In the manual mode, samples of liquids trapped in the well can be withdrawn and examined, or vapors detected using vapor detection equipment. In the automated mode, sensing equipment installed in the well transmits data about the presence of liquid or vapor to a monitoring station. Wells and well equipment need regular inspection and preventative maintenance to insure cleanliness, accurate calibration, and proper operation of equipment. The more complex the monitoring design, the more important the maintenance oversight.

Where monitoring has been performed within secondary containment barriers, in interstitial spaces of double-wall steel tanks, or in clean, new engineered backfill areas, vapor detection equipment has proven to be relatively effective and is being more widely used. In certain counties in central California, new tank installations are required to have vapor sensing devices transmitting their data to the local fire department. (Vapor monitoring of native soil or non-homogeneous backfill is of limited value due to the complex makeup of the soil conditions.) Vapor monitoring using adsorption technology is most frequently employed. Liquid sensing equipment is also normally installed in these spaces. As noted earlier, FRP double-wall tanks can be monitored in a hydrostatic pressure mode by using a brine solution, and need not use vapor sensing equipment to monitor the interstitial space.

Leak detection for piping is necessary where the product dispensing occurs under pressure. In this type of system, where there is no secondary containment installed, any break in the piping will allow product to escape into the surrounding environment as long as the pump is running. A pipeline leak detector is necessary to detect that this condition exists. This device is inserted into the piping system near the discharge side of the pump. It contains a pressure-sensing diaphragm bellows which is designed to sense that the proper backpressure in the pipeline has built up within seconds after the pump starts. Any opening (a leak, an open nozzle, or a loose connection) in the line will prevent normal pressurization as the product moves downstream and fills the piping. A valve segment in the detector will not open fully if the proper pressure is not detected. This will prevent product flow and signal an unusual condition that needs further investigation.

Statistical inventory reconciliation analysis (SIRA) is a specialized form of leak monitoring, relying on highly precise computer analysis of daily inventory information. Firms skilled in SIRA review are equipped to identify the causes of unexplained inventory variations as resulting from stick reading errors, over or under deliveries, tank tilt effects, or similar "normal" contributions to error. After screening for obvious non-leaker losses, these programs are capable of identifying when losses may be from other than "normal" data errors. When unexplainable aberrations are detected, SIRA reviewers alert the tank owner to the abnormal variations as possible leakage or theft, recommending further investigation. The value of this analysis lies in its ongoing assessment of tank condition from daily inventory data without having to rely on special equipment or complex mechanical procedures.

Leak detection technology will continue to improve, since a lively market has developed because of regulatory pressure. Major security is provided against undiscovered releases from UST tanks, which otherwise cannot be readily observed. Reliable monitoring methods allow the owner to utilize UST systems with minimum concern of experiencing undetected pollution events.

UST System Testing for Liquid-tight Integrity

A facet of UST system operation that has also received widespread market growth in recent years has been the development of credible methods for testing the condition of an underground tank. Obviously, since the tank cannot be examined physically, indirect methods of determining its liquid-tight integrity must be relied upon. For many years, it was a difficult goal to achieve. Even today, tank testing still remains an art, rather than a true science.

Tank integrity testing, unlike continuous monitoring, provides information only about the current condition of the tank at the time of testing. It is an investigatory tool to be applied primarily for seeking further information when suspicions of questionable tank condition have been aroused. It is also useful as a screening strategy to determine which tanks among a large population might be currently defective.

Prior to the mid-1970s, the early use of testing was primarily by fire officials who, when responding to a complaint of the presence of gasoline in a building or detected elsewhere, required that all known underground tanks in the vicinity of the complaint be tested in order to locate the leaker. A simple hydrostatic method, known as a **Standpipe Test,** was used. It was a crude procedure, limited in precision and often gave contradictory results, but, since the objective of the testing was to locate an already seriously leaking tank, it was useful as a screening tactic.

Hydrostatic testing was an essentially volumetric analysis: comparing the liquid level at the start of the test period with that at the end, and assuming any difference to be caused by leakage. The test procedure involved mounting a standpipe approximately 6 feet high on the fill pipe of the tank to be tested. The pipe was the same 4 inch diameter asthe fill. The entire UST system, tank and piping, was filled with liquid (usually the product already in the tank) to a scribe mark placed on the standpipe at approximately 4 feet above grade. At the start of the test, liquid level normally dropped a few inches, as the tank shell expanded under the hydrostatic pressure, and as initial temperature effects caused expansion or contraction of the body of liquid. Product was regularly added back into the standpipe to the scribe mark to maintain a constant pressure head. The test protocol was based on the theory that, after a few hours in a tight tank, product movement in the standpipe would stop, and no further liquid would need to be added. Where product continued to drop in the pipe and required constant replenishment, the tank could be considered a potential leaker.

While the operational theory of the standpipe test appeared straightforward, there were many methodology factors which, if not recognized and adjusted for, distorted the results and affected the precision and accuracy of the data. A liquid test medium such as gasoline introduced a high coefficient of expansion. Temperature changes during the test period could affect the apparent level of liquid in the system and produce an erroneous conclusion about the actual gain or loss in liquid height. The method did not include obtaining temperature data to correct for the apparent volumetric change from expansion or contraction of the body of liquid. Expansion of the tank shell under hydrostatic load tended to produce an apparent decrease in liquid level that might be mistaken for leakage. Product evaporation in the standpipe produced apparent shrinkage of liquid level. In addition, since the test required filling the tank and all the piping, any air trapped in the system could distort measurements.

The petroleum industry during the 1960s, in cooperation with general industry and fire officials, were instrumental in developing a more sophisticated and precise version of hydrostatic volumetric testing, originally known as Kent-Moore, later called the **Petrotite method.** This procedure continues to be widely used throughout the petroleum industry and by general manufacturing concerns using UST systems in their operations. Following introduction of this improved testing protocol, the National Fire Protection Association (NFPA) developed a recommended code for tank testing in its publication NFPA 329, which came to be adopted by most authorities and industry groups as the operating standard. The term "precision test," together with threshold test criteria, was introduced in these standards.

In the middle 1970s, as the market for testing services expanded, additional methods were introduced using different physical principles. The **Leak Lokator test procedure,** developed by Sun Oil Company, attempted to detect leakage by measuring the change in buoyant forces on a weight suspended in the body of liquid. If leakage occurred during the test period, a change in buoyant force could be measured, thus signaling the tank defect. Texaco developed a procedure using a manometer to measure small changes in level transmitted toa small orifice measuring tube. Other concepts involved a bubble tube principle, a method for accurately measuring temperature at multiple points in a sealed tank, laser beam technology, and test methods which incorporated hybrid measuring techniques adopted from the other procedures.

In the 1980s, following the new EPA regulations, a number of approaches, different from volumetric measurement, were introduced. The regulations changed the basic standards for testing methods from those of the previous NFPA protocols. As a result, the door was opened for innovative processes not available previously. One new procedure drew a partial vacuum in the head space of the tank, thus encouraging air flow through any openings that might be in the tank shell. By listening for air movement in the tank, one could detect leakage. Another technique utilized a non-hydrocarbon volatile "tracer" additive placed in the body of liquid. If there was leakage, the tracer element could be detected by vapor sensors placed externally around the tank.

Tank testing services and differing techniques have proliferated in the past few years and appear likely to continue to do so. In mid-1980s, EPA operated a testing evaluation program at Edison, NJ, to study testing technology and to evaluate various testing concepts found in the marketplace. Their data and conclusions formed the basis for the testing and test standards included in the UST regulations.

Upgrading Existing UST Systems

There are many thousands of existing UST systems with steel components that have years of serviceable life expectancy remaining. The regulations specify that these systems must be upgraded to acceptable performance standards by 1998 in order to remain in service. The requirements for upgrading include installing an approved form of leak detection (to be completed by 1993), providing a method of spill and overfill protection, and installing either corrosion protection or internal lining, or both. Metal piping systems must have corrosion protection. Tank systems which operate under pressurized pumping must have pipeline leak detectors in place.

Corrosion protection most suited for upgrading existing systems is the "impressed current" form of cathodic protection, although use of sacrificial anodes may be applicable under very specialized circumstances. As noted above, cathodic protection is a technique used to negate the normal corrosion circuitry present in the soil by changing the electrochemical forces in the vicinity of the tank system to be protected. An impressed current system operates by placing an anode bed around or near the protected system and impressing direct current voltage on this anode bed. An electrical circuit flows in a direction in which the tank system becomes "cathodic," or the non-corrosive element in the circuit. The installed anodes become the corrosion site. These systems can be adjusted for specific site conditions. They do, however, require close supervision and regular maintenance oversight to insure that no stray currentsescape to damage nearby unbonded metal structures.

Sacrificial anode systems rely on magnesium or zinc anodes placed in close proximity to the metal to be protected. No voltage is impressed on the system; current flow results from the electromotive forces that occur naturally between the adjacent dissimilar metal surfaces. The amount of protection is dependent entirely on the electrical forces derived from the amount and placement of the anodes and the area of surface metal to be protected. Sacrificial anodes are relatively low power, static systems, requiring limited oversight. But they should be inspected to determine the extent of anode degradation, and their replacement planned. Stray current damage is not a factor, but, depending on the backfill material and tank system design, adequate levels of protection for the tank system may be of concern. Sacrificial anodes are useful for piping protection, and in the case of existing tanks that are of FRP construction with steel piping, for both design simplicity and low cost.

While cathodic protection provides external protection against corrosion, it provides no protection against internal corrosion. A practical solution for this problem is lining the internal surface of the tank with a non-corrosive surface coating. Also, when properly applied, the lining should be of sufficient thickness to withstand the effects of external corrosion pits or holes reaching the inner surface of the tank. The lining materials used for this process are of polyester or epoxy formulations, either sprayed or troweled on the prepared inner surface of the tank. The procedure is highly specialized, performed by contractors trained and equipped for the operation. The process requires careful preparation of the surface by sandblasting. With an epoxy coating, the application can be either by spray gun or trowel. In either case, the coating is applied in one pass, about 1/8 inch thick, evenly spread, with no sags, holidays (skipped spots), or slumping. Epoxy curing takes from 24 to 48 hours; no product should be allowed in the tank until the coating is thoroughly dry and hard. Polyester coatings are quick-setting. They are sprayed on in very thin-film layers. Each layer will cure in minutes, so a succession of coats can be applied during a one-day application. The sandblasted etch pattern on the prepared steel surface is particularly important with polyester, since the first pass must adhere tightly to the metal surface without blistering. Polyester coated tanks may be placed back in service almost immediately after the final layer has been applied.

Since internal lining is a coating process, careful inspection of the tank shell is required before attempting to use the method. While some holes may be present in the shell (coatings have been used for years as repair technique for tanks already damaged), they must be widely separated to insure no weakening in the structural integrity of the tank wall. Seams or welded sections are particularly critical; any weakness in this area can rupture the coating as the tank flexes and breathes. Coatings used in combination with external cathodic protection provides effective internal andexternal protection in those cases where no prior damage to the shell already exists.

UST System Closure

Closing existing tank systems requires that processes be followed to ensure that tanks will not create future damage to the environment or be a hazard to the use of the property. Before the closure can proceed, local regulations should be reviewed, and approvals should be obtained from officials having jurisdiction. Definitions vary as to the time limits allowed for temporary closure, and local rules may dictate the approved procedure allowed for permanent closure. While EPA UST regulations specify that the site must be inspected for evidence of releases, most states direct how the inspection will be made, what tests and reports are required, and to whom the information must be submitted.

For temporary closing, the tank should be emptied, cleaned, electric service disconnected, and access openings capped with light concrete filler. Where the tank had been in flammable liquid service, such as with gasoline, the atmosphere in the tank should be rendered safe by inserting CO_2 to neutralize the explosive potential while the tank is out of service. Normally, about 15 pounds of CO_2 ice per 1,000 gallon capacity is most effective. To put the tank back in service, check it for cleanliness, remove the concrete caps, restore electrical service, and refill the tank. Check the product discharge for clean, dry liquid before attempting to use the tank system in regular operation.

For permanent closure, local rules may require that the tank be removed from the ground, the hole backfilled with clean fill, and the site be examined for evidence of a release of the product stored in the tank. Under certain conditions, the tank may be allowed to be closed in place, such as when the tank is under a building structure, or when its removal may result in damage to nearby facilities. Approval of local officials is usually required. In this case, the tank must be filled with an inert solid, normally sand or concrete slurry, and the site inspected for evidence of any release of the product stored. In either method of permanent closure, the registration information on the tank must be resubmitted to reflect the new condition.

Installation of UST Systems

In the past, many tank failures could be traced to poor project specifications, faulty installation practices, poor supervision during installation, or inept handling of equipment. Since the tank is out of sight after its installation, errors are hidden until failure occurs. In an attempt to insure that contractors exert greater care and attention to proper procedures, many states are requiring licensing or certification of installers. The EPA UST regulations require that written certification be provided that proper installation practices were followed. That documentation remains on file in the tank records.

There are a number of steps during the installation process that tank owners should oversee.

1. Tanks should be carefully examined upon arrival at the job site.
2. Awaiting placement, the tank should be secured against damage.
3. After the tank hole has been excavated, proper bedding and base backfill material should be set in place.
4. Due regard should be paid to potential water collecting in the hole, assessing the need for anchoring.
5. No product should be put in the tank until backfill completely surrounds the tank shell.
6. Backfill material must be clean, homogeneous, and free of trash, rocks, debris, or foreign matter.
7. While the tank hole is still open, the tanks should be ballasted to prevent floating in case of rain or flooding in the hole.
8. When the installation is completed, an impermeable cap of concrete or asphalt, to protect against surface water seepage, should cover the hole, extending well beyond the back-

filled area. The cap should be constructed with regard for the possibility of vehicle traffic over the tank.

Inspection of the tank upon arrival at the job site is important to ensure that no damage has occurred during delivery, and that, upon placement in the hole, the tank is sound and liquid-tight. FRP tanks are particularly vulnerable to shell damage during handling, and therefore require close examination. The tank should be inspected visually for any obvious signs of cracks or punctures. The external surface should be coated with soapy water, and then air pressure, no greater than 5 psi, should be impressed on the tank. Again, an inspector should visually examine for holes or punctures evident by bubbling. After ensuring proper tank condition, the tank should be carefully secured to prevent rolling, wind damage, and accidental impacts from equipment working at the site. The tank should be carefully inspected again immediately before being set in the hole.

Proper bedding under the tank to support it in place in the hole is essential for reliable long-term service. As the tank is subjected to variable hydrostatic loads from filling and withdrawing the product, the shell expands and contracts. A tank should NEVER rest on a hard, unyielding base. Such positioning causes all the weight in the tank to be supported by the narrow spine in contact with the rigid base. Extremely high pressures are exerted on that small area, and shell sections or seams can split under the high stress. The bedding should be at least 12 inches of backfill material, carefully leveled before placing the tank. This placement of bedding material will provide support across the entire tank diameter, distributing the weight load evenly over a large area and avoiding imposition of highly concentrated pressure loads.

Before setting the tank, the installer must consider the possibility of water in the hole, causing tank movement, flotation, or shifting of an empty or partially empty tank. If this is a potential problem, the tank must be anchored to prevent damage from the buoyant effect of water in the hole.Anchoring is accomplished by placing straps over the tank, which are then secured, or "anchored," to a concrete pad under the tank, or to concrete piers (called "deadmen") placed to the side of the tank bottom. When the tank has been secured in this fashion, the weight of backfill material on the pad or deadmen, acting in a downward direction, will more than offset the buoyant forces of water acting upward on the tank.

Backfill material should be clean, free flowing, and free of any contaminants, foreign matter or debris. For steel tanks, a clean, washed sand or crushed stone is acceptable. For FRP tanks, use pea gravel backfill; crushed stone may be acceptable as an alternative. Sand backfill should not be used with FRP tanks, except when the installation is under the close supervision of a factory representative. (Sand tends to slump more readily than stone, allowing voids to develop under the tank. This loss of support can be critical to the reliable service of an FRP tank. FRP tanks, with relatively thin wall shells, must have the full support of properly placed backfill. In steel tanks, structural strength will usually compensate for slumping problems.) Because of lack of support until the hole is filled, liquid should not be put in the tank until backfill support is in place up to the tank top.

However, when the hole has been backfilled to the tank top, and the installation awaits completion of piping and equipment placement, tanks should be filled with liquid for ballasting, to prevent flotation in the event of rain or surface water flooding the hole. This step may not be necessary with anchored tanks, but should be considered. The open tank hole is an efficient dry well; backfill material such as pea gravel is very permeable versus the surrounding native soil. Any water flow will be readily retained within the backfilled area. If the tank hole has been lined for secondary containment, liquid cannot escape by percolation, thus exacerbating the water condition.

The final consideration is capping the tank hole. Protection must be provided against surface water drainage into the hole, and for support of traffic movement over the tank. The cap should extend well beyond the backfilled area to prevent the surface water seepage through the interface area between native soil and backfill. In a traffic area, overburden on top of the tank is usually over 3 feet, with 6 inches of reinforced concrete or asphalt. Above an FRP tank, however, total overburden should not exceed 7 feet.

During the installation process, a record should be kept of each stage in the procedure. This may be documented on a form provided by the equipment manufacturer, or by a supervising engineer's check list. It is also recommended that photographs be taken or video tape be made during the installation, for future maintenance reference. The record of the installation should be signed off by the installing contractor or supervising engineer, and made part of the permanent record of the UST system.

UST System Management and Operations Planning

Operating underground storage tank systems has become a complicated exercise in logistics, oversight, and regulation compliance. Tank owners are faced with a bewildering array of rules enforced by numerous levels of officialdom. Federal, state, county, city, town, and village all have developed some form of UST regulation. In order to cope, the tank owner should prepare a logical management plan addressing the concerns of the various agencies, while at the same time allowing himself to operate his business at reasonable cost. At the same time, improvements in UST technology provide the owner with the means to virtually eliminate the potential for an undetected leak or spill to occur. An effective plan for managing the tank population will help take advantage of that opportunity to enjoy long-term security against environmental damage.

The plan is based on a careful survey and analysis of the present condition of each system in the tank population. The survey should include as many details as possible about each system. (This will not only assist in developing a compliance management plan, but will be invaluable in future maintenance and operations programs.) At the very least, in addition to the basic description data, the survey should cover information about existing protective equipment, about the environment surrounding the tank, about both the current service and future plans, and about the pathway/receptor setting to be impacted in the event of leakage (where will the leak travel, and who or what will be affected?).

With this information, it is possible to construct a profile of relative risk for each UST system. A simple arithmetic weighting of various key survey parameters: age, corrosive potential, protective equipment, soil aggressiveness, monitoring capability, etc., will provide a comparative scale on which to rate the various tank systems. Included in the analysis should be the future plans for specific tank systems and an explanation of how the regulations will impact the timetable for acting on a particular system. This rating can, in turn, provide an estimate of the order, extent, and priority of upgrading actions among the various tanks.

Working from this framework, one should construct a budget plan and timetable for upgrading. Specific upgrading actions can be identified. Estimates of costs for compliance with the regulations can be developed. Contingent possibilities should be evaluated. The time line by which to plan the program can be established. At this point, the plan can be documented in written form to provide for project planning, personnel training, a compliance program for enforcement authorities, and justification for budget approvals.

There are ten basic subject areas to be addressed in the plan, together with their specific costs and timetable for completion:

- Loading and Unloading Procedures
- Inventory Management Controls
- Maintenance Program
- Safety Program
- Files and Recordkeeping
- Leak Detection and Monitoring Program
- Tank Testing Program
- Upgrading Existing Tank Systems
- Replacing Existing Tank Systems
- Closing Tank Systems

Specific operating instructions should be documented in the plan, together with costs to be incurred and the timing for actions. Cost and timetable data can be developed in spreadsheet format for internal evaluation. The written instructions and descriptive information can be used in training programs and audit protocol.

Operating Practices

The material in the first section, Loading and Unloading Procedures, might include supervisory requirements when transferring product, spill prevention practices required, reports to be prepared, and training procedures for personnel involved in transfer operations.

Inventory Management documentation covers methods and precision of measurement, acceptable inventory variations, comparison of meter vs. physical measurements, and the precision required.

Maintenance and Safety are standard operating activities. Subjects should include inspections and inspection schedules, cleaning, painting, preventative maintenance, liquid handling procedures, product dangers, emergency response procedures, notices posted, notification, and follow-up.

Establishing useful Files and Records is extremely important for operating oversight as well as regulation compliance. This section of the plan should describe the types of records to be maintained, the form of record, and the purpose. Typical records should include purchase records, installation certification, instructional manuals, maintenance files, calibration records, corrosion protection records, SPCC plans, OSHA records, NPDES permits, Coast Guard operating manuals, copies of regulations, and similar documents or files. Of singular value in the plan would be a discussion of the purpose and form of the records, and their disposition over time.

The sections of the plan discussed above can be thought of as administrative, providing documentation for operating practices. They represent a nominal cost to implement and will require periodic updating, but will have ongoing value in facility operations for many years. They form the nucleus of material for specific training exercises or refresher courses to acquaint operating personnel with their duties and responsibilities. Auditors may find the material useful in reviewing actual field practices relating to the operation of UST systems. They can provide continuity and standardization in operating activities as personnel changes, and time goes by.

Regulatory Requirements

The next five sections of the suggested plan deal with engineering and upgrading efforts to be expended in meeting the requirements of the regulations. They incur the major costs, and are subject to mandatory deadlines for completionunder the UST rules.

The Leak Detection and Monitoring Program has, in most cases, been implemented to meet the deadline of 1993. However, as tank upgrading advances, the form and methods of monitoring may change. The plan should include discussion of both current monitoring practices and future modifications. Subjects to cover will be specific to the various installations, but might include a description of observation wells, their location and use, use of automatic sensing equipment, in-tank gauging practices, schedules for inspection of monitoring equipment (which may also appear in the maintenance section), and use of an inventory and testing program or SIRA, if involved.

Tank Testing may be included in the operating plan and should be carefully described in its section of the plan. Subjects might include approving methods for testing, supervising test programs, reporting test results, retesting where indicated, and compiling records and response to enforcement agencies relative to test results.

Engineering and Construction Program

The last three subjects are specific to the actual upgrading effort and constitute the engineering and construction program. The section on Upgrading Existing Tanks should include discussion of the various upgrading practices: cathodic protection, internal lining, spill and overfill protection as a retrofit, site specific monitoring equipment, and economics versus tank replacement. The locations selected for retrofit should be identified, and the budget and timing established.

Replacing Existing Tank Systems is covered in the next section. Tank designs, ancillary equipment, budget and timetable should be reviewed. Permits, site approvals, installation supervision, documentation, and facility records should also be addressed. This section might also include construction drawings and specifications, or provide an index to such information.

The final section of the plan addresses Closure, discussing temporary and permanent closure procedures, regulations pertaining to closure, and identifying the specific tank systems to be closed. This section should also address the procedures for assessing the site for evidence of product release, tests to be used, reports to be filed, and the records of site assessment to be maintained.

When the plan has been developed, its detailed documentation can be used to explain the overall program to a variety of interested parties to justify budget requirements and to be available to demonstrate to enforcement officials that a program has been put in place to comply with the rules. During plan development, logical expenditure estimates evolve, and various risk analyses can be evaluated. Once completed, the plan provides ongoing guidance for operating UST systems in a safe, reliable manner, with minimum concern for future undetected product releases.

ABOVEGROUND STORAGE TANK TECHNOLOGY

With the high costs and complex operating requirementsimposed on UST owners by the new regulations, there has been increasing interest in the use of aboveground storage tank (AST) systems to replace USTs. The incentives are: relief from regulation, lower cost of installation, lower insurance rates, and ability to readily detect leakage. All of these assumptions should be viewed with a degree of skepticism, however. While there is some general validity to these opinions, they may not hold true in specific situations, and there are other considerations that may outweigh the perceived benefits from the AST alternative.

While UST systems tend to be consistent in size and configuration (commonly found in sizes less than 20,000 gallons), and of a standard rounded horizontal shape, AST systems range widely in size, structure, and configuration. Historically, their general utility has been where high volumes of storage capacity has been required, as in refineries, oil fields, large chemical plants, large airports, and major liquid distribution centers. Sizes ranged from the small 20,000 gallon tank to multimillion gallon vessels, where storage is measured in barrels (42 gallon unit). For the purposes of the following discussion, comments will be confined to the low end of AST configurations. These are comparable to UST system sizes, described in standards UL 142 or API Specification 12F, and are generally smaller than 50,000 gallon capacity tanks.

AST Regulations and Standards

As of the early 1990s, AST systems have not been regulated in the same manner as UST systems. There have been no federal rules specifically addressing AST designs, upgrades, or leak detection. There have been, however, state regulations that vary widely in complexity and rigor in different localities. Many states dictate construction standards, and, to some degree, operating standards, but are relatively inconsistent from state to state. Since most AST operations occur at large physical plant facilities, usually situated within a single state, this regulatory condition poses little problem to the tank owner at that location. AST systems do not proliferate throughout the country in thousands of small settlements, as do UST systems. While a release from a large AST system could be a catastrophic event, it is largely a local problem, normally resolved at state level. Federal concern over releases from AST systems tend to be related to the Clean Water Act, and the possible impact of a spill on waterways. The federal rules more or less uniformly affecting facilities with AST systems are 40 CFR 112, the oil spill (SPCC) regulation for facilities handling oil products near navigable waterways, and 33 CFR 154, 155, and 156, which empower the Coast Guard to regulate waterside oil terminal operations. All affect AST operations indirectly, but no rule contained specific standards for storage tank design or operations. This lack of a specific federal rule for AST systems has led to interest in using AST equipment to replace USTs.

Two federal laws, however, provide EPA with de facto power to exercise close oversight of any size AST system: the CleanWater Act, requiring spill prevention, containment, and countermeasure (SPCC) planning, and the Oil Pollution Act, under which EPA can require planning for actions to be taken to cope with the "worst case" release scenario. Since the SPCC Plan requires specific contingency planning, including the updating of existing facilities, AST owners can be required to take upgrading actions similar to those required of UST owners. Further, Congress has been considering passing legislation to regulate AST systems to the same degree now required for USTs.

AST System Designs

Small AST systems appear in two different configurations: with the tank bottom resting on the ground, a so-called "onground" tank, and one fully supported above the ground on cradles or saddles, the "aboveground" tank. The standards describing these tank configurations are found in UL 142 and API Specification 12F. UL 142 discusses both onground and aboveground designs, to a maximum of 50,000 gallon capacity, while the API standard addresses onground tanks only with maximum volume of 31,500 gallons.

While the typical horizontal aboveground tank meeting UL Standard 142 resembles a similar tank for underground service described in UL Standard 48, they should not be confused as interchangeable systems. The UL 48 tank does not contain certain features required in the UL 142 Standard, specifically in their normal and emergency venting configurations. For aboveground service, a UL 48 tank will not meet code requirements and may not be substituted for a UL 142 tank.

API Specification 12F describes only onground vertical tanks which are shop-fabricated and shipped to the work site ready for installation. (The larger AST tanks are usually erected from steel plate at the site and referred to as "field-erected" tanks.)

Current AST design, construction, and operating guidelines evolved from industry standards and fire prevention codes. The industry standards have been largely the work of the American Petroleum Institute (API), Underwriters' Laboratories (UL), American National Standards Institute (ANSI), and the Steel Tank Institute (STI). The American Society of Mechanical Engineers (ASME) and the American Society for Testing and Materials (ASTM) also contributed. The fire codes have been developed by the National Fire Protection Association (NFPA) and the Western Fire Chiefs Association (WFCA) in their Uniform Fire Code (UFC). Where the recommended NFPA or UFC codes have been adopted as reference by certain states or municipalities, they have the force of law or regulation. In other jurisdictions, these recommended codes and industry standards are viewed by local officials as definitive, authoritative sources on which to base personal enforcement judgments. Approval authorities, primarily fire officials, invariably require tanks to have UL labels, referred to as "UL listed tanks." Most tank owners follow these codes and industry standards regardless of their regulatory status.

NFPA 30, NFPA 30A, NFPA 395, and UFC Article 79 address flammable and combustible liquid storage and handling, and are regularly updated to remain current with evolving practices and conditions. NFPA 30A, an extract from NFPA 30, is specific to service stations and marinas. NFPA 395 contains codes for farms and isolated construction projects. The codes describe detailed installation and design criteria to protect against fire or explosion damage. In addition to tank design and construction descriptions, code specifications include:

- installation standards; tank and building spacing
- overfill and spill control; diking requirements
- venting specifications
- corrosion protection
- support and foundation descriptions
- installations inside buildings
- controlling ignition sources
- testing
- closure, abandonment, or reuse

Underwriters Laboratories is a nationally recognized standard-setting organization which tests and produces design and operating standards for various devices, materials, and systems presenting potential hazards to life and property. UL 142 is the standard developed for shop-fabricated low pressure, cylindrical, horizontal or vertical, welded steel tanks, under 50,000 gallons. They are used as aboveground storage of stable, non-corrosive liquids with specific gravity not exceeding that of water. Specifications include:

- description of tank material for volumes
- shell plate thickness
- welding detail at shell joints
- compartment tanks and bulkhead bracing
- vent openings
- tank connections
- manholes
- fabrication details for vertical tanks
- heating coils and hot wells
- tests and markings

While no universal regulation currently exists, local officials normally require compliance with UL 142 when approving an AST installation.

API 12F is an industry standard for the manufacture of atmospheric, cylindrical, vertical, closed top, welded steel aboveground storage tanks in nominal capacities of 90 barrels (3,780 gallons) to 750 barrels (31,500 gallons). These tanks are shop-fabricated and delivered to the installation site ready for placement on prepared bases or foundations. The specification includes:

- material, shell thickness, design details
- dimensions, connections, and operating pressures
- welds
- bottom, shell, and deck design
- fabrication, testing, marking, and inspection

Shop-fabricated vertical tanks will usually require base plate corrosion protection not described in this standard. A vapor suppression feature, such as a floating internal pan, may be required if volatile liquid is stored.

The Steel Tank Institute has developed specification STI F911-91 for fabrication of an open top secondary containment steel dike structure to be used with a standard UL 142 primary storage tank in a unitized aboveground storage configuration. The specification describes the containment of a UL 142 tank with an open-top secondary outer shell, or dike, consisting of a solid steel floor, vertical steel sidewall diking, and tank supporting structures inside the dike. Primary tank capacities up to 20,000 gallons are included in the specification table. The external dike structure is capable of holding 110% of tank capacity.

The entire unitized structure is intended for aboveground, free-standing use. Specification details include:

- dike design standards
- dike buttress details
- welding dike sections
- foundation and tank supports
- testing and marking

While FRP tanks and ancillary equipment are widely used in UST installations, FRP tank manufacturers do not offer configurations for aboveground storage of petroleum products. There are a number of specialty FRP AST designs, in both onground and aboveground configurations. These are available primarily for chemical product storage, and are manufactured to meet special service specifications. There are presently no UL or API construction standards for aboveground FRP petroleum storage tank systems.

In specialized situations, portable tanks on skids, or a tank mounted on a truck chassis, may be encountered. These will be useful at small commercial operations, such as road construction sites, temporary job locations, or where mobile fueling capability is desired. NFPA 30, Chapter 4, describes the special standards for portable tanks of less than 660 gallon capacity. Portable tanks over 660 gallons must conform with the same standards applicable to fixed aboveground tanks.

To meet UL 142 requirements, horizontal AST configurations must be of new structural grade carbon steel, and with metal thickness of 3/16" or 1/4", depending upon the tank's diameter and volume. Flat bulkheads of compartmented tanks should be braced. Venting, specified by NFPA 30 and UL 142, must be designed for both normal and emergency pressure relief. Emergency venting may be either

1. a manhole designed to open upon pressure, insuring no greater pressure build-up in the tank than 2.5 psi; or
2. a manhole not less than 8" in diameter capable of rising 1 1/2" in emergency venting (for example, in a typical 20,000 gallon tank).

A single combination vent of 2" to 10" in diameter depending on tank volume, may be installed for both normal and emergency venting.

UL 142 vertical tanks do not exceed 35 feet in height, nor 50,000 gallons in volume. Minimum plate thicknesses is greatest at the bottom (base plate) and lower side walls, and thinner on upper sections. Tanks over 25 feet high must have side walls of at least 1/4" up to the 25 foot measurement. Venting requirements are the same as for horizontal tanks. The tank may be specially constructed with a weak shell-to-roof joint designed to rupture at excessive pressure.

Tanks meeting API 12F standards are nominally under 25 feet high. Shell plate thickness is 3/16" or 1/4" depending upon the purchaser's specification. Venting standards are as specified in NFPA 30. Based on the buyer's order, API tanks can include a cone-down bottom with internal drain line for water removal, can be either unskirted (external projection) or skirted (internally mounted), and can have a flat bottom without a drain.

All tanks over 5,000 gallons, and each compartment over 5,000 gallons in a horizontal tank, shall have a manhole of from 16 to 30 inches, depending upon tank size, to provide access to the tank. Specifications for manhole plate thickness and weld and gasket details are listed in UL 142. For smaller volume horizontal tanks, manholes may be best positioned on the tank top, together with vent connections and gauging openings. In larger volume tanks, horizontal, vertical onground, or shop-fabricated manholes may be either on top or in the side. Where manholes for access are located near the bottom, their lower part should be from 6 to 18 inches above the base. There may also be smaller manholes on top for emergency vent pressure release.

In addition to manholes, tanks require openings for product inlet and dispensing, venting, vapor recovery, level measurement, and water draw-off. These will vary in size and configuration from 2" to 10", depending upon tank size and connection use. Pipe and flange connections are described in detail in UL 142. Openings will be located in configurations most suited to the tank design and application. Delivery inlet and dispensing openings are normally located at tank bottom (on end plates for horizontal tanks). Product handling openings at the tank bottom should be situated 2 to 8 inches above the lowest section of the tank to preclude dispensing bottom sludge or waste with the product. Water draw-off openings may be at the lowest tank level. A series of openings on the tank top may be installed for special uses, such as pressure-vacuum vents, pressure relief flanges, vapor recovery connections, or product receipt.

At this writing, NFPA has published Tentative Interim Amendments to NFPA 30A, attempting to recognize the development of new technologies and tank designs, and to recommend acceptable standards to utilize innovative aboveground storage systems. In the meantime, new AST designs and storage concepts are rapidly being developed by industry in anticipation of the growing interest in aboveground storage.

One of the newest innovations has been the introduction of arectangular horizontal steel tank resident in a concrete container. The entire structure can rest on a concrete slab or horizontal supports above the ground to allow observation under the tank. A leak monitoring capability is provided with leak detection tubes inserted in the concrete wall. While the concrete outer containment is referred to as a **vault,** it actually forms a secondary outer shell directly contacting the tank surface, and thus functions as a composite tank design rather than a true vault system. (At present, "vaults," in storage tank system concepts, are deemed to be structures allowing either physical entry or unobstructed visual examination of the space between tank and the outer containment.) The concrete exterior meets most basic code requirements of NFPA 30 and UL 142 (6-inch reinforced wall thickness, 6-inch emergency vent, steel tank UL approved, etc.), and provides reliable aboveground storage, as well as excellent fire protection. The tank is relatively simple to place, and concrete pouring procedures, where performed at the site, are clearly described by the manufacturer. Tank sizes from 250 gallons to 6000 gallons are available .

When a tank is placed in an underground structure such as a "basement, cellar, mineworking, drift, shaft, or tunnel", and "is situated upon or above the surface of the floor" (RCRA Amendment, Nov. 8, 1984), it is exempt from regulation as an underground tank system. Tanks placed in underground vaults meet this qualification. For practical purposes, the vault in such an installation should meet the structural design standard of American Concrete Institute, ACI 350, Concrete Sanitary Engineering Structures, for the control of cracking. This standard is typically used for water retention requirements. Further, an elastomeric epoxy coating on the interior vault wall will prevent petroleum seepage into concrete pores. The vault must be ventilated, and must be of sufficient dimension to allow entry for tank inspection. The top of the vault should be under a cover to prevent surface water collection. The tank should be an aboveground horizontal design (UL 142), mounted on cradle supports for bottom inspection. (Economy and design considerations dictate that vertical tanks are impractical in closed vaults.) Ancillary equipment for overfill protection, vapor recovery, and spill collection should be installed. Fire fighting equipment should be available. Delivery and withdrawal equipment can be the same as that used with a typical UST installation. Tanks in sunken vaults will receive product by gravity flow, and dispense by either suction or submerged pump.

While very useful for UST installations, the double-walled tank is of limited value in an AST configuration. EPA and state agencies have not accepted the double-wall design as a secondarily contained storage system for aboveground service - as is a single-walled tank and dike - because the outer shell offers no containment of piping, valves, emergency vents, or inlet and outlet openings. The greatest environmental concern about an AST configuration is overfilling or spills; spillage at external appurtenancescannot be contained within the outer shell of the tank. The outer wall of the tank does not perform the same function as a diked area. A single-walled tank inside a dike can be regularly observed, and leakage from either the tank or ancillary equipment can be quickly detected and contained.

Aboveground tanks may be lined internally to protect against internal corrosion. Various processes are available: epoxy or polyester spray coatings, fiber glass lining materials, and flexible liners. Coatings or linings may be installed to cover the entire inner surface, the base plate and lower wall of vertical tanks, or only the lower portion of horizontal tanks. These have been used extensively in UST systems and are equally suitable for use with AST tanks. The surface area must be thoroughly cleaned and sand-blasted to bright metal. The premixed coating is applied by spray (or certain epoxies may be troweled) to a finished thickness of 100 mils (1/8 inch). Application must be made by a qualified contractor, properly trained and equipped for the work. Prior to putting the tank into service, an inspector should check the coating for thickness, continuity, and hardness, using a magnetrometer and a penetrometer.

Sheet lining material may be used as a corrosion barrier on the base plate and side walls of vertical tanks and in horizontal tanks, if desired. (Internal coatings are normally a more cost-effective choice for horizontal systems.) The metal surface must be carefully cleaned and dried before applying the adhesive material. Particular care must be taken at seams and right angles between the base plate and side wall to insure tight adhesion of the lining material. It is recommended that the material be applied only by qualified and experienced specialists.

A new innovation in internal lining concepts is the flexible liner, or **bladder insert.** This design is mainly suited for use with horizontal tanks, although may be utilized for vertical tanks that are to be completely lined on all internal surfaces. A vacuum pump is required. A geotextile lining material, custom-fit for the size of tank, is inserted to cover all surfaces. A partial vacuum is then drawn from the interstitial space between liner and tank surface, resulting in tight adhesion between the two materials. The vacuum must be maintained to insure continuing contact.

As noted earlier, Steel Tank Institute Standard STI F911-91 describes the combination tank and attached dike structure, a very recent design development in AST technology. The tank and integral dike are delivered as a unit to the job site, ready for placement with minimum setup effort required. The installer simply has to position the tank as specified on the plot layout and make piping and operational connections. Depending on project specifications, raised base supports may be installed to set the entire structure above the ground and to eliminate the need for corrosion protection of lower metal surfaces of the dike section.

SPCC Plans and AST Operations

A major consideration in the use of AST systems is protecting against spills and overfill releases. Except under veryunusual conditions, all aboveground tank systems receive product deliveries by pump under pressure; most USTs receive product by gravity flow. The possibility for overfilling an AST is far greater than in an UST system. The need for prior planning and close supervision of the transfer procedure is important.

The EPA, pursuant to the Clean Water Act of 1973, developed regulations (40 CFR 112) at that time to protect against uncontrolled releases from facilities handling oil products near navigable waters. The regulation required the facility to prepare a **Spill Prevention, Control** *and Countermeasures Plan,* designed to anticipate where spills might occur, and, as the title suggests, to have in place various strategies to mitigate their effect. Originally intended for petroleum facilities located on major waterways, the concept actually applied to any AST installation capable of impacting the waters of the U. S. This could include the great majority of any type of AST facility near rivers, streams, creeks, or any drainage network leading to waterways. In addition to this rule, Congress recently passed the Oil Pollution Act of 1990, giving EPA further authority to oversee the effectiveness of various spill countermeasures adopted by AST facility owners or operators. As a practical matter, these two federal laws empower EPA to require AST owners to perform many of the activities now specified in UST regulations, such as monitoring for leak detection, upgrading certain equipment to provide better SPCC performance, and maintaining detailed files and records about their activities in this regard.

SPCC planning already imposes a responsibility on the AST facility owner or operator to take proactive steps toward spill prevention. Planning frequently includes facility upgrading programs, personnel training routines, monitoring efforts, and revised transfer and handling procedures. The regulation requires that, if two minor spills, or one spill in excess of 1,000 gallons, occur within any 12 month period, the SPCC Plan must be reviewed by both the EPA and the state agency having jurisdiction. From that review, the owner or operator can be required by either agency to take whatever actions or to make revisions to the plan deemed effective in preventing further incidents. The Oil Pollution Act empowers EPA to approve spill preventive plans at the outset without prior spill history. The penalties for non-compliance are severe, including fines or revocation of operating licenses.

On the positive side, SPCC planning makes a great deal of sense. Most spill events occur because of poor oversight of the transfer operation (a personnel training problem), or lack of adequate facility design to protect against the consequences of spillage (an engineering problem). By applying due diligence to each of these elements, AST operators can make marked reductions in the potential for uncontrolled spills. A well-designed facility, maintained by effectively trained personnel with efficient transfer handling equipment, proper alarms, and adequate containment designs, is an excellent defense against serious spills. An effective maintenance program can protect against unexpectedequipment failure, further reducing spill potential. Since the consequences of spillage are so severe in expense and business interruption, the cost and effort in effective SPCC planning and implementation are nominal by comparison.

Comparison of Benefits of AST vs. UST Configurations

The rationale for preferring an AST system over a similar sized UST system is twofold: AST systems avoid the costs and regimentation of UST regulations, and AST systems are not regulated under federal authority. Insurance costs are lower for AST designs; there are no financial responsibility rules; and insurance carrier supervision is less stringent than with USTs. AST tanks are accessible, and pollution problems can be quickly discovered and corrected. AST systems do not require extensive leak and spill prevention equipment. Installation costs are relatively economical since no excavation is required.

Looking at these opinions in closer detail, one should consider the alternative view relative to USTs. While AST systems are not now regulated uniformly under federal law, they are quite widely regulated at many state and local levels under various fire, safety, and environmental codes. The National Fire Protection Association and Uniform Fire Code of the Western Fire Chiefs' Association have specified design and operating criteria for AST installations for many years. UL and API standards for AST designs have been adopted by reference in some local community codes. Given the current climate for stringent oversight of storage tank matters, most fire marshals now exercise tight control over the use of AST equipment in their jurisdictions. Obtaining their approval -- for which they have wide discretionary authority -- is an absolute requirement for the installation of any AST system. Local environmental agencies often must approve plans, concepts and layouts of any exposed structure potentially affecting the environment. Building officials, zoning officials, and certain other local boards or commissions may also exercise oversight of the proposed installation. Each of these official groups can, and usually will, suggest or require a variety of ancillary equipment to be added for safety or aesthetic reasons. The net cost of satisfying reviewing authorities to obtain permits can push the installation cost of an AST system beyond that of a similar capacity UST design. (The UST rules and limitations are now reasonably consistent, allowing very little room for costly interpretation or design adjustment by local officials.)

Community concerns and pressures may affect the choice. After installation, an AST system remains highly visible, subject to all manner of hazards due to its exposure. Vandalism, community criticism, product mishandling constraints and safety are heightened operating concerns. The aesthetic effect on the surrounding community of highly visible tankage may pose public relations problems. Because they are situated above the level of the delivering vehicle, tanks must always have product pumped into them under pressure. The potential for spills from overfilling is greater than with UST storage, which is normally served by gravity flow. AST overfill spillstend to be catastrophic events. When a spill occurs, an AST facility is immediately subject to all federal and state environmental and pollution regulations. The fine distinction between an AST and a UST installation, and their operational regulations, is no longer at issue. Where required by air quality regulations, methods for capturing vapors from the tank when filling it presents a fire hazard, and the proper vapor control equipment is often more expensive than that used with UST facilities. AST systems that store products affected by seasonal temperature changes may require insulation, heating, or cooling equipment, which are not necessary in UST systems.

There are operational matters to consider. Labor and maintenance costs are higher for AST systems. Where an AST system operates within a close range of navigable waterways, a Spill Prevention, Control and Countermeasures Plan (SPCC) must be prepared and implemented. This federal rule affects any facility with the potential to pollute waterways. An SPCC Plan can be expensive to develop, implement, and administer, and requires continual review and periodic updating. (UST facilities under 42,000 gallons capacity are exempt from SPCC requirements by virtue of the UST regulations.) Pursuant to the Oil Pollution Act of 1990, EPA is currently reviewing the existing SPCC rule and expects to have new regulations in place shortly. These rules can be expected to impose more stringent conditions for all manner of storage tank systems. AST systems will be required to be registered, undergo periodic inspection and testing, have effective corrosion protection and leak detection capability, and meet certain design and operating criteria like those specified by the UST rules. Of particular concern will be the requirement to have an approved discharge response plan in effect for responding to a "worst-case" discharge or threat of such a discharge. The costs of developing, implementing, and administering such planning can be substantial. Failure to have an approved plan will be grounds for denying a facility the ability to store oil products -- effectively putting the operation out of business.

Fire protection is a major concern in facilities storing flammable or combustible liquids. UST installations have a proven record of safe and effective defense against serious fire or explosion, while AST systems are clearly exposed to fire potential. Public safety officials naturally prefer storing dangerous liquids in the safest manner, which is underground. Environmental officials, on the other hand, concerned with the effects of pollution, usually prefer to remove storage from the ground, and tend to favor AST designs. The tank owner is caught in the middle. At the very least, approval for an AST system usually involves substantial expense for providing adequate fire protection. Modern UST designs avoid that problem. (They are, however, often required to have more elaborate leak detection capabilities, which can usually be incorporated into the basic UST design at a minimal added cost.)

An UST installation requires excavation, product containment, leak control, and properly engineered backfill and flotation protection. Likewise, AST systems include a number of added cost features. Secondary containment in the form of impervious layers under and around the tank area is required. Diking, which has always been required for fire protection, must now meet newer, more stringent environmental rules for spill containment. Emergency venting designs are required. The piping and handling controls must be designed to prevent product release while pumping. In some cases, depending on local approval criteria, roofing may be required over the transfer areas to reduce rainwash effects. Special pumping and handling equipment for surface water collection and small spill containment, retention, and storage may be necessary. A perceived cost advantage for installing an AST over a similar volume UST configuration may be deceptive.

An incentive frequently cited by tank owners for switching to AST systems is the lower cost of insurance and limited financial responsibility. This may have been true in the past, but there appears to be a narrowing of the cost difference between UST and AST coverage. First, as with pollution liability coverage for UST systems, insurers are requiring ground condition assessments at AST sites, at substantial cost to the tank owner for hiring qualified environmental consultants. Next, insurers are looking carefully at such matters as product flammability, exposure for damage to surrounding structures, compliance with NFPA and UFC standards, and pollution history of sites. Premium rates may well be affected by the findings. Having learned about environmental exposure from experience with UST installations, insurers are applying many of the same acceptance principles to AST configurations, and will be basing acceptance and premium costs accordingly. Also, proposed AST rules now under consideration include requiring financial responsibility plans similar to those in place for UST systems. As these events unfold and new UST state-of-the-art designs proliferate, with their vastly improved pollution safeguards, it is reasonable to assume insurance rates will reflect those changes, with an equalization of premium and attendant coverage costs between UST and AST systems.

With the UST regulations mandating upgrading of all existing UST facilities by 1998, tank owners have been searching for reasonably cost-effective, efficient avenues to operate their liquid storage facilities for the future. Certainly, aboveground designs offer one approach to solving that puzzle. However, the decision to utilize AST storage should not be made solely on the basis of avoiding UST regulations. AST systems are already regulated, but in a haphazard, patchwork manner likely to cause added cost and greater aggravation in certain jurisdictions. The UST rules now in place offer some uniformity and consistency. And, in any event, with Congress seriously considering regulating the AST population in a manner similar to the UST program, it is highly likely that there will be that type of federal AST regulation within two to five years. (With the Oil PollutionAct and new SPCC regulations, EPA effectively imposes universal "de facto" AST rules already.) If so, money spent now on AST configurations may require supplemental expense later to modify or retrofit for the new rules.

Today's rules require UST tank owners to take prompt action to upgrade tanks. The upgrade question is really how to get best long-term value from investment in equipment for storing bulk quantities of liquids. Whether spending for new UST facilities, or for a similar storage volume above the ground, the tank owner is faced with a major cost. Rather than simply choosing an AST system to escape UST rules, the tank owner should examine the benefits and liabilities of each method of storage to determine which may best suit his own long-term needs. Today's state-of-the-art UST system, properly designed and installed, offers a liquid storage design virtually free from risk of serious undetected pollution. Similarly, a properly designed and installed AST system also offers a high degree of protection against pollution risk. UST designs offer a number of benefits not available with AST: better fire protection, better temperature control and vapor suppression, reduced exposure to vandalism, better resistance to overfilling, minimum requirement for SPCC oversight, reduced public objection to undesirable aesthetics, and more efficient use of land area. ASTs, on the other hand, depending upon local conditions, may or may not offer savings in installation costs and lower insurance costs. Their supposed advantage in leak detection and reaction has not always held true; AST overflows and serious leakage have occurred suddenly and catastrophically. The costs of providing ancillary services such as fire protection, SPCC administration, operating oversight, product transfer supervision, and plant security may offset assumed savings from lower insurance costs or installation expense.

SUMMARY

Of course, there is no "correct" answer to the question of which method of storage offers the best solution. Each situation is unique, and affected by many other factors than those cited. At the moment, AST systems seem to be of interest to tank owners in the more rural areas, where land space is available and installation costs low. The interest is somewhat less in the more populated communities, where higher land costs and greater public concern over safety issues are encountered. Industrial firms tend to show more interest in AST designs than do service station operators. Anyone considering replacing an UST configuration with an AST system should review the proposed installation with the local fire and zoning officials before expending any appreciable sums for site review or preliminary engineering. Their approval must be obtained, and should be sought at the beginning, to avoid unnecessary design and planning expenditures for systems that may later not be permitted. In new installations with modern state-of-the-art designed equipment, and when properly operated, either method provides reliable protection against undetected releases. Cost differentials for insurance and financial responsibility mayshortly disappear. Other factors are subject to opinion and attitude. Considering the costs and consequences, tank owners should carefully weigh the relative long-term benefits of each method of operation before making a decision that will affect their facility for many years to come.

9

GEOLOGY AND GROUNDWATER HYDROLOGY

Porter-C. Knowles, P.E., P.G.
Ground-Water Hydrologist

OVERVIEW

The earth beneath us holds the fascinating story of geology. As we go from mountains to the coastline, from desert to the wetlands, we see around us the surface of the earth being shaped in many different forms by a wide variety of external and sometimes internal forces. It is the weathering of rocks that provides us with an immense diversity of landscape. It is also the rocks that provide the medium through which water flows -- a water which we call ground water, providing 90 percent of man's water supply. To understand the presence and the movement of ground waters, we need first to have a basic understanding of geology, the study of the material through which ground water moves.

To set this stage, we will look at the different fields of geology. We will then move from this broader picture to look at the building blocks of our environment--the chemical composition of minerals and the rock types they form. These rock types, over time, are subjected to forces that may change their location, their character, and their physical characteristics. Although some of these changes may occur deep within the earth's surface, many changes also occur on the surface. These surface changes we call **weathering**. The weathering of the earth's surface is visible all around us as landscape--the mountains, the valleys, and the seashore with material transported by the wind and water.

A large part of our understanding of geology, and subsequently ground water, also lies in an understanding of the many tests, analytical techniques, and methods that the geologist and engineer use to solve the "geologic puzzle." This puzzle is just one step in the process of understanding groundwater hydrology and contamination of our groundwater resources. If contamination exists and is a threat to our environment and public safety, an understanding of geology is also critical to addressing cleanup and practical remediation. Let's start with some basic terminology and background material.

GEOLOGY

Geology is the science of the earth--its composition, its structure, its history, and its past animal and plant life. A number of other sciences are also involved in the study of geology. Mathematics is, of course, fundamental to the study of all physical sciences. Following close behind are physics, chemistry, biology, and atmospheric sciences. Why so many sciences? Let's think about the bigger picture.

To understand the earth, we need knowledge of the earth's origin. We therefore need to add the science of astronomy to our list. The evolution of the earth, from its molten state, to the creation of atmosphere and the changing mantle or crust, is the start of the geologic process--a process which has been continually changing over billions of years. Physics and chemistry study both the composition and structure of the earth. Biology is also important since rocks have recorded the history of life over geologic time by leaving a fossil record of key or index species in various geologic time periods. The rocks can be age-dated by using sophisticated measurements of radioactivity. For recent material, Carbon-14 isotopes are very effective for dating rocks and sediments that are less than 40,000 years old. Water canalso be dated using similar techniques. Finally, the atmospheric sciences involve the weathering and remolding of surface rocks, sediments, and soils.

The study of geology can be broken down into the fields of physical and historical geology. In **physical geology,** we deal with the materials that comprise the earth, the structure and surface features of the earth, and the processes that make these features change and evolve. In the field of **historical geology,** emphasis is on the changes in the

Geology and Groundwater Hydrology / 251

earth's surface over time and the living things whose remains are found in those rocks.

Geology also can be broken down into branches such as:

- **Geochemistry**-- the study of the chemistry of rocks, waters, and atmosphere;
- **Geomorphology**-- the study of land forms related to rocks, minerals, geologic structure, and weathering by wind, water, and other forces;
- **Paleontology**-- the study of life in past geologic periods and the evolution of animals and plants as part of the fossil record;
- **Geophysics**-- the study of physics related to the earth and its properties.
- **Petrology**-- the study of rocks and their origin;
- **Stratigraphy**-- the study of structures, particularly sedimentary rocks;
- **Petroleum geology**-- the study of the origin and occurrence of oil and gas;
- **Mining geology**-- the study of mineral deposits and nonmetallic resources; and
- **Geohydrology**-- the study of ground water and surface water resources

The above list is not complete but is intended to illustrate that there are a great many specialties in the field of geology. Nevertheless, the reader can gain considerable insight into our multidisciplined world through a basic understanding of certain principles and concepts from many sciences. The following sections are designed to provide the overview and insight which will allow us to focus on one of the above mentioned specialty areas--geohydrology or ground water hydrology.

THE EARTH

The earth has a diameter around the equator of about 7,927 miles. The circumference around the equator is approximately 25,000 miles. In looking at the earth's surface, a little over 70 percent is covered by oceans. The earth's gravity is also important. Gravity not only plays a role in the formation of different minerals and rocks because of lighter and heavier materials but also influences rock formation by erosion and deposition. Different gravities on the many planets in the solar system create different forces with likely different end results.

The major divisions of the earth are the lithosphere, hydrosphere, and atmosphere. Each division is important in understanding the field of geology (Figure 1).

The **lithosphere** is essentially the earth's crust comprised of continents and oceans. Whether on the surface of the continents or beneath the ocean surface, physical features such as mountains, plains, and canyons are typical of the earth's surface. These larger features are the result of erosion, deposition of material, and volcanism, as well as faulting and folding of materials.

The **hydrosphere**, also an important part of our geological process, is comprised of oceans, lakes, and streams. The hydrosphere also includes ground waters in the subsurface soils and rock.

The **atmosphere** consists of various gases, water vapor, and dust. It is in the atmosphere that we consider the start and origin of **the hydrologic cycle,** which we will discuss in more detail later in the chapter.

Before discussing minerals, which are the building blocks of rocks in the lithosphere, let's consider the elements that make up the earth. These elements may combine together to form various compounds. The elements and the compounds may be gaseous, liquid, or solid at normal atmosphere, temperature, and pressure.

More than 99 percent of the earth's crust is composed of nine elements. These elements are as follows:

oxygen (O)	46.60%
silicon (Si)	27.72%
aluminum (Al)	8.13%
iron (Fe)	5.00%
calcium (Ca)	3.63%
sodium (Na)	2.83%
potassium (K)	2.59%
magnesium (Mg)	2.09%
titanium (Ti)	0.44%

Six other elements added to this first list bring the composition of the crust to 99.5 percent, with all other elements representing only .50 percent. This secondary list of six is as follows:

hydrogen (H)	0.14%
phosphorous (P)	0.12%

Geology and Groundwater Hydrology / 253

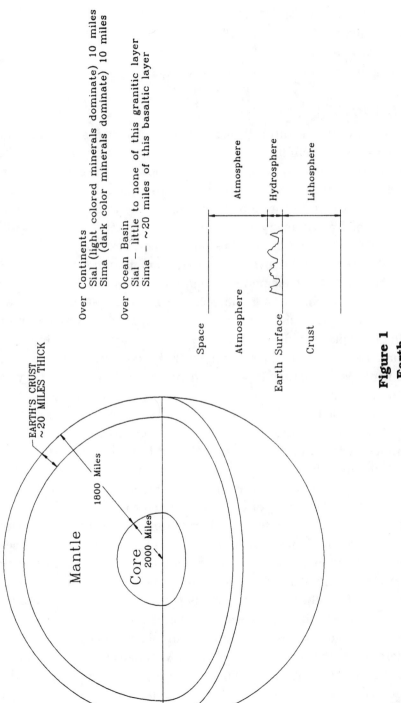

Figure 1
Earth

manganese (Mn) 0.10%
sulphur (S)   0.05%
carbon (C)   0.03%
chlorine (Cl)   0.03%

MINERALS

Minerals are naturally occurring solid elements or compounds. A mineral has a definite composition, or range of composition, and an orderly internal arrangement of atoms known as the **crystalline structure.** This structure gives the mineral unique physical and chemical properties, including a tendency to assume certain geometrical shapes and forms known as crystals.

The mineral diamond is simple in composition and is composed of only one element, carbon. Another common mineral called halite is a compound of two elements, sodium and chlorine. Halite is common table salt. Limestone is, in fact, calcium carbonate--the combination of the elements calcium, carbon and oxygen. As you can see, minerals can be simple in composition, or they can be increasingly complex combinations of elements. But whatever the combination, the atoms or elements join together in arrangements which we call the crystalline structure.

We have been talking about the crystalline structure of solid elements or compounds. This crystalline structure refers to the orderly arrangement of atoms in a mineral, and the arrangement of these atoms distinguishes one mineral from another. Earlier we mentioned carbon as the element in the mineral diamond. Graphite is also a mineral composed entirely of carbon. What's the difference? Well, in a diamond, the carbon atoms arebonded in a compact framework, with each atom surrounded by four others, all at the same distance from one another. In graphite, the carbon atoms are arranged in layers.

Then why is one mineral of the same composition formed at the expense of another? The answer is that it is under tremendous heat and pressure, and over a period of time, the diamond crystalline structure is formed. In nature, diamonds are formed deep within the earth where those specific conditions exist. Man has even learned to imitate these same conditions, to a limited degree, with machines, and hence, industrial diamonds have been made in large quantities for many industrial uses. Even some gem quality diamonds have been made, but these have been small in size and of lesser quality than those found in nature. But remember, nature has created a lot of flawed diamonds for every flawless one. How do we tell minerals apart and distinguish one from the other? We know from the diamond/graphite example that diamond is very hard and graphite is soft and slippery. Here is a list of some of the physical properties of minerals that are used to distinguish differences:

- crystal form
- hardness
- specific gravity
- cleavage
- color
- streak
- striations (parallel thread-like lines or narrow bands on the face of a mineral)

Let's discuss some of these properties because they are important tools that a geologist uses in describing rocks. These rocks may, in turn, be various combinations of a number of minerals. Each rock also has its own classification or properties.

Most minerals have a three dimensional pattern because of their atomic structure. Under favorable conditions of temperature and pressure, most crystalline minerals will grow as crystals that will have faces of the crystal oriented in accordance with their internal structures. Rock crystals are a source of wonder and value when they achieve any size in nature. Crystals occur in a great variety of forms, and the facets of crystals are helpful, in many cases, for identifying the specific mineral involved. The discipline of mineralogy focuses on mineral detection and evaluation sometimes using sophisticated electron microscopes to identify the chemical composition of an unknown rock or mineral. Today, with the help of modern technology, various minerals and elements are created synthetically in the laboratory. Whether these elements or compounds are used to create synthetic gemstones or industrial grade materials used in chemical processes, common mineral concentration and processing is an important part in our industrialized world.

Typical mineral identification involves the ability to describe minerals and compounds using their many properties. A number of descriptors are as follows:

Color. Specific minerals may vary in color due to the presence of impurities or slight changes in chemical composition. Color is one of the least helpful physical properties in identifying specific minerals. Certain metallic minerals, for example, will tarnish in the presence of oxygen so that only a fresh scratched surface will reveal the true color of the mineral.

Streak. When crushed, a few minerals will show color in powdered form that is different from their color in larger fragments. Sometimes this color is distinctive enough to help in their identification. Unglazed porcelain piece known as a "streak plate" is used to rub on the specimen to obtain a resultant color. As one might expect, most minerals turn white as a powder or have a paler tint than their ordinary color.

Luster. Luster is normally divided into a metallic or non-metallic luster. Some other terms used in describing non-metallic luster might be: pearly, greasy, silky, glassy, resinous, or brilliant "diamond like."

Hardness. The hardness of a mineral is measured by its ability to scratch or be scratched by other minerals in a scale. Over a century ago, a scale was developed as a standard. The scale, called **Mohs' scale**, was named after a mineralogist and is as follows:

1. talc
2. gypsum
3. calcite
4. fluorite
5. apatite
6. orthoclase
7. quartz
8. topaz
9. corundum
10. diamond

On the scale, talc is the softest, while diamond is the hardest. To compare common objects which might be utilized to evaluate hardness, the following list is helpful:

2½--fingernail
3--copper penny
5½--window glass
6--knife blade of good quality steel

Specific gravity. The relative weight of a mineral compared to water as a standard is termed its specific gravity. With water representing 1.00, specific gravities of all types of compounds cover a gamut of less than 1.00 to 5 or more. Non-metallics generally range between 2.5 and 3. Metallic minerals are usually over 5.

Magnetism. Iron represents 5 percent of the earth's crust. Identification of iron varying compounds is most helpful in identifying minerals. An ordinary pocket magnet is helpful in this case.

Rock-Forming Minerals

In looking at rock-forming minerals, it is important to understand that although there are many thousands of known minerals, only about 25 minerals are considered abundant in the earth's crust. The importance of identifying minerals, particularly the rock-forming minerals, relates to insight into the origin of the rock or rock type. This knowledge, in turn, provides other useful information related to physical properties of the rock or mineral which impacts the resulting movement and occurrence of ground water. The ground water is also influenced by both the mineral composition or solubility. Dependent on the subsurface environment, preferential pathways for movement of water may be developed. The subsurface environment may also provide minerals that can chemically combine with"contaminants" in ground water to immobilize them or to reduce their solubility. Some "contaminants," when present at less than federal primary and secondary drinking water standards, provide water with that good "taste" that we appreciate and look forward to when we're thirsty. Examples would be sulfate, chloride, and fluoride.

Several groups of rock forming minerals deserve mention. The groups are: silicates, oxides, carbonates and sulphates, and ore minerals.

Silicates. A first group of aluminum silicates called "feldspars" constitutes almost half of the earth's crust. Orthoclase feldspar is a potassium aluminum silicate. Plagioclase feldspar has a chemical constituency of sodium and calcium aluminum silicate. In general, the orthoclase feldspars are plate gray, while the plagioclase feldspars range from gray to pink. The feldspars are generally light in color and are non-ferromagnesian silicates. For classifying igneous rocks, identifying feldspar is the key.

The second group of silicates, called the ferromagnesian silicates, are much darker in color, ranging from green to blacks. This color is a product of the iron and magnesium elements in the mineral makeup. Such minerals as olivine, "greenish" hornblende, and the shiny and flaky biotite mica are good examples of these ferromagnesian silicates.

Oxides. Silicon dioxide or quartz is an excellent example of a very common oxide and mineral found in many rock types. Quartz can range from its crystalline form to a more micro-crystalline form such as flint, chert, or jasper. The color variance of gray to colorless is provided by the addition of chemical impurities to the pure colorless quartz (silicon dioxide). Iron dioxides are also very important. Limonite(Fe_3O_4) has generally a rusty or blackish color with a yellow-brown streak. Hematite, another dioxide(Fe_2O_3), has a reddish to brown color. Finally, a most important oxide is hydrogen dioxide, or water.

Carbonates and sulphates. Probably the most important carbonate is calcite, a calcium carbonate. This mineral is the primary constituent of limestone, and when subjected to additional heat and pressure may change into marble. Dolomite or magnesium carbonate is also a common rock-forming mineral.

Carbonates are found as cementing material in rocks precipitated from ground water. These rock types may be found in massive rock "beds" formed in deep seawater. Sulphate are also an important group. One of the most common sulphates is gypsum, a hydrated calcium sulphate that occurs as a sedimentary evaporite deposit. Gypsum is quite soluble and is a major source of sulphate in ground waters.

Ore minerals. These minerals are defined as mineral deposits that are mined on a commercial basis. Several hundred minerals are mined throughout the world. A number of these ores are metal sulphites, combining a metal with sulphur. Common iron ores are magnetite, hematite, and limonite, all oxides of iron. Pyrite is an iron sulphite that has a distinctive gold color and crystal form. Pyrite is known as "fools' gold" with a distinctive cubic structure. Lead sulphite is also a major ore and is called **galena**. Zinc sulphite is known as **sphalerite**. Lead and zinc deposits are commonly found together.

There are also a number of non-metallic ore minerals. The most important ones are coal, oil, and gas. Building stone and road aggregate are also important parts of our industrialized society.

ROCK TYPES

Now that we have briefly discussed minerals, it is appropriate to look at rock types. Again, most rocks are the combination of two or more individual minerals. But remember, most of the earth's crust consists of only the most abundant 25 minerals. Now we are going to move from the minerals that make up the rocks to looking at the different kinds of rock bodies and types of rocks.

To begin with, there are three basic types of rocks. We also have a large number of soil types resulting from the weathering of these rocks.

Igneous Rocks

The first rock type is *igneous*, a rock material which at one time was in a molten or liquid stage. Typical igneous rocks are the many types of volcanic material that all of us have seen in pictures or in person from volcanic eruptions or spewing from a volcano, flowing down a hillside, and later solidifying into rock. A less obvious type of igneous rock occurs when molten material solidifies **beneath** the surface of the earth rather than on top of it. This type of rock can be found beneath the volcano and beneath the earth's crust where there is a continuing melting and solidifying occurring at depth. These changes are caused by the heating and cooling of the earth itself from its core. A good example of this rock type is "granite" found in the core of many mountain areas. Granite is the light speckled rock with dark and light minerals seen in many mountain road cuts.

By examination, we are able at times to tell the origin and history of a rock. The geologist normally uses a ten power hand lens or looks at sections of rock through a microscope. Dependent upon the chemical composition of the molten lava or magma, igneous rocks are categorized by coarse or fine textured crystals or grains. This texture is a function of cooling time--the faster the cooling, the finer the texture or grain. The chemical composition of the materials may also typify the source of the rock.

A simplified classification of various igneous rocks is based upon two factors: texture and mineral content. Igneous rocks are normally classified by sight identification of hand specimens usually under a ten power magnifying lens.

Texture refers to the size, shape, and pattern of the mineral grains. This texture is a function of the rate of cooling, chemical composition, and pressure. Coarse-grained or coarse interlocking textures usually result from slow cooling. Fine textures usually result from rapid cooling and in some cases they occur so fast that only a "glass" is formed. In some cases, a mixed texture occurs that can be generally explained as resulting from multiple stages of cooling. An order of crystallization occurs in a magma as it cools. Minerals are derived from the chemical reaction of a solidified mineral with the remaining liquid magma. In 1922, N.L. Bowen proposed a series of minerals based on this principle (Figure 2). The chemical or mineral content of igneous rocks is generally broken down into the light colored versus dark colored minerals. As described in the previous section, the light colored minerals have high concentration of quartz and feldspar. Again, granite is an excellent example of this type of rock. In many cases, this type of rock is identified as being "acidic." The chemical composition sometimes is a function of the area of the earth's surface and the depth where the rock solidified. The dark-colored minerals have high concentrations of iron and magnesium compounds with a lower content of silica. These rocks are identified as being "basic."

With the help of microscopes and rock thin sections, the experienced mineralogist can develop a history of the rock formation. This history takes into account the volatility of various chemicals and minerals, along with the texture of crystal, indicative of cooling rates over time. This information may also serve as a help in locating similar rock or ore bodies. Radioactive age-dating may also help to identify the point in geologic time when the rock was formed. In addition, the direction of magnetism found in the rock may also indicate the previous position or location where the rock was initially solidified.

Prior to moving on to the second rock type, sedimentary rocks, here is an appropriate time to discuss weathering processes. Weathering creates sedimentary material and rock.

WEATHERING

When rocks are exposed to the atmosphere they slowly break down. This "weathering" is the result of the rock coming into contact with air, water, and physical organisms. This breakdown or disintegration can be either physical or chemical, or both. Chemical weathering or decomposition and physical weathering may occur at the same time or separately.

Geology and Groundwater Hydrology / 261

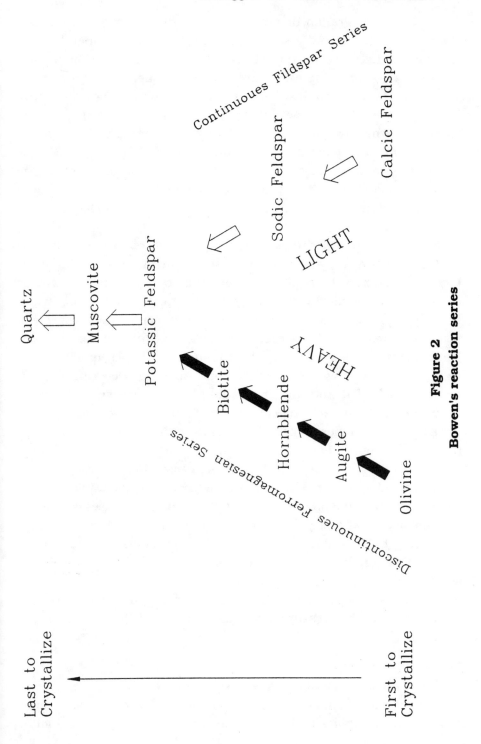

Figure 2
Bowen's reaction series

The disintegration or physical weathering can be caused by a number of different forces working on the rock and/or mineral. Temperature changes causing rocks to expand and contract over time cause cracks to develop in the rock and/or mineral usually along a plane of weakness. Sometimes this plane is a crystal interface, or it may have been caused by the manner in which the rock was deposited. Small cracks allow moisture to enter and plant growth to begin. Plant growth and roots are a very effective physical weathering process.

Once water or moisture can enter into a rock, temperature change causing freezing or melting can exert tremendous pressure into these zones of weakness. Frost heavage as well as spalling or exfoliation can occur.

Chemical weathering is a process that can remove soluble constituents from rock by leaching, or gases in the atmosphere can combine with the rock to yield new compounds or minerals. Water, for example, can chemically weather feldspar into clay minerals. Oxidation can also occur and in the presence of iron compounds, especially pyrite, can result in the release of acids as part of a chain reaction. Pyrite, in the rock matrix, brought to the surface of mines, quickly turns to brown and yellow-brown with intense staining occurring down a mountainside. The iron sulphate oxidizes and releases sulphuric acid.

The rate of weathering is also important, and differential weathering of different types of rocks provides us with the varying landscape that we see around us on the surface of the earth. The more resistant rocks form the top of ridges, while the less resistant are eroded easily away by wind and water. In areas where there is considerable rainfall, chemical weathering proceeds more rapidly, and some cases can result in extensive depths of soil above unweathered rock. Whereas limestones may be dissolved more quickly in a moist climate, granites tend to weather more rapidly in an arid climate where their coarse grain texture and complex minerals make them much more susceptible to the processes of physical disintegration. Formation of the rock at higher temperatures and pressures also makes it more unstable at the earth's surface.

Sedimentary Rocks

The second type of rocks are **sedimentary**. Sedimentary rocks have one thing in common: they are composed of small units, ranging in size from molecules to dust particles to pebblesand large boulders brought together and deposited on the surface of the earth's crust (above or below water). The components were transported by water, or by wind, glaciers, or gravity. All of the mineral matter composing these rocks were once part of other rocks--igneous, previously existing sedimentary rocks, and metamorphic (to be discussed next). It should also be pointed out that some of these materials may have passed from solution in water prior to their becoming part of rock material. The consolidation and/or cementing of these particles is sometimes described by the term **lithification**.

There are several processes by which sedimentary materials become compact rocks. These processes are: compaction, cementation, crystallization, and chemical alterations.

Compaction occurs when older sediments are overlain and buried by younger ones, subjecting the individual particles to the pressure generated by the weight of the overlying sediments. In some cases, compaction squeezes the water out of sediments unless the sediments remain in a saturated condition.

Cementation occurs when substances soluble in water such as calcium carbonate, silica, iron oxide, clay, and gypsum are deposited out of solutions in between the grains of the rock matrix. This deposition binds the grains together by creating a cement.

Recrystalization may occur in different chemical environments with ground waters. Sometimes crystals grow in open spaces within the matrix material or may replace earlier minerals with more stable mineral forms.

Chemical alterations occur under acid and reducing environments. The reduction of iron compounds by organic matter is a good example of this type of chemical alteration.

The matrix material of sedimentary rocks is generally classified based upon a size scale of the fragments. The following modified **Wentworth scale** uses part of that classification process:

Name	Diameter in millimeters
Boulder	Larger than 256
Cobble	64 to 256
Pebble	2 to 64
Sand	1/16 to 2
Silt	1/256 to 1/16
Clay	Size less than 1/256

Most, but not all sedimentary rocks are stratified, having been deposited in layers or beds. However, not all stratified rocks are sedimentary (volcanic lavas are often stratified). The physical processes causing the time of initial weathering of rocks also have impact on how well the particles may later be cemented together. Again, it is a cementation of the materials that is the difference between rock and soil. Some soils (soil particles) become cemented together after their deposition due to chemicals left by ground waters.

As you can imagine, the texture of sedimentary rocks is dependent upon the various mixtures of the different sizes of particles. Why is this important? A well-sorted material with particles very nearly the same size will have a much larger space between particles (called **porosity**) and will also allow better connection of those pores. When we measure that connection it is called **permeability**, the ability of rock to transmit liquids or air (Figure 3).

Let us examine for a minute a piece of sandstone. Sandstone is one of the most frequently observed sedimentary rocks, not because it is more abundant than other types, but because it has a tendency to resist weathering and to appear in prominent cliffs, forming the walls of canyons, ledges and shorelines. Generally, you can determine the bedding planes that separate one layer or several of sand as part of the deposition process. These planes, or parallel surfaces, are weaker than the rest of the rock and are usually etched more deeply by weathering and erosion. Using a magnifying glass, the sand grains that make up the bulk of the rock may vary in size and show evidence of cementation by calcium carbonate, or by iron oxide, if it exhibits a rusty or yellowish color. Calcium carbonate is a typical cementation, caused by the deposition in fresh or salt waters, or by the movement of some ground waters. The test for this cement or calcite is to dab a sample with weak acid and look for a bubbling reaction.

Another common sedimentary rock that is composed mainly of calcium carbonate is limestone. Carbonates dissolved in fresh or salt water may be precipitated in solid form in many ways. In the most common processes, aquatic animals and plants secrete the calcareous material to construct their shells, bones, teeth, etc. This biochemical precipitation may sometimes result in limestone beds consisting almost wholly of fragments of sea shells that are more or less broken and worn by waves and currents. Other limestones are obviously ancient

(a)
poorly sorted matireal
low porosity
low permeability

(b)
Well sorted sand
high porosity
high permeability

(c)
Well sorted clay
high porosity
low permeability

(d)
Solutioned limestone
medium to low porosity
high permeability

**Figure 3
Porosity and Permabililty**

reefs built by corals and other associated animals and plants. In chemically analyzing the rock, many limestones contain a significant amount of calcium-magnesium carbonate rather than just calcium carbonate. When magnesium carbonate is the primary constituent, the rock is called dolomite. Dolomite reacts less strongly to weak acid than limestone.

The most frequently occurring sedimentary rock on all of the continents is shale. Shale is comprised of silt and clay sized particles. Together, shale, sandstone, and limestone account for 99 percent of all sedimentary rocks.

The three most common types of sedimentary deposits are: by water, by wind, and by ice. These vehicles for transportation take particles of various sizes of rock during erosion and transport them to another location (Figure 4). In our later study of ground water, sedimentary rocks play the most predominant roles as aquifer units and sources of ground water for industrial and drinking supplies.

There are several types of transportation by water of rock material. Through the force of gravity and velocity, streams carry material to rivers and oceans. The finer materials of clay and silt size are carried much further than the larger sized material or the materials that are heavier with a greater specific gravity. You can appreciate that water transportation is a dynamic and changing process. Water can erode material and deposit material. Stream beds and channels are constantly eroding, filling and widening. The deposits in these valleys are called alluvial and are highly variable in composition, grain size, and thickness. Over time, many of the deposits may become cemented and quite extensive. In the geologic past, extensive deposits of sandstone, siltstone, shale, and limestone are evident throughout the world. Again, this deposition may be with fresh water inland sources, or it may be on the continental shelf or deep ocean under salt water marine conditions.

Marine sedimentary rocks are widespread and have thick sequences. If these beds have not been disturbed by mountain building activity causing faults and folds, the units are generallystratified and in a horizontal position. Because of the manner in which these particles were deposited, ground water tends to flow at a greater velocity in a horizontal direction parallel to the bedding plains than in a vertical one.

Windborne deposits are much more rare but are created particularly in desert areas. In some cases, these deposits represent sand

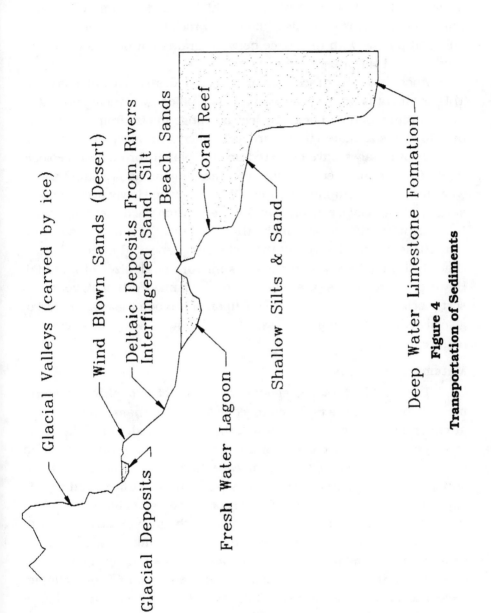

**Figure 4
Transportation of Sediments**

dunes formed along the beaches of large water bodies. A common characteristic of wind borne deposits is a high of degree of sorting. This uniform size means that this sedimentary rock tends to be very permeable and porous with the space between grains not normally filled by clay-size and silt-size particles.

Another windborn sedimentary rock that deserves mention is loess. This sediment is primary silt-size and is typified by vertical jointing.

Sediments transported by ice can range in size from huge boulders to clay-size material and are associated with glacial deposits. A number of ice ages have occurred on this planet throughout geologic time. Glaciers have cut through mountainous areas providing somewhat unique topographic features. While moving slowly over the land surface, the glacial ice incorporates material from the underlying rocks. It is only when the ice later melts that this material is deposited. **Glacial till** is a term describing the sediments deposited directly by ice. Glacial till may have a heterogenous mixture of all sizes of material. However, in some cases, streams and water may further rework and sort glacial materials. Dependent upon the type of glacial material, glacial sediments can provide substantial ground water as a resource (Figure 5).

Metamorphic Rocks

The third and final category of rocks is ***metamorphic***. The term metamorphic means simply "changed in form." The name concentrates attention on the processes by which the rock evolved. All of the rocks in this class were once either igneous or sedimentary, but have been changed by pressure, heat, or chemical action of liquids or gases. Pressure is caused by burial of rock types beneath the surface, and heat is applied either as a function of depth or in close proximity to molten rock from magma beneath the surface or with igneous intrusions into the upper rock layers. Depending upon the chemistry of the rocks involved, new crystals and mineral compositions may occur that are only created under such high temperature or pressure. For example, rocks buried deep beneath the Gulf of Mexico, over eight miles below the earth's crust, represent enough pressure forces to start the process of metamorphism. Some typical metamorphic rocks and the parent material which you will easily recognize are limestone turning into marble, or shale (silt-sized sedimentary material) changing into slate. Sand-

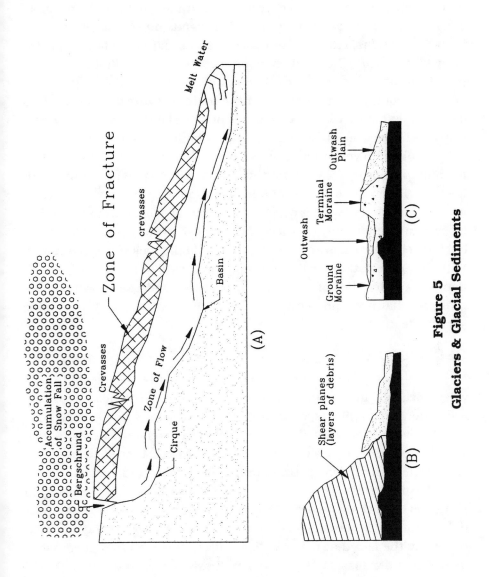

**Figure 5
Glaciers & Glacial Sediments**

stone, in many cases, is turned into quartzite, if it has a silica-rich water environment with increased pressure and temperature.

Metamorphic rocks are important because the processes of metamorphism sometimes concentrate high-grade, metal-bearing ores. Magma, at depths adjacent to iron-rich sandstones, may, under the proper conditions, result in the concentrations of iron sulfides and other metal sulfides. The most common metal ores are sulfides, and they solidify at lower temperatures than many other minerals.

The concept in geology called **the metamorphic or rock cycle** is described in Figure 6. A given particle of material may travel completely around the cycle, starting as a fluid, solidifying as an igneous rock, being broken down into a sedimentary rock which then may undergo metamorphic processes, later returning as molten magma or other sediments. Thiscycle is a continuous operation over the earth's surface with an incredible number of variations. The geologist, from his studying and understanding of rock types, can evaluate at which point in the circle/cycle the rock is and by what route did the rock reach that point. With an appreciation for the erosional processes of rocks and depositional characteristics of rocks, a geologist can understand why topography is what it is and conversely can map rock types by topographic surficial features.

GEOLOGIC TIME

The **geologic time scale** has been created by geologists to provide a means of identifying a date when an event took place, whether it was thousands, millions, or almost billions of years ago. Geologic time also involves the concept of "relative time" which relates whether a rock was created before or after another rock. A geologic time scale has been constructed which divides times into eras, periods and epochs.

Absolute time age dating by measurement of radioactive decay in materials has been incredibly helpful in giving an age to particular rock units. However, there are limitations. Radioactive age dating has been helped by the fact that many geologic ages have been identified based upon key index fossils. A **key index fossil** is the fossil remains of a plant or animal that is characteristic of a particular period in geologic time. When radioactive age dating is not possible, fossil dating is an accepted alternative. For example, the trilobites were common and prolific in the Cambrian era. Finding a trilobite fossil with other fossil

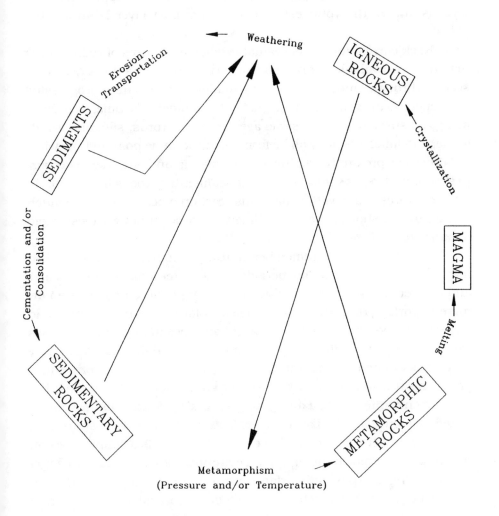

**Figure 6
The Metamorphic Cycle**

remains may provide enough information to age date a particular rock type to the Cambrian era. The following geologic column provides a time scale associated with some fossil records. In general terms, the Precambrian era contains little or no fossil record (Figure 7).

There are several additional basic principles which have been helpful in dealing with both relative and absolute time scales.

The **Law of Superposition** is very important. The law states that if a series of sedimentary rocks has not been overturned, the topmost layer is always the youngest, and the lowermost layer is always the oldest.

Rock correlation is also valuable when exposures of sedimentary material are available to draw such a correlation. Let's think of a cross section through land, a beach, and shallow ocean waters progressing to deeper ocean waters. At a particular moment in time, a particle along the surface is of the same age. In other words, siltstone, sandstone, and limestone may all be forming at the same point in time within fairly close proximity. As the ocean moves in and out, there is **transgression** and **regression** of different sedimentary rocks (Figure 8). In addition, rates of deposition may cause certain types of rock to be much thicker over a shorter period of time than other units with less rapid deposition.

It is important to remember that a particular rock unit or bed may extend over several time periods. As we later look at ground water and aquifer units, the physical water bearing characteristics are now more important than the geologic age and formation name. It should be remembered that there can be large variances in physical properties within a singular sedimentary unit such as a sandstone. Variances may be associated with rate of deposition and sorting; therefore, only a part of a sandstone unit may represent an aquifer. An additional law worth mentioning is the **Law of Cross Cutting Relationships,** illustrated by Figure 9. The Law of Cross Cutting Relationships states that a rock is younger than any rock it cuts. Again, this is a simple concept, but in the dynamic changing earth environmentit can be easy to forget when looking at a past subsurface record that may only be partly visible. Some sediments may have been eroded before deposition occurred again (Figure 9).

Geology and Groundwater Hydrology / 273

Era	Period	Epoch	Began Millions of Years Ago	Dominant Life Form
Cenozoic	Quaternery	Recent		Man
		Pleistocene	2	
	Tertiary	Pliocene	13	Grasses Become Abundant
		Miocene	25	
		Oligocene	36	Horses First Appear
		Eocene	58	
		Paleocene	63	
Mesozoic	Cretaceous		135	Extinction of Dinosaurs
	Jurassic		181	Birds First Appear
	Triassic		230	Dinosaurs First Appear
Paleozoic	Permian		280	
	Carboniferous: Pennsylvanian and Mississippian		310	Coil-Forming Swamps
	Devonian		345	
	Silurian		405	First Vertebrates Appears (Fish)
	Ordovician		425	
	Cambrian		500	First Abundant Fossil Record (Marine Invertebrates)
			600	
Pre-Cambrian			1,500	Scanty Fossil Record Primitive Marine Plants and Invertebrates
				One-Celled Organisms

Dominant life form groupings (right side): Mammals (Cenozoic); Reptiles, Amphibia, Fish (Mesozoic–Paleozoic); Marine Invertebrates (Paleozoic); Flowering Plants (Cenozoic); Conifer and Cycad Plants, Spore-Bearing Land Plants, Marine Plants.

Figure 7
Geologic Column

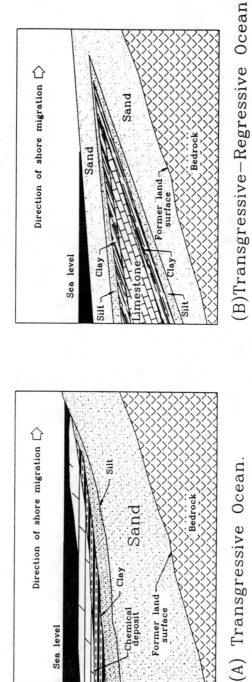

**Figure 8
Typical Cross Section of Transgressive and
Transgressive-Regressive Ocean**

Geology and Groundwater Hydrology / 275

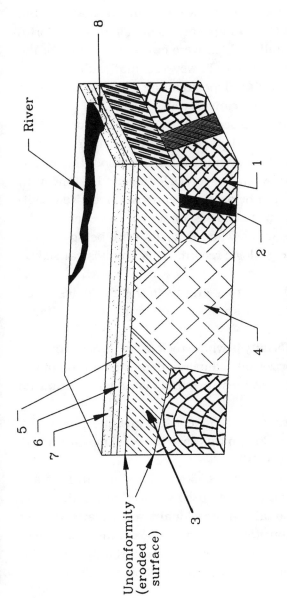

In decreasing age
1.) Oldest—Folded Limestone
2.) Dike (igneous rock)
3.) Siltstone (tilted from uplifting forces)
4.) Batholith (Solidified magma)
5.) Sandstone
6.) Sandstone
7.) Sandstone
8.) Alluvial deposits from river

Figure 9

Law of Superposition and Law of Cross-cutting Relationships

STRUCTURAL GEOLOGY

The easiest examples of geologic structure can be seen in massive road cuts along our highways or in the walls of the Grand Canyon. From a distance, the Grand Canyon appears to be a series of colorful rock beds laid down horizontally, which were cut by the action of the Colorado River. On closer inspection, however, a multitude of minor structural disturbance erosion and deposition are in those walls that make up the total sequence on the canyon wall.

Consider the fact that some sedimentary rocks laid down in a horizontal fashion in the ocean are then elevated by mountain building processes above the water surface. While in that condition, the formations are eroded and changed by various forces for a period of time until the sediments are again lowered below the ocean's surface. New sediments now are laid down upon the older sediments. However, for the period of time above water surface when deposits were being eroded, there is a gap or break in the geologic record. This break is caused by the cessation in deposition.

The surface between the two rock groups in the above described sequence represents an erosional surface. This surface is called a **disconformity**.

A variation of a disconformity is where rock units have been tilted and eroded with younger horizontal sediments deposited on top of the older rocks. The contact between these two units is called an **angular unconformity.**

There is also another case where igneous or metamorphic rocks are overlain by sedimentary rocks. In this case the erosional surface is called a **nonconformity** (Figure 10).

We mentioned tilting of rocks. Rocks and sediments that are below the earth's surface, being compressed by overlying sediments, may become folded by additional compressional forces created during the process of **orogeny** or mountain building. Folds come in all sizes, shapes, and angles. Again, if you have ever driven through one of the mountain ranges in this country, rock cuts will show excellent examples of folds and faults.

In structural geology, a number of terms are used to describe various structures. If an arch is created, the arch is called **anticline**. Conversely, a valley is called a **syncline**. The term **monocline** is used

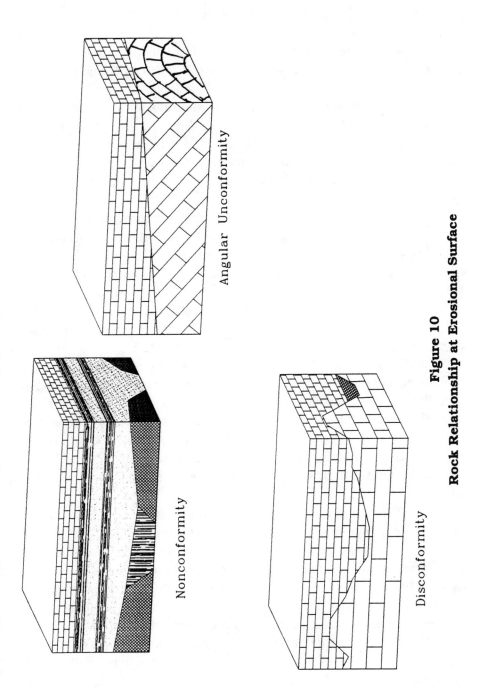

Figure 10
Rock Relationship at Erosional Surface

to describe rocks that are horizontal on either side of a flexpoint (Figure 11).

The structure in an area can range from gentle to severe. Some geologic structures extend over hundreds of miles whereas others are in millimeters or inches. Structures and contacts between beds or formations have particular importance in evaluating ground water. These contacts may represent a change in flow direction and in some cases may provide a surface where more dense constituents of ground water such as contaminants may move downward by the force of gravity. On occasion, this downward movement may be opposite from general groundwater flow.

Fractures or faults in rock are evidence that movement has taken place along a surface. Joints are also fractures, but movement has not taken place along a joint surface. Joints may be the result of differential rates of cooling, of different minerals or rock constituents, or the result of compaction or deposition forces of sedimentary material. Joints, faults, andfractures provide zones of weakness whereby more volatile constituents such as ground water may move at a much faster rate than through the adjacent rock matrix material.

Remember that a fault is not a movement in a particular straight line direction but may be a movement in a combination of directions at different angles, and may even be a curved surface (Figure 12). Extensive faulting and folding is associated with the mountain building processes throughout geologic time and also along our coastlines where the crust is undergoing deformation and dynamic change. This change from **continental drift** is the source of many earthquakes around the world.

Faulting usually occurs most commonly in areas of extreme deformation or uplift such as from mountain building forces. Depending upon whether these forces are in compression or tension, the resulting movement has been categorized and described for purposes of drawing maps of both the surface and subsurface. Faulting, of course, is occurring on a continuous basis throughout the planet as evidenced by earthquakes of various magnitudes which occur on a daily basis.

During deformation, both hills (anticlines) and valleys (synclines) are created. Synclines also occur in ocean areas where sedimentary deposits build up, causing downward compression of underlying soils and rocks. One can also imagine that these hills and valleys are not

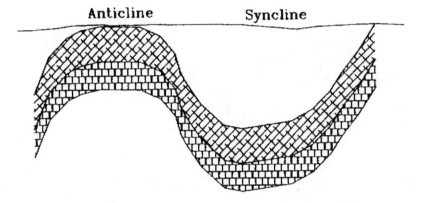

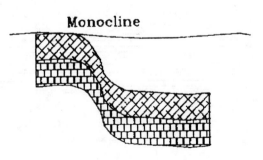

**Figure 11
Rock Cross Section**

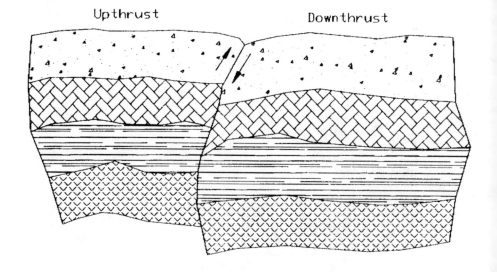

**Figure 12
Fault Cross Section**

necessary symmetrical but can have many varying types of dips and physical orientations from the horizontal, depending on the weight of the overlying sediments. Some dips may be very slight, whereas other dips may be almost vertical.

It is important to understand that the geologic structure of the rocks beneath the surface may not necessarily reflect the topographic surface. Let's think about this for a second. As erosion occurs, there is a leveling process of the rocks and rock structure. After some extensive uplifting, erosion can occur that will result in a flat horizontal surface after many, many millions of years. Nevertheless, the structure of the rocks below the surface may be highly deformed from the original uplifting forces. In understanding the subsurface, there is tremendous value in appreciating that surface features generally reflect the resistance to weathering of rock and soil types and may be indicative of the subsurface rocks. However, subsurface structure needs to be evaluated based upon primary and secondary exploration methods to confirm our subsurface understanding.

Primary methods include drilling into the subsurface and taking actual samples. Secondary information may be in the form of geophysical sounding techniques that might help identify subsurface formation contact indicative of the change between one rock type and another. Drilling into the subsurface is always one of more expensive investigation techniques, is costly, is time consuming, and may be representative of only a small area. It is for this reason that many different geophysical techniques measuring conductivity, magnetism, density, and other physical features are used to explore large areas quickly and to help us confirm our understanding of the subsurface geology and structure. However, geophysics without drilling is not the answer.

Understanding geologic structure is important in preparing maps of the subsurface. It is these maps that provide visual images of subsurface materials, their physical nature, and also, when rocks are porous and permeable, the nature and extent of ground water resources.

GEOLOGIC MAPPING AND INTERPRETATION

Geologic surveys of large and small sites have many common approaches to solving the geologic puzzle. The presence of topographic maps, area photos, and geologic maps, even if regional in nature, are extremely important tools in the process of a geologic survey. From both the topographic and aerial photos, visible surface features and forms provide some insight with respect to rock outcrops and identification of formations. The plotting of rock outcrops in highway cuts is also important in preparing a map of subsurface conditions. Field investigations and ground proofing are critically important to this process. A common instrument used by the geologist in the field is a **Brunton Compass.** This particular compass helps not only in the location of features on a topographical map but also allows the geologist to measure both the strike and the dip of formations (Figure 13). The **strike of the formation** is the intersection of that formation contact with a horizontal plane. The true dip of a formation is perpendicular to the strike and measures the angle between the formation contact and the horizontal plane. These fairly simple measurements are critical in identifying subsurface geologic structure.

In looking for the best places where bedrock might be exposed, stream beds and hillsides are helpful. Cuts for railroad access also yield important information.

Fortunately for the geologist and groundwater hydrologist, the United States Geological Survey has prepared detailed geologic and hydrologic maps for most of the country. Providing extremely high quality work since its inception, the U.S. Geological Survey information is usually a critical starting point for investigations of small and large areas alike. The U.S. Geological Survey has created an immense database of both geologic and hydrologic information. This information has allowed the United States to be able to characterize and develop resources to the degree not found in many other countries in the world.

Geophysical Surveys

In helping the geologist or groundwater hydrologist look beneath the surface, geophysical surveys may play an extremely important role. The primary advantage of using geophysics to help identify specific subsurface material and structure is the relatively cheap cost of coverage for large site areas. Geophysical surveys measure by the use of sensitive instruments the variations in a number of physical properties. These variations or anomalies may be helpful if known structure or formations create a characteristic anomaly. Known correlation is an important and critical part of this process. These properties are: (1) density, (2) electroconductivity, (3) elasticity, (4) magnetic intensity, (5) temperature, and (6) radioactivity. Preparation of maps using these various

Geology and Groundwater Hydrology / 283

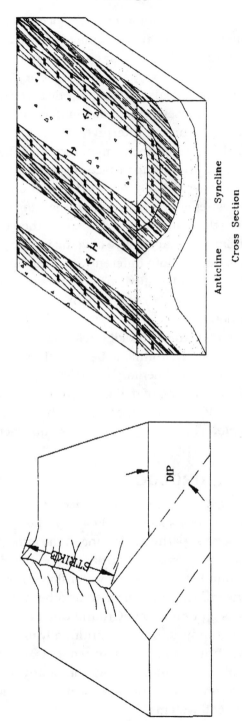

(A) Strike is the intersection of formation contact with the horizontal plane.

(B) The arrow indicates the direction of the dip.

**Figure 13
Strike and Dip of Formation**

types of surveys and used in conjunction with specific known subsurface data from drilling cores or outcrops may help identify what has happened to various formations, beds, or contacts at depths below the earth's surface.

The field of geophysics deals with a great deal of secondarily developed information that by scientific reasoning or deduction is then used to make specific subsurface interpretations. Geophysical data can be collected on the surface using monitors at different spacing lengths to obtain "pictures" of the subsurface at specific depths or zones. In addition, some geophysical tools may be lowered down borings or drill holes to provide cost-effective information regarding rock types and contacts. Sometimes, this information measures the conductivity of water in the rock formations that may be of value in tracing both aquifer andconfining units that contain different water quality. Once a pattern or marker bed can be identified, new or unsampled borings or wells can be geographically logged, and subsurface data can be correlated to the new location.

In our context today, the hydrology relating to ground water is controlled by the physical environment beneath the earth's surface. The extent of pores between sedimentary particles, the fractures caused by rock stress, or the solutioning of material provide not only space for water but a mechanism for water to move beneath the surface as part of the **hydrologic cycle.** This new cycle leads us into discussion of the science of ground water.

GROUNDWATER HYDROLOGY

The **hydrologic cycle** can be just as varied as the previously discussed metamorphic and/or rock cycle. Basically, the hydrologic cycle provides a simply illustrated pathway starting in the atmosphere where water is then deposited on land and surface by either rain or snow (Figure 14). At the ground surface, some water flows in surface runoff into lakes, streams, rivers, and oceans, whereas other waters move into the earth beneath us as ground water. Ground water movement for the most part reflects topography with water continually moving down gradient to a point of equilibrium which in many cases is the ocean. Water returns to the atmosphere through evaporation of any surface waters and through transpiration from plants that bring ground waters to the surface through their root systems.

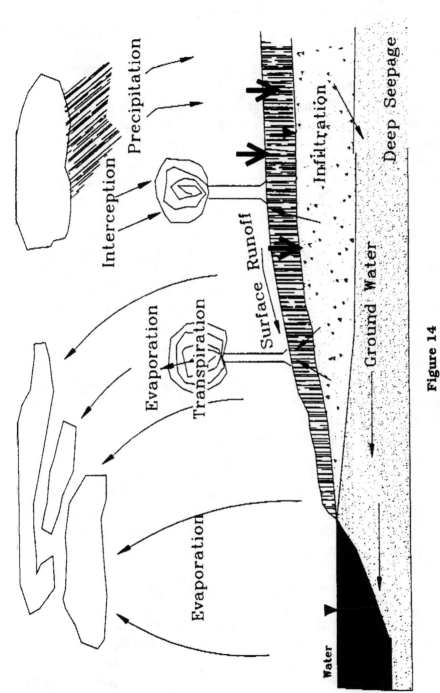

Figure 14
The Hydrologic Cycle

In looking at the distribution of water in the United States, for example, well over 80 percent of the water storage is in ground water. However, if you look at the amount of water circulating as part of a hydrologic cycle on an annual basis, ground water represents only slightly more than 25 percent.

In looking at a simple picture of a runoff cycle, water infiltrates through surficial soils to a water table. The **water table** is the surface below which soils are saturated. Usually above the water table there is an intermediate or **vados zone** which is not completely saturated. Once into the water table, subsurface formations may be designated as **aquifers.** Some saturated formations are so impermeable they are called **confining units.** For confining units, the formation may be saturated but represents a change in the ability of the soil to transmit water. For example, clays and silts will provide a barrier to ground water movement and are usually considered confining units. In contrast, more permeable material such as sands, sandstone, gravels, etc. comprise good aquifer units. Aquifers store and transmit water in significant amounts to adjacent beds or bedrock or to the surface as springs. For ground water supply, wells are installed in aquifer units to retrieve good quality water in sufficient volume to meet demand.

Before we discuss groundwater hydraulics, it may be helpful to evaluate various formations and rock types as to their **permeability** and **porosity.** Their ability to store (porosity) and to transmit ground waters (permeability) determines if a hydrologic unit may be used for water supply. A water supply may be of high quality for drinking or of lesser quality for industrial use or for cooling.

In looking at the question of the amount of ground water that can be stored in a rock or sediment, several definitions and equations become important. Porosity can be defined as the percent of openings or pore space that when saturated can contain water.

$$\text{Porosity} = \frac{\text{volume of pore space}}{\text{volume of bulk solid}} \times 100$$

Another important definition of even more concern to a ground water hydrologist is how much water can be actually released from storage in the aquifer. In other words, when water is drained from some saturated material under the force of gravity, only part of the volume of water stored in its pore space is released. This quantity of water is called **specific yield.** Specific yield is expressed like porosity in percent. If the specific yield of a rock or sediment represents the amount of water that is released, the amount that remains is called **specific retention.**

Porosity = specific yield + specific retention.

The specific yields of unconfined aquifers may range from 0.01 to 0.30. Note that the specific yield in this case is equivalent to the unconfined aquifer's storage coefficient. In the case of confined aquifers (those aquifers under pressure), specific yields are not able to be determined because the aquifer materials are not actually de-watered during pumpage (Figure 15). The storage coefficient is much lower in confined aquifers because of this fact. Water released from storage is obtained primarily by compression of the aquifer and expansion of the water when it is pumped. The yield from an unconfined aquifer is much greater than that from a confined aquifer under an equal decline and head.

For unconfined material, it is almost intuitive that the smaller the average grain size, the greater is the percent of water that is retained. Conversely, the coarser the sediment, the greater will be the specific yield when compared to the porosity. In other words, the finer sediments have lower specific yields than compared to coarser sediments even if both have the same porosity. The water that is held in pores is held by surface tension and other adhesive forces. The following table gives some examples of porosity for various sediments and rock materials as well as typical specific yields. Note that specific yields of unconfined aquifers range from 0.01 to 0.30. Typical storage coefficients for confined aquifers range from 10^{-5} to 10^{-3}.

Material	n(%)	Specific Yield (%)
Clay	45-55	1-10
Silt	35-50	3-15
Sand	25-40	10-30
Gravel	25-40	15-30
Shale	0-10	0.5-5
Sandstone	5-30	5-15
Limestone	1-20	0.5-5

One last important term involves what is called **effective porosity.** The definition for effective porosity (n_e) is the amount of intercon-

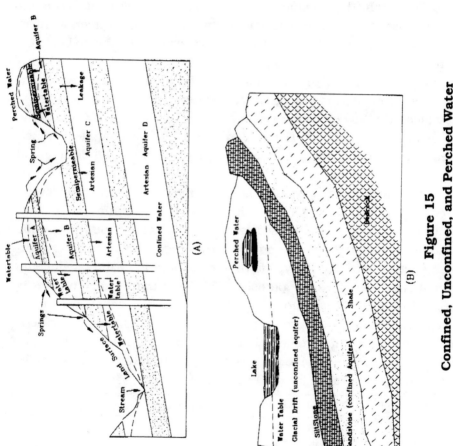

Figure 15
Confined, Unconfined, and Perched Water

nected pore space through which fluids can pass, which is expressed as a percent of bulk volume. Effective porosity will be less than the total porosity because there are, in the pore space, static fluids being held to various mineral and rock surfaces by surface tension.

The following table illustrates some water-bearing properties of common rocks showing their permeability and porosity. Permeability or **hydraulic conductivity (K)** depends on properties of the fluid as well as the characteristics of the medium.

Hydraulic conductivity is governed by the size and shape of the pores, the effectiveness of the interconnection between the pores, as well as various physical properties of the fluid media. When we are dealing with contaminants in ground water, the hydraulic conductivitytimes the dynamic viscosity of the fluid divided by the fluid density and acceleration of gravity equals the intrinsic permeability. Intrinsic permeability is used in the petroleum industry because there are movements through rock media that need to be analyzed involving petroleum and water as three phased gas, oil, and water (the same concepts are applicable with contaminant vapor (gas), contaminant (liquid), and water.

Average Values of K

Soil Class	K, cm/sec
Gravel	1 to 10^2
Clean sands (good aquifers)	10^{-3} to 1
Clayish sands, fine sands (poor aquifers)	10^{-6} to 10^{-3}
Silts and clays (confining units or cap material)	10^{-4} to 10^{-9}

Note the orders of magnitude between the various types of soil. For example, a gravel aquifer may be 100 to 1,000 times more permeable than a clean sand aquifer. Also note that a confining unit to an aquifer may have a sharply different lower conductivity and be "relative" to the aquifer unit.

Again, porosity is the pore space available for water storage between rock or material particles, whereas permeability relates to the connection of the pores or the ability of water to flow through rock material. In general, sands, sandstones, and gravels will have much higher permeability than silts, clays, or crystalline rock.

Very much like the gradient on the surface of the earth, the steeper the slope, the faster the movement of ground water through rock or soil material. **Gradient** as well as permeability/porosity are important parameters to know in evaluating ground water movement. Although gravelly riverine materials adjacent to the river may have very high permeabilities and porosities, the gradient or slope may be very flat. This combination may result in very slow movement of ground water adjacent to those rivers and streams. On the other hand, ground water as well as surface water in mountainous terrain will move much more quickly to streams or the water table with high gradients or slope and lower permeabilities. Water moving through soils and/or rocks above the water table(saturated zone) move at a gradient of one (1). This gradient is gravity movement of liquid water through saturated sediment. In studying moving water through the subsurface soils, you can understand that water may run into layers of less permeable materials such as silts or clays. If these layers are above the water table, the water will "pond." Water will then flow horizontally until it reaches more permeable material and can proceed downward to a lower water table elevation or equilibrium. This particular situation, common in many localities,is called a **perched water table.** The perched water table may be of significant thickness or may be so thin as to be marginal with respect to water withdrawals (Figure 15).

If one single concept in ground water hydraulics needs to be placed first on a list of importance, Darcy's law describing the flow of ground water is that basic equation. Henri Darcy was a French engineer who was very deeply involved in water flow concepts in the mid 1800s. In 1856, he conducted a series of experiments on a vertical pipe filled with sand recognizing that the flow of water through the ground is analogous to pipe flow. Darcy concluded that the rate of flow through the column of saturated sand was proportional to the difference in hydraulic head at the ends of the column and inversely proportional to the length of the column. This famous law and basic equation is as follows:

$$V = K \frac{h_1 - h_2}{L}$$

where V is the velocity of specific discharge
(h_1-h_2) is the difference in hydraulic head
K is the hydraulic conductivity, and
L is the distance along the flow path between the points where h_1 and h_2 are measured.

In the above equation, the difference in hydraulic head over the distance of the flow path is the same as the **hydraulic gradient**(I).

By substitution, the following equation is another version of Darcy's Law:

$$Q = KIA$$

In this equation, Q is the discharge per unit time, and A is the cross-sectional area.

The hydraulic gradient under which ground water movement takes place is the slope of the water table, or under confining pressures, the **potentiometric surface.** The total flow, in another variation of Darcy's Law through any vertical section of an aquifer, can be calculated if we know several parameters. We need to know the thickness of the aquifer(b), the width of the aquifer, the average hydraulic conductivity over the height of the aquifer, and the hydraulic gradient. The result is the following equation:

$$Q = KbI$$

Another important term is **transmissivity.** Transmissivity is or can be made part of Darcy's equation. Transmissivity is the capability of an entire thickness of the aquifer to transmit flow. In other words:

$$Q = TIw$$

The width of the vertical section through which the flow occurs is w (Figure 16).

The concept of transmissivity is important because transmissivity can be determined by three general methods. T can be determined by collecting data during a pumping test, (2) by analyzing the hydraulic property of aquifer material, and (3) by measuring the decline in ground water levels during a pumping test. A number of equations are available to describe flow near wells, and these equations give the transmissivity value if the declines are known. Theis developed the now-famous **non-equilibrium well equation** in 1935. This equation was the first to take into account the effect of pumping time on well yield and was a major advance in ground water hydraulics. By using this equation, the drawdown can be predicted at any time after pumping begins. The earlier terms of transmissivity and average hydraulicconductivity can be determined during the early stages of a pumping test rather than

292 / Environmental Science and Technology Handbook

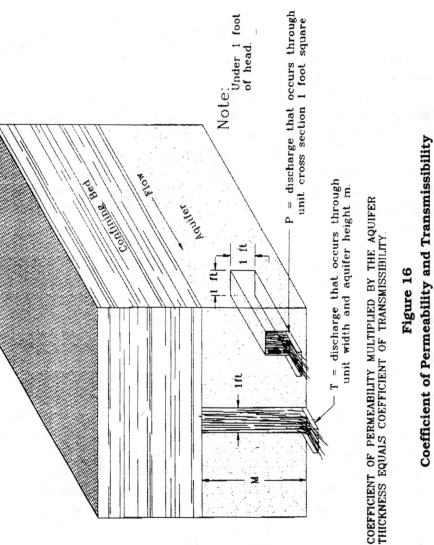

Figure 16
Coefficient of Permeability and Transmissibility

after observation wells have been stabilized, (under equilibrium conditions).

Aquifer coefficients, which again are terms used to describe physical parameters, can be determined from time drawdown measurements in a single observation well. Under equilibrium conditions, equations were developed to evaluate both unconfined aquifers (water table) and confined aquifers (under pressure). For an unconfined aquifer, the equation for well yield is (Figure 17):

$$Q = \frac{K(H^2 - h^2)}{1,055 \log R/r}$$

where, Q is pumping rate in gallons per minute
K is hydraulic conductivity in gpd/ft^2
H is head from the bottom of the aquifer in ft.
h is the depth of water(ft) in the well while pumping

For confined aquifers (Figure 18), the equation is:

$$Q = \frac{Kb(H-h)}{528 \log R/r}$$

where, b is thickness of the aquifer (ft).

For tests where two observation wells are available, the following equation applies for unconfined conditions (Figure 19):

$$K = \frac{1055 \, Q \log r_2/r_1}{h_2^2 - h_1^2}$$

where, r_1 is the distance to the nearest observation well(ft.)
r_2 is the distance to the farthest observation well(ft.)
h_2 is the saturated thickness at the farthest observation well(ft.)
h_1 is the saturated thickness at the nearest observation well(ft.)
Q is the pumping rate during the test
K is the hydraulic conductivity(gpd/ft^2)

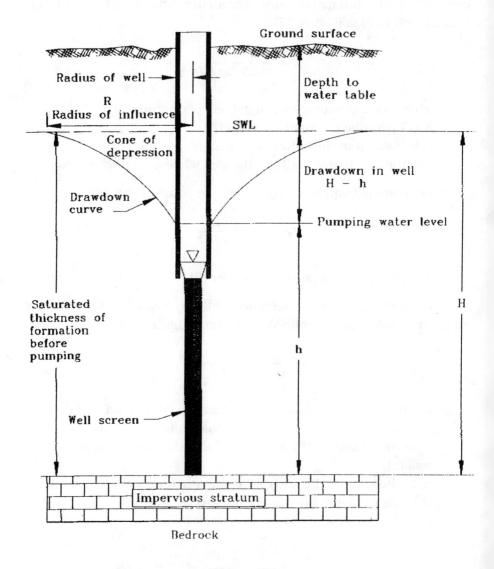

**Figure 17
Unconfined Aquifer**

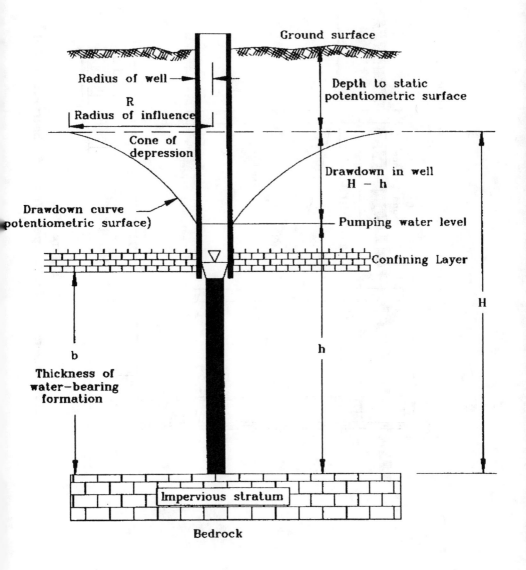

**Figure 18
Confined Aquifer**

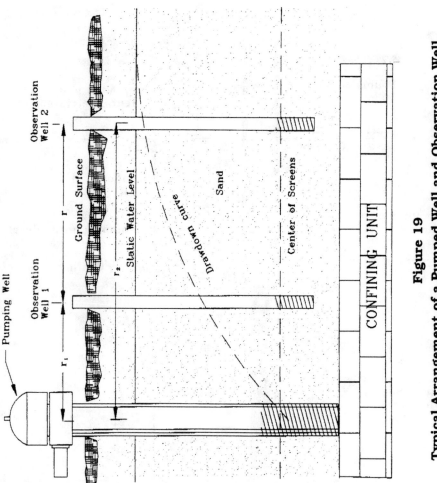

Figure 19
Typical Arrangement of a Pumped Well and Observation Well

If there are confined conditions as compared to a water table aquifer, the hydraulic conductivity from a test installation is:

$$K = \frac{528 \, Q \, \log r_2/r_1}{b(h_2-h_1)}$$

where b is equal to the thickness of the aquifer.
h_2 is the head (ft) at the farthest observation well measured from the bottom of the aquifer
h_1 is the head (ft) at the nearest observation well measured from the bottom of the aquifer

A second method to determine transmissivity involves determining the hydraulic conductivities of the materials in an aquifer. An estimate is made based upon grain size analysis of cuttings from a subsurface boring or other exploratory well drilled in the aquifer. The hydraulic conductivity characterizing the aquifer layer is multiplied by the thickness of the layer to obtain transmissivity. If there are different hydraulic conductivities for different portions of the aquifers, the series of hydraulic conductivities and strata thickness are added together to obtain the overall aquifer transmissivity.

The third method involves testing in the laboratory with relatively undisturbed samples of material by measuring quantities of water that are forced through a column of material. The head loss is determined for the field sample and the hydraulic conductivity can then be calculated.

The best aquifer values are obtained during pumping tests and reflect hydraulic conditions within the aquifer. The problem with laboratory testing and sieve analysis of cuttings or samples lies in the fact that these latter methods provide specific values only from a specific small area and may not be representative of the aquifer unit.

Pump testing provides the best values. Although expensive, pump testing is extremely important to obtain good aquifer parameters. These values may, in turn, be used in computer modeling for predictive calculations of water flow and also for evaluation of contaminant transport.

GROUNDWATER FLOW VELOCITY

As mentioned earlier with Darcy's equation, flow velocities become more important when related to underground contamination problems. Clearly, the ability to be able to predict the rate of contaminant flow from a contaminant source is a major objective of regulations enforced by United States and by the EPA. It is important to know the rate of flow and also to determine the general direction of flow. The direction of flow calculation is made in a similar procedure as the direction of dip in geologic surveys. The potentiometric levels of water in at least three wells are measured to provide contours of water level. The groundwater flow is perpendicular to the contour of equal water elevation (Figure 20).

In some circumstances, groundwater flow can be physically measured in the field by utilizing a type of tracer or dye. The use of tracers to measure velocity requires that the tracer must be detectable in extremely low concentrations and must not react chemically or physically with the ground water or aquifer material. Sophisticated testing devices are now capable of detecting extremely low concentrations in a series of wells at a distance from the injection point or source. A word of caution, however. Sometimes the fact that a tracer is not identified at some distance downgradient from the injection source may mean that the receiving point or well does not pick up a representative flow. The concentrations of tracers also may have become diluted by milking over even short distances, and the most sophisticated machines cannot pick up the downgradient concentrations.

Tracers are generally only effective over reasonably short distances. Since aquifer materials may vary over distances, tracer information may be of limited value. There are a number of common types of tracers. These include rhodamine, a number of strong chloride electrolytes, radioactive isotopes of water and iodine, as well as some detergents. It should further be noted that tracer studies are particularly important in indicating dispersion rates encountered in the aquifer. Dispersion involves both mechanical mixing and molecular diffusion.

Although this discussion has primarily emphasized sediments and sedimentary rock materials, water can move through the subsurface through faults, joints, and zones of weakness in the rock or material, as well as fracture patterns. Water can also move from the water table via a spring to the surface and then back into the groundwater table again.

AQUIFER SYSTEMS

In general, there are four major types of aquifer systems. Three of the four types involve sedimentary material and the last is a combina-

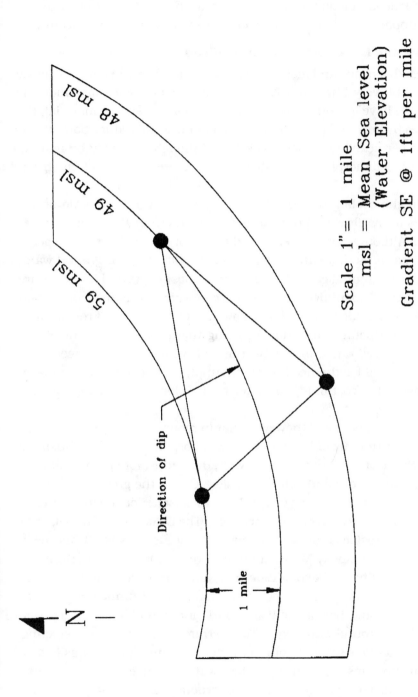

Figure 20
Determination of Groundwater Gradient

tion of igneous-metamorphic rock. From the standpoint of large volume supplies, igneous-metamorphic rocks are the least important.

Igneous and Metamorphic Rock Aquifers

Both igneous and metamorphic rocks, for the most part, are dense and exhibit very little porosity and permeability. Since metamorphic rocks can have igneous or sedimentary rocks as a source material, the changes by heat and pressure tend to minimize and reduce both porosity and permeability. These rock types, though they may be marble, shale, or metamorphose sandstone, do not store or transmit much water.

In the cooling process, however, joints and fractures do occur, and it is primarily through these joints and fractures that ground water moves. Although some igneous and metamorphic rocks exhibit specific joint patterns that can provide a general direction for ground water flow in a substantial geologic unit of rock, most joints and fractures are not that easily identified nor are they connected. It is this connection that would provide any volume of flow or even flow at all. From a water supply standpoint, many houses along the front range of the Rocky Mountains are located for the beautiful view and without concern and understanding for the need of water supply and septic tank drainage. As a result, the cost of piping water to such locations is extremely expensive.

It should be noted, however, that in mountain ranges and bodies comprised primarily of igneous and metamorphic rocks, a surprisingly large volume of water can move downgradient into more permeable sedimentary rocks found in valley deposits. One of the greatest concerns in tunneling for both hard rock and soft rock (softer sedimentary materials) is the occurrence of water seepage. The Eisenhower Tunnel, constructed to provide an all-weather passage through Loveland Pass near the center of the Rocky Mountain Range west of Denver, underwent an extensive drilling and exploration program to provide guidance to potential problems that might be encountered in the tunneling process. Several thousand borings indicated little if any problem with joint or fault related ground water flow. To illustrate the difficulty of predicting ground water flow through joint patterns, the tunneling process in fact encountered substantial delays when water seepage dramatically exceeded predicted volumes. The construction of necessary drainage systems and the control of flows during tunneling caused delays in the tunnel construction process.

From a practical standpoint, metamorphic and igneous rocks rarely provide more than a small amount of water to a single water well. One major exception involves igneous rocks such as lava flows. Many times these lava flows contain permeable zones created by the migration of gas bubbles during cooling and rock solidification. In some flows, these rocks can provide substantial ground water resources. In the Hawaiian Islands, for example, substantial flow in various lava zones provides many areas with substantial water supply. On Oahu, the extremely permeable lava flows in the center of the Island provide more than adequate water for Honolulu and the Island. In this case, horizontal drilled chambers collect water flow for storage and treatment. Water meets the needs of people on both sides of the Island. It is an interesting sight to watch heavy rainfall occurring on these rock formations without hardly a drop of that rainfall becoming surface water flow.

The three other major types of aquifer systems that are sedimentary in nature are sedimentary, alluvial, and glacial. Basically these three types present the effects of water erosion, wind erosion, and glacial ice erosion.

Sedimentary rock aquifers. Although sedimentary rocks make up only a small portion of the earth's crust (5 percent), these rocks contain about 95 percent of the ground water. Inthis section, this aquifer type is considered to be in a deposit caused by erosion from water. The water could be on land or it could be beneath the ocean.

Regardless of the type of sedimentary rock, there are certain major characteristics of these rocks that make them either good or poor aquifers. A number of different factors are important: the grain size, the grain shape, and the sorting of the grains, as well as the stratification. One additional factor also important to note is the gain in porosity and permeability caused by **secondary** weathering or solutioning of a rock already deposited. As you would expect, cementation can affect both the porosity and permeability of a rock and its ability to be a source of water supply. Grain size is important, but sorting of the grains is also a factor in the increase or decrease of rock permeability. A well-sorted material is a material that has uniform grain size. This type of sedimentary rock tends to be much better at yielding water than a material that has a wide variety of grain sizes which can fill the spaces between grains with smaller grains. A rock may also have extremely small grain size such as silt and clay and can become compacted very tightly to prevent the movement or flow of water.

Alluvial aquifers. Rivers and streams build alluvial deposits. Many major aquifer systems consist of these deposits, and they deserve mention. Rivers and streams constantly change the landscape as the water tries to reach equilibrium conditions. River valleys develop through a number of specific stages easily recognized in today's topography. In the initial stages, a stream may have lakes, waterfalls, and rapids. This situation particularly occurs in areas with high relief and where, of course, the rainfall is high enough to cause the eroding surface flows. As this initial stage starts to mature, the lakes slowly disappear, but the river cuts into the topography with continuing rapids and waterfalls. Normally the pathway for the river is a fairly narrowly incised cut. In early maturity, the profile begins to show a smoother grade without rapids and falls and starts the initial stages of a flood plain. When approaching full maturity the stream has a flood plain almost wide enough to accommodate all the meandering of the stream bed within the flood plain. At full maturity, the flood plain is broad with freely developed meanders. In the flooding during the spring of 1993, graphic pictures of the Missouri River provided evidence that many areas are at full maturity for the river while in other areas, the river is approaching full maturity. Even though man has attempted to control streams and rivers within the flood plain, massive rains of Spring 1993 overflowed almost all of those restraints.

With all of these stages, it is easy to understand that river-deposited sediments can vary considerably in type and character. Streams are constantly filling and eroding and widening their channels, leaving sorted materials in some areas while in others providing a high variability in grain size composition and thickness. With the river constantly moving, there is transgression and regression of various beds of different soil and rock types as the river moves and meanders in the flood plain deposits.

Now let's consider the same deposits in geologic time over many millions of years and it is easy to understand the complexity of the subsurface materials that we find in ancient river beds. To predict bedding thickness and direction is difficult, even in small areas.

From a hydrologic standpoint, the boundaries between different formations within an alluvial deposit strongly influence ground water flow. Since streams and rivers can meander within their flood plain and create a flood plain consisting of many different types of materials(going from very fine grained confining materials to the coarser grained aquifer units), solving the geologic puzzle based upon a series of borings or wells is sometimes difficult to evaluate. Nevertheless, in spite of these potential numerous changes in permeability, porosity, and ground water flow, pumping tests can evaluate aquifer parameters that may be representative of some distance in the aquifer. This test information may take into account the aerial changes and thicknesses of material. The test number may represent an extremely complex series of units with one number that is representative. Sometimes this type of testing averages the parameters adequately to model both groundwater movement and contaminant transport.

Glacial Deposits

In the midwestern and northern United States, glacial aquifers are an extremely important category of aquifer systems. Glacial aquifers occur throughout much of the northern areas and Canada. In many cases these aquifers are the only aquifers present in these geographic areas. Beneath the aquifers we commonly find igneous and metamorphic bedrock, which as earlier described is lacking in water or contains water of poor quality.

Glaciers erode and transport sediments ranging in size from boulders to clay. In the United States, there are valley glaciers as well as evidence of the more extensive continental glaciers. Both types result in similar types of deposits. In general, ice will flow at a uniform rate over flat topography. However, when the topography is not flat, the ice moves more quickly in the valleys or depressions and picks up rock debris more easily as a result of the faster flow. The Great Lakes of North America, for example, are deep depressions formed from glacial activity. The different sands and gravels, called **till,** are gradually deposited as the glacier melts. Some of this glacial till is then later reworked by melt water streams and further reworked by the environment.

The tills, which are generally clay rich, have little pore space and serve more as confining units than aquifer units. In a more sandy and gravelly material, water can be found in substantial volume and can be a major source of water supply. Reworking of deposits may occur many times; therefore, the geology and hydrology of glacial material may be extremely complex just like the earlier described alluvial materials. Many wells in glacial moraines and outwash can provide many thousands of gallons per minute. In many cases these moraines can be up to a thousand feet in thickness.

We have mentioned gravity and slope or gradient as factors in the movement of ground waters. Another factor related to contamination of ground waters and impacting movement is density. For example, saline waters or salt water/sea water, being more dense than fresh water, will underlie fresh water when both are present under equilibrium conditions. Islands in the ocean, for example, normally will have a fresh water "lens" resulting from rainfall that basically "floats" above the saltwater surrounding and underlying the island environment.

GROUND WATER CLASSIFICATION

One of the most unusual properties of water is its ability to dissolve a great range of substances, more so than any other liquid. As you might expect, ground water moving slowly through different minerals in soil and bedrock may dissolve many constituents before reaching some type of chemical equilibrium.

Although silicates, as a mineral group, comprises about 90 percent of the earth's crust by volume, silicates are only slightly soluble in water. The same is also true of aluminum oxideand iron oxide--again, common substances in the earth's crust. However, such constituents as sodium, calcium, magnesium, and potassium combine with oxides and are quite soluble. These major cations (positively charged atoms) are found commonly in ground waters. The primary anions (negatively charged molecules) in ground water are bicarbonates (HCO_3^-), sulphates (SO_4^{-2}), and chlorides (Cl^-).

The concentration of minerals in water is referred to as **total dissolved solids (TDS)**. Units used are generally milligrams per liter (MG/ℓ).

There are a number of important properties of water. Hardness basically describes the property of water to produce suds from soap. Rainwater, for example, requires little soap to produce suds and is therefore considered "soft" water. If the water requires considerable soap to produce suds, then it is considered "hard" water. This property is a function of primarily the calcium and magnesium cations as well as some heavy metals such as iron and magnesia that consume soap.

Hardness of water may be divided into two types: carbonate and non-carbonate. Carbonate hardness can be removed by boiling that precipitates the calcium and magnesium carbonate and sulphate minerals. Non-carbonate hardness is the difference between total hardness and carbonate hardness. The non-carbonate hardness cannot be removed by boiling. Hardness is usually expressed in terms of calcium carbonate. Hardness of less than 50 mg/l is considered soft. Water with 100 to 150 mg/l will deposit scale in steam boilers and at greater concentrations usually requires softening for household use. Municipal water supplies generally reduce hardness to about 85 mg/l.

Another important property of water is conductance, which is the ability of water to conduct an electrical current. Current flows in mineralized and ionized water because the ions are electrically charged and move toward a current source that neutralizes them.

The **hydrogen ion concentration (pH)** is also important as a property. Neutral is considered to be a value of 7 with pH values less than 7 being considered acid and those greater than 7 indicating alkaline solutions. Most ground waters in the United States have pH values ranging from 6 to 8.5. The more alkaline values represent ground water occurring in limestone (calcium carbonate) and dolomite. Natural acid waters occur in areas of volcanic activity that are usually high in sulphate ions. Acid and alkaline ground water influence the ability of that water to dissolve many constituents and is an important property related to water quality.

Fresh water is usually considered to have total dissolved solids between 0 to 1,000 mg/l, brackish water from 1,000 to 10,000 mg/l, saline water from 1,000 to 100,000 mg/l, and brine water above 100,000 mg/l. Depending upon the constituents found in ground water, a wide variety of treatment practices take objectionable levels of minerals down to levels that are acceptable for the particular use--whether it be potable consumption for household use or industrial water for steam boilers. A set of National Primary Drinking Water Standards has been established by the EPA. In more recent years a number of organic chemicals have been added to the Primary Drinking Water Standards. In addition, there are a number of secondary drinking water standards for public drinking water supplies. These standards are for aesthetic and taste purposes.

GROUNDWATER REGULATIONS

The first major federal law that recognized the importance of ground water was the **Safe Drinking Water Act of 1974.** This particular act established a number of standards forinsuring the safety of drinking water for injecting waste into the subsurface and for ground water quality. Two other important laws are significant relating to ground water quality. They are the Resource Conservation and Recovery Act of 1976 (RCRA), which established guidelines for managing solid and hazardous waste and the Comprehensive and Environmental Response Compensation and Liability Act of 1983 (CERCLA), which was created to help clean up waste sites where ownership could not be determined. The 1983 law established a trust fund to finance the cleanup of spills and sites. This fund has become known as Superfund.

CONTAMINATION

When we look at contamination in ground waters, an important factor is whether the chemical constituents are (1) soluble or insoluble in water; or (2) are more dense than the water itself or are significantly lighter. The light fractions will be found at the upper surface of the water table, known as "floaters." The more dense contaminants will tend to move through the ground water to a confining unit or barrier (they may move more quickly vertically than horizontally) and are known as "sinkers" (Figure 21).

An important aspect of physical and chemical properties of water relates to using chemical evaluation of waters to determine the source or origin of the water. Where water came from may be of value in solving a contamination problem. There are several means of evaluating water origin. One technique uses radioactive constituents to age date the water and provide a perspective as to whether waters of different ages are mixing. Sometimes, waters in different aquifers beneath the surface may be at different ages and sources. Another technique also uses the chemical evaluation of anions and cations plotted in a triangle to evaluate a "signature" or "fingerprint." If a question is raised as to whether waters from different aquifers are mixing together through zones of weakness or connection, the use of **trilinear diagrams, or Stiff diagrams** facilitates rapid comparison of water analyses from different depths to evaluate their origin (Figure 22).

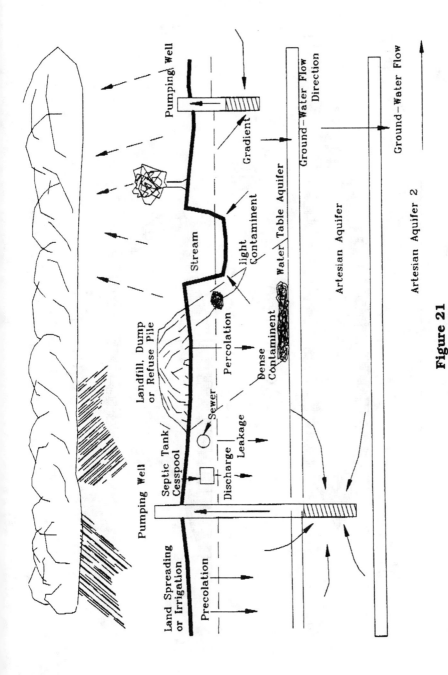

Figure 21
Typical Groundwater Contamination Sources

308 / *Environmental Science and Technology Handbook*

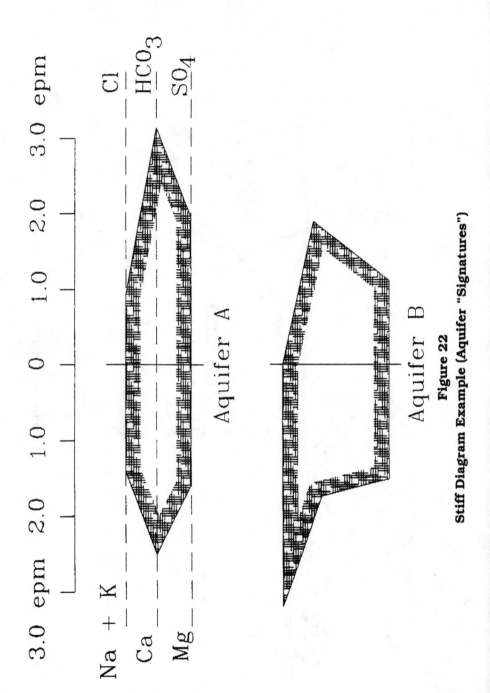

Figure 22
Stiff Diagram Example (Aquifer "Signatures")

As you can determine, ground water movement in the subsurface is primarily controlled by the permeability of the soils/rock (ability of the rock to move water) and the gradient or slope of the water table (in the case of artisan waters, the potentiometric surface). Ground water velocities can range from several feet per day to several feet per year. In the cases where there are very tight confining units, vertical movement may be tenths of a foot per day to a tenth of a foot over ten thousand years. The point should be made, however, that clay liners or caps, although highly impermeable if placed properly, will still accommodate movement of water through the clay. However, this movement it is at a very, very slow rate. It is only with synthetic liners of impermeable materials that no movement can occur across such a barrier. It is also understandable that some chemical constituents involving molecules much smaller than water can move through somewhat impermeable soil or rock material much faster than water. Decades ago, the installation of natural clay liner material or slurry walls at times did not take this physical consequence into account. It was assumed that these liners would prevent **all** movement. They do not.

After discussing some general geology and hydrology, we are at a point to briefly discuss groundwater contamination--the identification, the evaluation, and the clean-up. Why was all of the preceding background information important? For us to understandcontamination and its movement, we need to understand the physical environment where that contamination can be found.

In fairly simple terms, an evaluation of a contamination problem always starts with evaluating the physical characteristics of the subsurface. Then a more cost-effective program of drilling and sampling can be undertaken.

Some surface geophysical techniques, which may be able to identify a containment "signature" and the general extent of both horizontal and vertical contamination, should be considered when looking at large areas. We can use that information to provide more site specific testing programs. Some geophysical techniques also help to define the characteristics of subsurface materials such as clays and sands that may provide either barriers or pathways to groundwater flow. Examples of geophysical techniques used on the surface, which cover large areas at a low cost, are the measurement of magneticism, conductivity, resistivity, sound waves, and ground penetrating radar data. Each one of these methods provides a different and indirect means, through anomalies, of evaluating subsurface physical conditions. Techniques usually use combination surveys and always have physical correlation. Such techniques are effective tools. It cannot be emphasized strongly enough that the value of geophysical tests is in proportion to the direct correlation with borings and samples.

It should also be mentioned that down-hole geophysics employed in borings and wells provides a similar tool to evaluate borings where samples were not taken. Control wells can be minimized when looking at a large geographic area since drilling and sampling are at a much greater cost than the geophysics. As mentioned, geophysics may provide the framework for qualitative information on which to make other decisions and to allocate funding effectively in investigating the subsurface and in taking water quality samples.

Drilling the subsurface, collecting samples of both soil/rock and water, and then testing the aquifer or confining layer to measure specific permeability and other physical parameters are required for contamination studies. To adequately characterize subsurface aquifers and confining units and to evaluate a contamination problem may involve a wide range of costs. Costs can range from thousands to millions of dollars. Understandably, the complexity of the subsurface conditions, both geologically and hydrologically, greatly impacts costs of investigation and remediation. Whenever drilling and testing are required, invariably costs are into and above five figures. As mentioned earlier, RCRA and Superfund sites have cost millions of dollars or more.

WELL HYDRAULICS

An important part of hydraulics is conducting the proper tests and making the right assumptions to evaluate aquifer characteristics. The information obtained is used to design additional wells for potable or industrial use. Hydraulics also can provide significant information that can be utilized in ground water contamination investigations and cleanup-remediation activities. The purpose of a pumping test is to provide basic data indicating aquifer performance. These data are usually transmissivity and storage coefficients. The data are calculated from test information which includes measurement of withdrawal from a pumping well and the drawdowns of additional wells located at some distance from that pumping well. Formulas were discussed earlier in the chapter. Well placement in a pumptest is particularly important as it is necessary to be within the radius of the cone of influence of the pumped well (Figure 23).

The most basic tests involve constant rate pumping and step drawdown pumping. In the first case, a well is pumped for a period of time at a single rate until stability or equilibrium is obtained in the pumping well and observation wells. In the second case, a well is pumped at successive rates for relative short periods of time. Information from both types of aquifer pumping tests is then analyzed to obtain important aquifer characteristics as well as specific well characteristics. It is important to understand that the results of properly conducted pump tests provide additional techniques to be used. Aquifer data may then be utilized in predictive groundwater modeling to evaluate water flow and transport direction. This additional important tool is critical in providing information for ground water withdrawals for well fields and remediation design of contamination planes.

The techniques can provide information to create hydraulic barriers to future contaminant flow and movement (Figure 24). In addition, well and aquifer testing information provide valuable data to optimize pumps and withdrawal rates for water supply well fields. Well fields can then be pumped cost effectively and safely without long term damage to the aquifer from either salt water intrusion or "mixing" of the water in storage.

One of the purposes of a pump test is to stress the aquifer. Therefore, from a planning standpoint, appropriate large volumes of water need to be pumped from the aquifer and discharged. For aquifers that are highly permeable and transmissible, monitoring wells need to be fairly closely placed to the pumped well. In the case of tighter formations with less permeability in porosity, observation wells can be at some distance in order to obtain the proper test results.

The cost of pump tests can vary from as little as several thousand dollars to several hundred thousand dollars. Cost depends upon the aquifer thickness, depth, and the area of investigation.

Once the subsurface hydrology and geology have been defined, as well as a contaminant plume, state-of-the-art computer modeling techniques are available to predict longer term migration of various contaminant levels and the effectiveness of different clean-up schemes. This modeling may be important in evaluating risk assessment relating

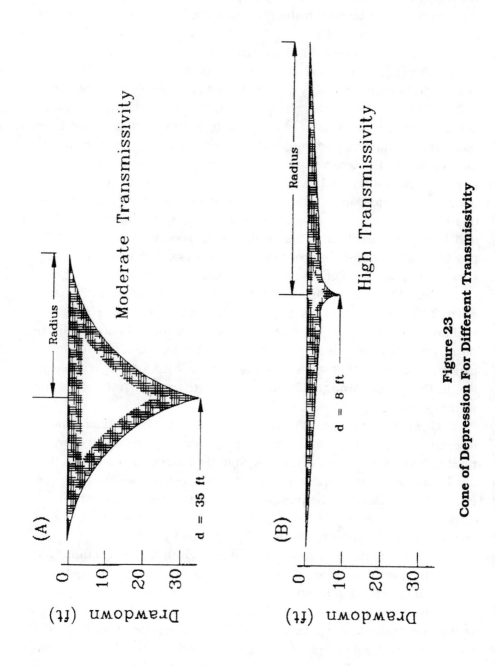

Figure 23
Cone of Depression For Different Transmissivity

Geology and Groundwater Hydrology / 313

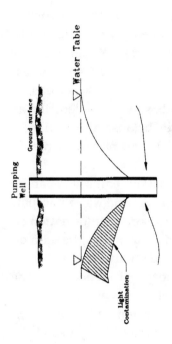
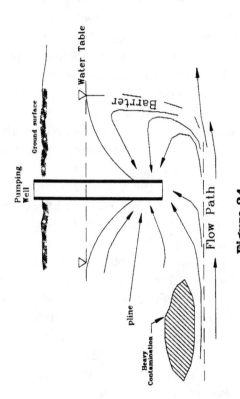

**Figure 24
Hydraulic Barrier**

to both human pathways and environmental pathways. Dependent upon the quantity and quality of subsurface information, it is possible today to utilize three-dimensional modeling techniques for these predictive efforts. Modeling, however, is also not cheap. Five figure modeling efforts are common, with the more complex and three-dimensional models approaching six figures.

GROUNDWATER MODELING

Computer modeling of ground water movement in the subsurface has reached new sophistication over the last ten years. The model itself is an attempt to provide a means of predicting past performance in the hope of predicting future change. With the availability of extremely fast personal computers, the groundwater hydrologist has, at his disposal, numerous mathematical methods which can describe all types of groundwater situations in different dimensions. Although most models still are two dimensional, many models can provide three dimensional output.

Based upon the available aquifer information, equations can be written to describe boundary conditions and the ground water flow pattern. We mentioned earlier that pump tests can provide information on actual aquifer performance, which in many cases averages the data on many different formations, boundaries, and flow patterns. It is this simplification that enables us to create conceptual models and the mathematical framework for additional analysis.

Two dimensional models remain the most common. Information collected in most investigations does not and cannot justify use of the more sophisticated three dimensional model. The cost of obtaining adequate subsurface information (quality and quantity) is prohibitively expensive and time consuming.

Analytical equations to model conditions are also often used when the data are insufficient to use computer models. However, in the number of situations where the data are adequate, numerical techniques effectively to analyze and predict subsurface conditions that have complexity and irregular boundaries.

In general, two basic types of models are utilized in numerical analysis. They are finite difference and finite element models. In both cases, different types of grids are used to dissect an area with values for the grids serving as the basis for numerical solutions that approximate continuous partial differential equations.

Again, where there is enough information to justify computer modeling, this tool has become extremely important in helping to make predictive conclusions and to support other conceptual or analytical models. An important point to make is that the accuracy or precision of computer data output is a function of the information input and conceptual knowledge of the problem solution.

Computer modeling techniques may provide additional insight into a situation but, for the most part, should never provide anything but support for a conceptual model already developed. If you do not have an idea as to what will happen physically, do not use a computer model for that purpose.

SUMMARY

In conclusion, it should be clear that to understand the earth beneath us is a challenging, complex, and expensive effort. Nevertheless, modern technology involving a number of primary and secondary measuring techniques from satellite imagery to surface geophysics to subsurface drilling and testing can provide us with information with which to evaluate, use, and preserve our environment and ground water resources. Whether the search is for a long term, potable water supply; a search for non-potable cooling water for industrial plants; or for preservation of water levels, water quality, and flows in the environment impacting surface ecology: these problems can be addressed with an appropriate technical and economic level of solution.

For those who wish to look further into the fields of geology and hydrology, there are a number of excellent references and basic texts on the market that should satisfy readers of varying technical backgrounds. An excellent source of elementary materials may be found from the U.S. Geological Survey with offices in Reston, Virginia; Denver, Colorado; and many other locations. Your local EPA Regional Headquarters also has available a number of primers to help understand these fields, and your local state geological survey is an excellent source. Many state geological surveys also have roadside handbooks which explainin detail geology and sometimes hydrology across a particular state. In addition, the following references may be of value.

REFERENCES

1993, Hydrology, An Environmental Approach, Ian Watson and Alister D, Burnett, Buchanan Books, ISBN 0-9636039-0-6.

1990, 1991 EPA Ground Water Handbook. Government Institutes. 1992, ISBN 0-86587-279-1.

1986 Ground Water and Wells, Johnson Division, Fletcher C. Driscoll, ISBN 0-9616456-0-1.

1980, Applied Hydrogeology, C.W. Fetter, Jr., Bell & Howell Company, ISBN 0-675-08126-2.

10

GROUNDWATER POLLUTION CONTROL TECHNOLOGIES

Stanley G. Puszcz and Michael V. Tumulty, P.E., CGWP
H2M Group

OVERVIEW

Groundwater is a vital resource that needs to be protected. As a resource in the United States, groundwater:

- provides drinking water to 50 percent of the population;
- makes up 95 percent of the fresh water reserves;
- is used by 75 percent of the cities in some way; and
- is used by 95 percent of the population in rural areas.

During the last decade there has been increasing awareness of how sensitive this resource is to our daily activities. Out of this awareness has come a demand to clean up contamination in our groundwater resources.

Further, the degree of cleanup or "remediation" at any given location is often a function of several factors. Philosophically, it may be our goal to restore groundwater to its natural, pristine state. But, while the technology is generally available to achieve that goal, the cost of cleanup can be so high that such a goal is not practical in every case. However, there is rarely only one way to restore groundwater at a site that has been degraded by pollution. In addition, depending on the specific case, one technology may be more effective, workable or less costly than another. The following sections of this chapter will discuss and evaluate the technologies that are most commonly used to clean up the environment.

EVALUATION CRITERIA

The U.S. Environmental Protection Agency (U.S. EPA) provides guidance to the scientists and engineers working on Superfund projects for evaluation of remedial alternatives. This guidance, commonly referred to as Feasibility Studies, has been used extensively in the Superfund Program and, in a conceptual way, with most large groundwater cleanup projects.

An evaluation of alternative methods of remediation follows the investigative stage of a program, during which sampling is performed and analytical data is developed. With that investigative data from the site in hand, a decision about the various alternatives can then be made.

In general, the protocol calls for screening technologies, assembling alternatives and evaluating them using the following three criteria:

1. Effectiveness Evaluation
   - performance
   - reliability
   - safety
2. Implementability Evaluation
   - ability to be permitted
   - workability at the site
   - availability of necessary program components
3. Cost Evaluation
   - capital costs
   - operation and maintenance costs

REMEDIAL TECHNOLOGIES

There are many alternative technologies considered for groundwater cleanup projects. They include, among others:

- Pumping and treatment (e.g., air stripping, carbon adsorption, and bioremediation for volatiles and precipitation for metals),
- Subsurface barriers (e.g., slurry walls and grout curtains),
- Subsurface drains (e.g., french drain),

- Bioremediation, and
- Capping.

A successful remediation project may include one or several of these technologies done together.

Take, for example, an industrial site in New York that was recently remediated. Spent solvents and cutting oils had been disposed of on the ground in the back of the plant some 15 to 20 years ago. The groundwater was contaminated; the waste product, although well broken down, was still on the surface, and the soil beneath the waste pile was contaminated down to the water table, 50 feet below.

At this site, the cleanup program involved several technologies. First, the waste pile and the soil immediately surrounding the pile was excavated for off-site incineration, thereby eliminating a continuing source of groundwater pollution. There was no attempt made to excavate the deeper soils that were impacted to a lesser extent. For these soils, a soil vapor extraction system was constructed to provide fresh air flow through the in-place soils, stripping the solvent contaminants from the soil. The groundwater was remediated by pumping it up through extraction wells and through an air stripping tower to remove the volatile contaminants. The contaminated air from the stripping tower and the soil vapor extraction system were passed through carbon, so that only clean air was released to the environment.

In addition to these measures, a bioremediation aspect was incorporated. Adding oxygen and nutrients to the treated groundwater and reinjecting it so that it would flow through the contaminated area enhanced nature's own biological treatment of the contaminated groundwater by sustaining the natural microbes that were degrading the contaminants.

DEVELOPMENT OF A REMEDIAL PROGRAM

The most significant part of a groundwater remediation program entails the pumping, treatment and recharge of treated water. However, other technologies such as subsurface barriers (e.g., slurry walls), capping, and subsurface drains often play a significant part in a remedial program. These technologies employ geotechnical engineering expertise, beyond the scope of this chapter. The focus of the following sections will be on the treatment methods employed under a pumping and treatment scenario.

The goal of a remedial groundwater pumping scheme is to modify the "hydraulic gradients" in the area of contaminated groundwater to prevent any further migration of the **plume.** A plume is the mass of contaminated groundwater that moves in the direction of groundwater flow. The hydraulic gradient is the slope, or change in water levels over a given distance.

Remedial pumping should create groundwater flow patterns toward the extraction sites to remove the contaminated water in the shortest possible time with the least possible dilution, thereby minimizing operational costs and providing for effective treatment of the groundwater. Figure 1 presents a section of a typical remediation well.

The necessary modification of groundwater gradients to capture the plume is accomplished by the installation of **extraction wells,** which are wells used to pump out groundwater for treatment. Strategic placement of the extraction wells will allow for a greater lateral influence on groundwater gradients while minimizing the rate of pumping at each well necessary to capture the plume. Optimizing pumping rates would further minimize dilution of the plume, enhancing the rate of contaminant removal.

The design pumping rates of the extraction wells are based on a calculation of the estimated change in hydraulic gradient and subsequent **zone of capture.** The zone of capture is the area that is affected by pumping to the extent that flow is toward the well. Estimates of the zone of capture for extraction wells can be calculated. Once the treatment system is in operation and sustained pumping has taken place, groundwater elevations can be measured within adjacent monitoring wells for development of groundwater contour maps. The groundwater contour maps will provide definitive demonstrations of the extent of hydraulic control. Any modifications required to affect the groundwater contours can often be accomplished by modifying one or more of the extraction well pumping rates.

Theis Equation

Evaluation of the theoretical response of an aquifer to pumping can be based upon the application of the Theis non-equilibrium well function equation with conservative estimates of hydrogeologic variables. The Theis equation, first developed in 1935, was the first equation to take into account pumping time on yields. Application of this simplified prediction method is a conservative approach to estimating

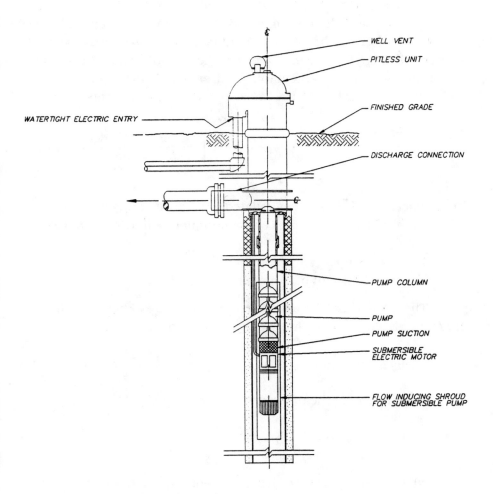

**Figure 1
Cross Section of Remediation Well**

the effect of pumping on groundwater contours. The derivation of the Theis equation is based on the following assumptions:

- The water-bearing formation is uniform in character, and the hydraulic conductivity is the same in all directions.
- The formation is uniform in thickness and infinite in areal extent,
- The pumped well penetrates and receives water from the full thickness of the water-bearing formation, and
- The water table and potentiometric surface has no slope.

To account for the application of this idealized equation, and deviations anticipated therefrom, a contingency should be applied to the result. In reverse form, if the equation predicts sufficient hydraulic control with significant contingency at the specified flow rate, then the flow rate is considered acceptable for design and start-up purposes.

In its simplest form, the Theis equation can be expressed as:

$$ho - h_{(r,t)} = \frac{114.6\, Q}{T} \int_u^\infty \frac{e^{-u}\, du}{u} \qquad (1)$$

where, $h_o - h_{(r,t)}$ is the calculated drawdown at any point a distance from the center of the pumping well, r, at pumping time, t [ft];
Q is the pumping rate [gpm]; and,
T is the coefficient of transmissivity [gpm/ft].

The integral can be expressed as W(u) and is known as the well function. Evaluating the integral by series produces:

$$W(u) = -0.5772 - \log_e u + u - \frac{u^2}{2*2!} + \frac{u^3}{3*3!} - \frac{u^4}{4*4!} + \cdots \qquad (2)$$

In determining the dimensionless value of u:

$$u = \frac{1.87\, r^2\, S}{T\, t} \qquad (3)$$

where, r is the distance from the center of the pumping well to a point where drawdown is calculated [ft];

S is the coefficient of storage [dimensionless]; and,

t is the time since pumping started [days].

In order to calibrate the Theis non-equilibrium well function to the site, theoretical results from applying the equation are compared to actual results from pump tests performed on the aquifer. A "correction factor" is developed for each observation well utilized. This "correction factor" is then applied to the results of the Theis calculations for determining design pumping rates at the proposed extraction wells. The "correction factors" generally result in a more conservative prediction.

Design pumping rates are arrived at by performing iterative calculations using varying pumping rates within each proposed extraction well. The objective is to affect groundwater contours such that flow patterns are created toward the extraction wells from all areas that exceed the groundwater quality criteria.

To account for multiple wells being pumped simultaneously, each drawdown curve is superimposed at an offset equal to the distance between the centerpoints of the two wells.

DESIGN FLOW AND CONTAMINANT LOADING

In order to accurately size the treatment system, the range of expected influent flows and contaminant concentrations must be known.

The design flow of the treatment system is based on the maximum potential pumping rates that may be required to provide hydraulic control. The design of the treatment process should take into account the potential need for increasing the combined extraction rate to the treatment system, as well as the potential need for decreasing the rate.

The contaminant loading rates for the design basis are developed by applying a conservative increase factor (e.g., 50 percent) to the highest concentrations reported during groundwater monitoring. While it is expected that contaminant concentrations will decrease over time during the operational life of the treatment facility, initial concentrations may likely be greater during start-up. Further, if **Dense Non-aqueous Phased Liquid (DNAPL)** is likely to exist for solvents with specific gravities greater than that of water, it may prove to be a continuing source of higher concentrations.

Contaminant Migration And Characteristics

Depending on characteristics (e.g. solubility and density), the pollutant can do four things when it percolates down to groundwater: it can float on top of the aquifer; dissolve into the aquifer; sink to the bottom of the aquifer; or adsorb onto sediments. It is probable that the contaminants will do more than one of these. When developing a cleanup program the characteristics of the pollutant should be considered. For organic contaminants, the following properties of the pollutant should be considered:

1. solubility
2. density
3. affinity to volatilize
4. adsorptability
5. biodegradability

Table 1, "Volatile Organic Pollutant Characteristics", presents values for these properties for various volatile organic pollutants.

Pure compound recovery is possible only when the contaminant is not generally soluble in water. Table 1 shows that compounds such as acetone and phenol have high solubilities and therefore, pure product recovery is not generally possible. Conversely, compounds such as napthalene may be recoverable as a pure product in the aquifer system, in part, as a result of its low solubility.

If a compound is not soluble, then it will generally either float on top of the aquifer or sink to the bottom as well as adhere to sediment. Table 1 also provides specific gravities for compounds. Compounds with specific gravities less than 1.0 would be lighter than water and tend to float, while compounds with specific gravities greater than 1.0 would be heavier and tend to sink. Petroleum products with volatile organic components such as benzene, ethylbenzene and toluene have a specific gravity less than 1.0 and will float on the water table as well as allow compounds to dissolve at varying rates into the aquifer. Figure 2 presents a cross-section schematic of a pumping system for oil recovery showing the recovery of oil floating on the water table.

Once dissolved in the groundwater, compounds will move with groundwater flow.

Table 1
Volatile Organic Pollutant Characteristics

Compound	Solubility [a] (mg compound/L Water) (@ deg. C Temp)	Density [b] (Specific Gravity)	Ability to Strip [c] (Henry's Law Constant)	Adsorptive Capacity [a,d] (mg compound/ g carbon at 500 ppb)	Biodegradability [a,e]
Acetone	Infinite	0.79	0.0017	43	degradable
Benzene	1780(20)	0.88	0.23	80	degradable
Carbon tetrachloride	800(20), 1160(25)	1.59	1.0	6.2	nondegradable
Chloroform	8000(20), 9300(25)	1.48	0.14	1.6	nondegradable
Methylene chloride	20,000(20), 16,700(25)	1.33	0.10	0.8	-
Chloro benzene	500(20), 488(10)	1.11	0.17	45	degradable
Ethyl benzene	140(15), 152(20)	0.87	0.28	18	degradable
Hexachloro benzene	0.11(24)	1.60	0.072	42	nondegradable
Ethylene chloride	9200(0), 8690(20)	1.24	0.038	2	refractory
1,1,1-Trichloroethane	4400(20)	1.34	0.77	2	refractory
1,1,2-Trichloroethane	4500(20)	1.44	0.31	3.6	refractory
Trichlorethylene	1100(25)	1.46	0.42	18.2	refractory
Tetrachloroethylene	150(25)	1.62	0.65	34.5	refractory
Phenol	82,000(15)	1.07	0.000011	161	degradable
2-Chlorophenol	28,500(20)	1.26	0.00037	38	degradable
Pentachlorophenol	5 (0), 14(20)	1.98	0.00014	100	refractory
Toluene	470(16), 515(20)	0.87	0.29	50	degradable
Methyl ethyl ketone	353(10)	0.81	0.002	94	degradable
Napthalene	32(25)	1.03	0.02	5.6	degradable
Vinyl chloride	1.1(25)	0.91	120	Trace	-

Notes:
Source: Nyer, Evan K. "Groundwater Treatment Technology" New York, Van Nostrand Reinhold, 1985
a - From Verschueren, Karel "Handbook of Environmental Data on Organic Chemicals" New York, Van Nostrand Reinhold, 1983
b - From Weast, Robert "Handbook of Chemistry and Physics," 60th Ed.: CRC Press, Inc., 1979, 1980.
c - Calculated from Montgomery, J.H. et. al. "Groundwater Chemicals Desk Reference" Lewis Publishers, 1990
d - From Uhler, R.E. "Treatment Alternatives for Groundwater Contamination" J.M. Montgomery Engineers
e - From Tabak, H.H., et. al. "Biodegradability Studies with Organic Priority Pollutant Compounds" Journal of Water Pollution Control Federation, 53:10 -1503, 1981

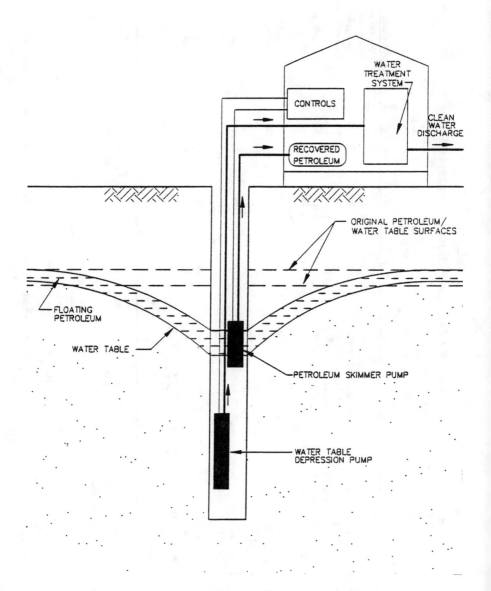

**Figure 2
Cross Section Showing Removal of Floating Petroleum**

BIOLOGICAL TREATMENT

Biological treatment of groundwater uses naturally occurring bacteria to break down complicated organic compounds into carbon dioxide and water resulting in the growth of new bacteria.

Bacteria can grow in either **aerobic** (i.e., uses oxygen) or **anaerobic** (i.e., oxygen not used) environments. Aerobic environments are most commonly chosen when treating for organic compound removal. The necessary inputs to the system are oxygen and nutrients (e.g., nitrogen and phosphorous) for the bacteria. Additionally, dissolved oxygen levels should be kept above 1 part per million (ppm), pH between 6 and 9, and temperatures above 50°F.

Certain organic compounds, such as most petroleum-related compounds, degrade very easily in a biological system; others are harder to degrade **(refractory)**; and others still will not biodegrade. A list of compounds and their relative biodegradabilities are presented in Table 1.

In-situ biological treatment of groundwater is growing in popularity. In relation to other technologies, it is sometimes considered very cost effective. Oxygen is provided by pumping in hydrogen peroxide, ozone, or by blowing in air (i.e., aeration). Nutrients are also pumped into the aquifer. However, if the oxygen or the nutrient supply is eliminated, the bacteria will die and stop degrading the contaminants.

Bioremediation entails the inoculation of microbial strains into the contaminated media to degrade a variety of organic compounds. To identify which bacteria will successfully biodegrade the contaminants of concern, representative samples are collected from the site and subjected to a consortium of bacteria during a laboratory pilot study. Microorganisms which use the specifically targeted compounds (i.e., site contaminants) as a source of carbon and energy for growth and metabolism are isolated during the pilot study and selected for field application. Upon completion of pilot testing, selected bacteria are introduced into the contaminated media in a solid, liquid, or slurry phase process. Oxygen, water, and nutrients such as nitrogen, phosphorous and potassium, are added to stimulate microbial biodegradation. To further enhance degradation rates of the various organic compounds in the soil, the bacteria, soil, and water are exposed to a unipolar magnetic field.

As this technology has only recently been developed, a limited number of full-scale bioremediation applications have been performed and completed. An evaluation of several case studies indicates that an immediate increase in the rate of contaminant degradation occurs when a bioremediation process has been applied under appropriate environmental conditions. There have been several instances, however, when microbial strains have had difficulty becoming established due to the presence of indigenous strains of microorganisms, or contaminant concentrations which are vastly different than those observed in the laboratory.

AIR STRIPPING

Air stripping is a pollution control technology that is employed to remove volatile organic pollutants that are dissolved in groundwater. It involves intimate contact between contaminated groundwater and air, so that the volatile compounds are transferred from the liquid phase to the air phase. The air then carries off the contamination.

The **packed tower,** or air stripping tower, is the most common method of air stripping for both potable water supplies and remediation programs. Figure 3 provides a schematic of the primary components of an air stripping tower. The stripping tower has water introduced at the top of the tower which flows down through an inert "packing" material while air is blown in from the bottom and exits on top. The packing material is generally made up of small inert plastic units that are typically one to six inches in diameter. The cleaned water is collected at the bottom and flows out to its final destination. The upward flow of air with the downward flow of water over the packing material is referred to as **counter current packed column air stripping.** Counter-current packed towers have been utilized in the chemical process industry for decades as a standard unit operation to affect mass transfer, both in adsorption (e.g., air pollution control) and desorption (e.g., groundwater treatment via stripping). The **adsorption** process is typified by the mass transfer of material from the air phase to the liquid phase, while **desorption** involves the mass transfer of material from the liquid phase to the air phase. For that reason, the physical chemistry and mass kinetics are well understood and documented. Essentially, thepacked tower is a chemical engineering process unit which promotes intimate contact between a gas phase and a liquid phase so as to enhance the establishment of equilibrium between phases.

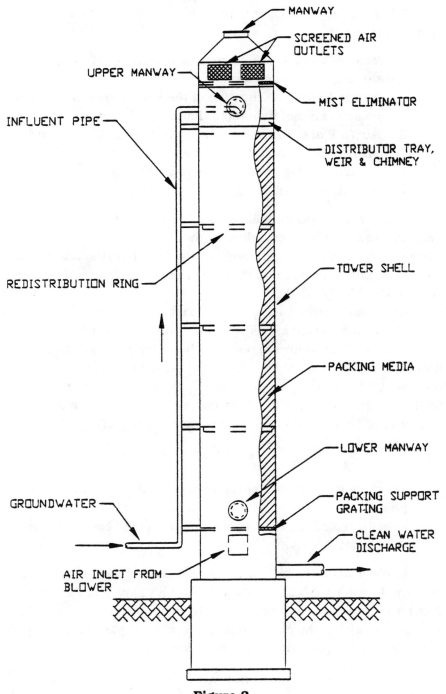

Figure 3
Air Stripping Components

When sizing a stripping tower, the designer has three basic parameters to define:

1. **Tower Diameter.** A function of flow rate (the higher the flow, the larger the diameter).
2. **Tower Height.** A function of the level of contaminants (the greater the contamination, the higher the tower).
3. **Air to Water Ratio.** A function of the specific contaminant being removed and the specified packing material (the more volatile the contaminant and greater the surface area of the packing material, the less air required).

Different compounds will be transferred from the water to the air phase at different rates depending on the Henry's Law constant for the particular compound. Compounds with a high Henry's Law constant will transfer from the water to the air phase more easily, and hence are more strippable (Table 1).

The organic compounds in water that are to be removed must be at concentrations which exceed an equilibrium concentration. By outlining some basic principles of physical chemistry, the vapor-liquid equilibrium is better understood. The relationships can most clearly be presented by starting with a principle of physical chemistry relating the total pressure of a system to the partial pressure of one contaminant in that system. This relationship can be defined by the following equation:

$$\bar{p}_A = \tilde{y}_A P \qquad (4)$$

where, $\bar{p}_A$ is the partial pressure of contaminant A in the gas phase;
$\tilde{y}_A$ is the mole fraction of contaminant A in the gas phase; and
P is the total pressure.

In order to determine the partial pressure of a contaminant in a solution, Raoult's Law can be utilized. Raoult's Law describes the relationship between the partial pressure of a component in the vapor phase with the product of the mole fraction in the liquid phase and the pure component vapor pressure. **Raoult's Law** is as follows:

$$\bar{p}_A = \tilde{x}_A P_A \qquad (5)$$

where, $\bar{P}_A$ is the vapor pressure of contaminant A at the appropriate temperature; and
$\tilde{x}_A$ is the mole fraction of contaminant A in the liquid phase.

A more appropriate form of Raoult's law applicable to low concentration scenarios is available and is known as Henry's law:

$$\bar{p}_A = \tilde{x}_A H_A \qquad (6)$$

where, H_A is a constant (i.e., **Henry's Constant**) and is found experimentally for each particular contaminant. Henry's Law has been seen to work well at low concentrations and provides for a linear relationship between partial pressure and concentration. By combining equations (4) and (6), an equilibrium relationship can be developed for a contaminant A in the vapor and liquid phases:

$$\tilde{y}_A = \tilde{x}_A H_A / P \qquad (7)$$

From the above equation, it can be seen that the greater the magnitude of the Henry's law constant, the higher $\tilde{y}_A$ becomes, and therefore, the equilibrium mole fraction of contaminant A increases in the gas phase. For any given contaminant in water, a higher Henry's law constant will mean that the contaminant would be more easily removed from water by stripping. Values of Henry's constant are available for selected compounds in literature and provided in Table 1.

In addition, the liquid phase resistance is the limiting factor in the transfer of volatile organic compounds from the liquid phase to the gas phase. The percent liquid film resistance is a function of Henry's Constant for each particular compound. For low solubility contaminants (i.e., where Henry's Constant is greater than 0.1), the liquid film resistance will control the stripping rate. Table 2 lists percent liquid film resistances for selected contaminant

Table 2

Liquid Film Resistance Percentages for Volatile Organic Pollutants

Compound	Liquid Film Resistance
tetrachloroethene	99.0%
trichlorethene	97.7%
cis/trans-1,2-dichloroethene	99.8%
1,1-dichloroethene	99.9%
1,1-dichloroethene	95.5%
methylene chloride	90.0%
vinyl chloride	99.996%

In general, the higher the value of the Henry's Constant for a particular contaminant, the easier it will be to remove by air stripping. The removal efficiency of a given size air stripping tower for a given contaminant is dependent upon the overall transfer coefficient, usually denoted "$K_{Lx}\ a$", where "K_L" is the overall mass transfer coefficient, and "a" is the specific surface of the packing media, expressed as wetted surface area per tower volume. The overall transfer coefficient can be determined empirically through the results of pilot studies, which can be performed prior to formal design efforts by utilizing the packaging material that will most likely be specified.

The design of an appropriate air stripping tower is based upon selection of reasonable overall mass transfer coefficients, liquid loading rates and packing height matched to the design conditions. Conventional practice in sizing the packing height for a tower of known cross-sectional area is based upon determination of the height of a theoretical transfer unit (i.e., HTU) and the number of theoretical transfer units (i.e., NTU) required to achieve the desired removal. The extent of mass transfer (or removal efficiency) can be estimated for a stripping tower with a specified cross-sectional area and packing height.

The mass balance equations can be expressed as:

$$Z = Required\ Packing\ Height = (HTU) \times (NTU) \tag{8}$$

$$HTU = \frac{Q/A}{K_L a} \quad (9)$$

where, Q = Flow
A = Cross-Sectional Area
$K_L a$ = Overall Transfer Coefficient

$$NTU = \int_{C_e}^{C_o} \frac{dC}{C - Cg/H} \quad ; \text{ where, } Cg = f(C) = Cge - \frac{Co-C}{G/Q} \quad (10)$$

Performing a mass balance across the tower and substituting it into equation (10) followed by integration yields:

$$NTU = \frac{1}{1-A} \ln \left[\frac{Co(1-A) + ACe - Cgo/H}{Ce - Cgo/H} \right] \quad (11)$$

When the influent gas phase contains no volatile organic compounds being stripped from the liquid phase (i.e., Cgo = 0), then:

$$NTU = \frac{1}{1-A} \ln (A + (1-A) Co/Ce) \quad (12)$$

For system analysis and design, the steady state equation for the liquid film is utilized. From this equation, the tower packing height can be calculated:

$$Z = \frac{\overline{Q}}{(1-A) K_L a} \ln (A + (1-A) Co/Ce) \quad (13)$$

where, $\overline{Q} = Q/A$, and the removal efficiency of a given packed tower is given by:

$$\%R = 1 - Ce/Co = 1 - \left[\frac{1-A}{e^{[K_L a Z (1-A/Q)]} - A} \right] = (1 - e^B)/(A - e^B) \quad (14)$$

where, $B = K_L a Z (1-A)/Q$

An evaluation of an air stripping installation must include consideration of the rate of contaminant discharge to the atmosphere. Conservative estimates of the emission rate potential can be calculated from the design influent flows and contaminant concentrations with the assumption of complete mass transfer to the gas phase.

CARBON ADSORPTION

Adsorption is a natural process in which molecules of a liquid or gas are attracted to and then held at the surface of a solid. Activated carbon is an excellent adsorbent due to the large degree of surface area contained within the carbon particle that is accessible for the adsorption process. In addition to the "outer" surface area on the carbon particle, "inner" cavities allow for significant surface area per mass of particle.

Evidence of the high degree of surface area and internal cavities is the difference between the activated carbon's density versus the density of solid carbon. Solid carbon densities range from 125 to 130 pounds per cubic feet, whereas the density of activated carbon ranges from 25 to 31 pounds per cubic feet. The difference is related to this surface area - inner cavity phenomenon.

The adsorption process consists of three steps:

1. diffusion of the contaminants through the fluid (gas or liquid phase) to the carbon particle;
2. diffusion of the contaminant through the "inner" cavities to the adsorption site; and
3. adsorption of the contaminant to the carbon particle.

Granular activated carbon (GAC) adsorption is most typically implemented through the use of downward flowing pressure vessels installed between the well pump and the point of discharge (for liquid treatment), or an upward flowing pressure vessel between the source of off-gas and the point of emission (for air pollution control).

Contaminants in the raw water or off-gas adsorb onto the GAC. The adsorptive capacity of the carbon varies with the nature and concentration of the contaminants. As the contaminant loading on the carbon reaches the adsorptive capacity of the carbon near the top of the filter (in the case of wastewater treatment) or the bottom of the filter (in the case of air treatment), the interface between the saturated and the "clean" carbon moves downward (wastewater) or upward (air) through the carbon bed inside the pressure vessel.

When the carbon in the filter vessel is fully loaded with contaminants (i.e., at its adsorptive capacity), no further removal will take place and contaminants will begin to be found in the filter effluent. Effluent monitoring and estimates of the adsorptive capacity of the carbon enable the carbon in the filter to be replaced prior to occurrence of contaminant breakthrough.

The GAC removed from the pressure vessel, after adsorptive capacities have been reached, can be regenerated by heating at high temperatures. On-site carbon regeneration facilities only prove economical for a facility having a very high rate of consumption. Off-site carbon regeneration is usually preferred.

The frequency with which the carbon must be regenerated or replaced depends on several factors:

- The nature and concentration of the contaminants to be removed;
- The total flow through the pressure vessel; and
- The total amount of carbon contained within the pressure vessel.

In groundwater treatment, adsorption of contaminants from groundwater would be accomplished via liquid phase carbon. As an option or possibly as required by regulation, vapor phase carbon can be utilized for the adsorption of the contaminants from the air stream following air stripping.

AIR STRIPPING WITH CARBON ADSORPTION

Vapor phase granular activated carbon (GAC) is the most common form of treatment for the control of volatile organic contaminants in offgasses from air stripping operations (as well as for soil vapor extraction technologies). This unit operation provides for the transfer of contaminants onto the GAC after it has volatilized from the liquid.

The advantage of using vapor phase GAC after an air stripper, as opposed to using liquid phase GAC and foregoing the air stripping operation, is the greatly increased adsorption capacity of the GAC in the vapor phase. By transferring the contaminants to the vapor phase during air stripping, the vapor phase carbon can adsorb more contaminant, and the carbon can last longer. Thus, the amount of spent carbon generated per unit operating time (or per unit mass volatile contaminant treated) will be reduced, with an increase in contaminant concentration per unit mass of GAC. Depending on the contaminants in question, the relative humidity and temperature of the air stream, and the approach velocity of the air stream, the vapor phase adsorption capacity can be from three to 50 times higher than the liquid phase capacity.

For vapor phase GAC to be properly utilized, the offgas relative humidity must be reduced to below 50 percent. This can be done by using desiccants, pressure condensate vessels, or by heating the air stream. If the relative humidity is not reduced, the capacity of the carbon is significantly reduced because the water molecules occupy adsorption sites preferentially. Another factor in the design of vapor phase GAC systems is the approach velocity, which must be maintained below 100 feet per minute for effective adsorption.

For a given application, where liquid phase GAC alone is compared to air stripping with vapor phase GAC, the increase in operating costs are typically at least twice that of air stripping in combination with vapor phase GAC. This considers the cost of energy as well as waste generation.

ULTRAVIOLET (UV) OXIDATION

This relatively new technology for treating groundwater contaminated with organic compounds provides for destruction of the contaminant in the treatment process rather than transferring of the contaminant to the air phase (air stripping) or onto another media (carbon adsorption). This process utilizes ultraviolet (UV) light as a catalyst to the chemical oxidation process where organic compounds oxidize to form primarily carbon dioxide and water. Hydrogen peroxide or ozone are commonly used as the source of oxygen for the reaction.

Inorganic compounds, including metals (i.e., iron) and minerals (i.e. calcium carbonate) will also be subject to the oxidation reaction. Scaling on the UV lamps will result from those reactions which decrease efficiency and therefore requires maintenance to clean the lamps.

EX-SITU CHEMICAL AND BIOLOGICAL OXIDATION

Three chemical oxidizers have been widely used for industrial treatment: chlorine, hydrogen peroxide, and ozone. Use of chlorine as an oxidizing agent will result in chlorinated compounds or higher levels of chlorine substitution which is unacceptable in groundwater treatment. Hydrogen peroxide is readily available and works very well on organic compounds with double and triple bonds. The costs and handling of peroxide, however, will limit its use to small flows and short durations. Ozone is the strongest of the oxidizing agents and, with sufficient time, can eliminate any organic compound. However, because of high capital and operating costs, ozone will be limited in its applicability.

Biological oxidation is the lowest costing remedial technology for removal of organic contaminants from groundwater. However, there are restrictions on when a biological method may be used. The main restriction is that it must be run continuously and cannot be cycled on and off. The reactor must have a continuous source of food (organic compounds) and nutrients. It must be maintained at optimum temperatures and pH. Another restriction is that the bacteria must be grown to a sufficient concentration in order to effectively remove the contaminants. This start-up period can take two to eight weeks and is labor intensive. Also, the fact that standard biological reactors are not designed for influent concentrations below 50-75 mg/l limits its applicability and effectiveness.

LIQUID PHASE GRANULAR ACTIVATED CARBON (GAC) TREATMENT

The adsorptive capacity of activated carbon for organic contaminants can be estimated from an adsorption isotherm, which relates the concentration of a contaminant laden wastewater to that which is adsorbed by the GAC. This is derived by fitting the results of multiple-batch equilibrium isotherm tests to the Freundlich equation to obtain values of empirical coefficients which characterize the properties of the carbon used and the contaminant adsorbed. Carbon adsorption isotherms are available in literature for many common groundwater contaminants.

The Freundlich equation can be expressed as:

$$\frac{Co - Cf}{M} = K\,(Cf)^{1/n} \qquad (15)$$

where, Co is the contaminant concentration of the influent;
Cf is the contaminant concentration of the effluent;
M is the total weight of carbon; and

K and $1/n$ are empirical constants unique to the contaminant and carbon.

The ultimate capacity of the carbon can be estimated by defining carbon as reaching saturation when the contaminant influent concentration equals the effluent concentration. Choosing the point on the isotherm where $Cf = Co$ will yield a value of carbon adsorption capacity $(Co - Cf)/M$ at that contaminant concentration.

In a similar manner, obtaining the empirical constants allows for a calculated estimate of adsorptive capacity $(K (Cf)^{1/n})$.

At the contaminant concentrations utilized as design criteria, the adsorption of different compounds can be approximated as being independent of the other contaminants present.

Changes in the contaminant concentrations found at each extraction well will alter the consumption rate of activated carbon. The design influent concentrations are conservative themselves in that they should be based on the highest reported concentrations with a contingency factor applied. Actual feed concentrations would typically be considerably less. The isotherms, along with the empirical constants, can be utilized to estimate the adsorptive capacity of the carbon at predicted future contaminant levels and subsequently for design concentrations. The impact of increasing contaminant levels on carbon usage would greatly increase carbon costs per gallon of water treated.

VAPOR PHASE GRANULAR ACTIVATED CARBON (GAC) TREATMENT

Vapor phase granular activated carbon (GAC) is the most common form of treatment for the control of volatile organic contaminants in offgasses from air stripping operations. This unit operation provides for the transfer of contaminants onto the GAC after it has volatilized from the liquid.

Insofar as vapor phase adsorption isotherms are considered proprietary by the manufacturers and venders of vapor phase GAC and are generally unavailable, the design of such a system is based on the performance specification of individual units. The mass balance (considering accumulation of the contaminants within the vapor phase GAC vessel), temperature, and relative humidity are specified, as well as the air flow rates. The vender typically provides a guarantee of performance based on the operational data specified.

INORGANICS REMOVAL

Inorganic contaminants (e.g., metals) in groundwater are associated with a wide variety of sources. In addition to occurring naturally, they also result from human activities. Common naturally occurring inorganics include arsenic, fluoride, and selenium. Inorganic compounds such as chromium, cadmium, and nickel are often associated with groundwater impacts from industrial activities, such as metal plating operations.

Some of the principal factors considered in the selection of an inorganics treatment technology are:

- Contaminant type and valence;
- Influent contaminant concentrations, (because many technologies remove fixed percentages of contamination);
- Desired effluent contaminant concentrations; and
- Influent levels of dissolved solids and pH.

While no single treatment technology is suited for all inorganic contaminants, certain treatment technologies are more favorable for treating the relatively low inorganic concentrations found in groundwater.

Treatment technologies include, among others: metal hydroxide precipitation, reverse osmosis, and ion exchange. **Precipitation** occurs when dissolved metals in groundwater change out of solution into an insoluble form as the result of a chemical reaction. The precipitates first formed by the chemical reactions are crystals of molecular size. The initial growth or increase in size of these colloidal crystals and coagulated solids is caused by ion charge reductions and Brownian forces. Additional growth is also a result of gentle mechanical stirring of the suspension (flocculation). This results in a sludge that is removed, leaving the water with lower concentrations of dissolved metals. The remaining concentrations are a function of the new chemical characteristics of the water and the new solubility of the contaminants.

Reverse osmosis uses a semipermeable membrane to selectively remove contaminants from solutions. Water with a contaminant concentration is placed on one side of the membrane. The water travels through the membrane by hydrostatic pressure to the other side, resulting in a lower concentration of contaminants. Since the membrane permits only water, and not dissolved ions, to pass through itself, contaminants are left behind in a brine solution. The brine solution must be periodically removed and disposed of.

With **ion exchange,** metals removal is accomplished through adsorption of contaminant ions onto an exchange medium. One ion is substituted for another on the charged surface of the medium. This surface is designed to be either cationic (positively charged) or anionic (negatively charged). The exchange medium is saturated with the exchangeable ion before treatment operations. During ion exchange, the contaminant ions replace the regenerant ions because they are preferred by the exchange medium. After the exchange medium reaches equilibrium with the contaminant ions, the medium is regenerated with a suitable solution, which then resaturates the medium with the appropriate ions. Because of the required "down time," the shortest economical regeneration cycles are once per day. Similar to reverse osmosis, a brine solution is generated during regeneration.

SYSTEM OF UNIT OPERATIONS (TYPICAL GROUNDWATER TREATMENT SCENARIO)

Generally, the primary objectives of a groundwater treatment facility are to process a specified flow rate of contaminated groundwater, provide for the removal of settleable and suspended solids, remove contaminants from the groundwater to specified levels, and discharge the treated water.

To accomplish these objectives, a groundwater treatment facility is often proposed to be constructed. A facility often includes various unit operations and associated piping, pumps and controls. A generalized schematic for a typical treatment system process for treatment of volatile organic compounds and suspended/settleable solids is provided in Figure 4. In this section we highlight certain aspects and features of the design which would be more fully described within formal design plans and specifications for the treatment facility. The core of this example of a generalized groundwater treatment system is an air stripping tower with off gas controls. The core of the system is supplemented with additional unit operations for the removal of suspended and settleable solids.

This example treatment system includes five major unit operations, namely: (1) gravity differential separation, (2) surge tank/surge pump; (3) cartridge filtration, (4) countercurrent packed bed air strip-

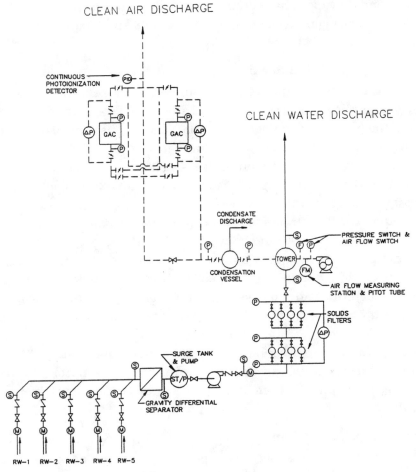

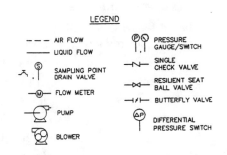

**Figure 4
Treatment Process Schematic**

ping, and (5) vapor phase granular activated carbon (GAC) adsorption. A brief discussion on the units that could be specified within the system are presented here.

Gravity Differential Separation

The high turbidity often encountered in extended groundwater may require control of settleable and suspended solids in the first stages of the process. A gravity differential separator is designed, based on the principles of Stokes Law, to allow particles greater than a specified mass and volume to drop out of the solution while experiencing laminar flow through the vessel. Stokes Law is an expression which describes the rate at which solids settle out or rise to the surface in a fluid based on the flow resistance between the particle and the fluid, under the influence of gravity and buoyancy. The driving force in this process is that of gravity acting on the particles with a resistance of the particles to settling due to their respective drag coefficients. Allowing the pumped groundwater to pass through the gravity differential separator upon entering the treatment process allows for the greatest removal of settleable solids. Further, removal of the majority of settleable solids prevents a buildup of such material within subsequent units. The separator is designed to provide for settled solids removal from the vessel by pumping off the bottom of the vessel or by "vacuuming" the bottom of the vessel through the access fittings.

Surge Tank/Surge Pump

Insofar as the initial design pumping rates are often insufficient to provide hydraulic control of the contaminant plume, provisions for allowing increased extraction rates are provided. However, the air stripping tower requires a constant flow rate (i.e., air to water ratio). Therefore, a flow equalization system is sometimes necessary. This system accommodates variations in the extraction pumping rates, while maintaining a constant flow rate to the stripping tower, which is necessary for removal efficiencies to be maintained. The surge tank and pump operate on a cycle using level controls to energize the pump. When the extraction pumping system is operating below the stripping tower pumping rate, the surge tank will fill without the feed pump to the tower operating. When the tank reaches its high level, the pump is energized and operates at the tower's designed influent flow rate, slowly emptying the tank until it reaches the low level mark, when the pump is deenergized.

Solids Filtration

Suspended solids within the groundwater as well as residual settleable solids not removed within the gravity differential separator are removed in the solids filtration units. The solids filtration system may consist of cartridge filters, bag filters, or plate filters. The simplest by far is a cartridge filter assembly. The cartridge filtration assembly is typically comprised of two or more parallel filters having a specified pore size filter fabric, followed in series by an additional two or more parallel filters having smaller pore size filter fabrics. Therefore, the cartridge filtration battery consists of primary filtration followed by polishing filtration. The efficiency of each filter will increase once online due to the nature of the filtered particles layering on the filter fabric around each pore opening, thereby decreasing the pore size. The cartridge filtration unit operation is designed with initially specified porosity filter fabrics. Based on data developed during the initial stages of treatment and data developed over the life of the treatment system, filter fabric porosities may be modified to suit the nature of groundwater being treated. The cartridge filtration units provide for this to be easily accomplished at any point during system operation.

Counter-Current Packed Bed Air Stripping

The air stripping tower is the first of two unit operations that make up the core of the treatment system. The tower is designed to allow the effluent to drain by gravity to the discharge location. With the two stages of solids removal preceding the air stripping tower, tower maintenance associated with fouling of the packing media is kept to a minimum.

Vapor Phase GAC Adsorption

To control the emissions from the air stripping tower, vapor phase granular activated carbon (GAC) is often utilized. The emission control typically comprises two pressure vessels in series preceded by a pressure condensation vessel. The pressure condensation vessel is designed to reduce the relative humidity of the offgas to below 50 percent. This is accomplished by passing the gas through the pressure vessel containing high surface area material. Under increased pressure, the humidity will condense on the high surface area material thereby reducing the relative humidity of the off gas. The condensate is removed by an underdrain and transferred to the surge tank or managed as waste.

The two GAC vessels, arranged in series, are valved to allow either vessel to be the lead vessel or the lag vessel. The operation typically requires the lead vessel to be taken off line for carbon replacement when it is spent. The second vessel, which previously operated as a lag vessel, then becomes the lead vessel. When the first vessel has had its carbon replenished, it becomes the lag vessel. This type of arrangement has two primary benefits: (1) that the vapor phase GAC can remain on line during carbon replacement, and (2) that the second (lag) vessel provides assurances that emission control will be maintained, even in the event of breakthrough on the lead vessel.

SUMMARY

Our current technologies for restoring the environment rest on the most basic natural processes and physical laws. Nature has enormous restorative capacities, although at many sites contamination so far exceeds these capacities that we must intervene to accelerate, focus, and assist the work of nature if we are to see any benefit in a realistic human time frame. We can expect the coming decades to accelerate both the pace of cleanups and the number of technological operations we have in designing them. Technologies previously utilized in other fields, as well as new emerging technologies, have added practicality and cost effectiveness to many remediation programs.

REFERENCES

Driscoll, F.G. "Groundwater and Wells -- Second Edition," Johnson Division, St. Paul, MN, 1987.

Montgomery, J.H., et al. "Groundwater Chemicals Desk Reference," Lewis Publishers, Chelsea, MI, 1990.

Mueller, James A. "Air Stripping of Toxic Organics from Water A-O Wastewater," Manhattan College, College of Engineering, Lecture Notes, 1987.

Nyer, Evan K. "Groundwater Treatment Technology," New York, Van Nostrand Reinhold, 1985.

Tabak, H.H., et al. "Biodegradability Studies with Organic Priority Pollutant Compounds," Journal of Water Pollution Control Federation, 53:10-1503, 1981.

Uhler, R.E. "Treatment Alternatives for Groundwater Contamination," J.M. Montgomery Engineers.

Verschueren, Karel. "Handbook of Environmental Data on Organic Chemicals," New York, Van Nostrand Reinhold, 1983.

Water Pollution Control Federation. "Groundwater, Why Should You Care?," 1987.

Weast, Robert. "Handbook of Chemistry and Physics," 60th Ed.: CRC Press, Inc., 1979, 1980.

U.S. EPA. "Carbon Adsorption Isotherms for Toxic Organics," Cincinnati, OH, 1980.

U.S. EPA. "Guidance for Conducting Remedial Investigations and Feasibility Studies Under CERCLA," Washington, D.C. 1988.

U.S. EPA. "Technologies for Upgrading Existing or Designing New Drinking Water Treatment Facilities," Cincinnati, OH, 1989.

11

POLLUTION PREVENTION THROUGH TOTAL QUALITY MANAGEMENT

Gary Vajda
Dames & Moore

ROLE OF POLLUTION PREVENTION

As we enter the 1990s, the development and implementation of an effective pollution prevention program will likely become the single most essential component of the successful corporate environmental program. "Success" will be measured in terms of both compliance and costs. Furthermore, the most successful pollution prevention programs will be those that comprehend the entire manufacturing process, not just the wastes that are generated.

The concept of "pollution prevention" has been heralded as the environmental focus of the 1990s, but the idea is neither new nor unique. Whether it's called "zero discharge," "waste minimization," or "emission reduction," the premise of prevention has been part of the environmental world since at least 1972 and the enactment of the Clean Water Act.

However, there is little doubt that it took the stiff regulations of the Resource Conservation and Recovery Act (RCRA), its 1984 Amendments, and even the more onerous liabilities of the Comprehensive Environmental Responsibility, Compensation and Liability Act (CERCLA) to make waste minimization, and now pollution prevention, ideas whose time had come. The Pollution Prevention Act of 1990, the recent stormwater regulations under the Clean Water Act, and the Clean Air Act Amendments of 1990 completed the statutory basis for a multimedia approach.

Another factor has played a role, however, and that is the concept of **Total Quality Management (TQM)** as a means of optimizing the productivity and efficiency of industry. When waste is recognized as a quality defect which can be reduced or eliminated, the circle is closed. By adapting traditional manufacturing quality control analytical methods to environmental factors, pollution prevention can be accomplished in the same manner as any other operational improvement.

The role of pollution prevention is clearly not just to achieve an environmentally desirable state of affairs, but to do so by increasing quality and productivity, and eventually, profits. This can be accomplished by integrating manufacturing and environmental issues.

DEFINITIONS

In its broadest context, pollution prevention can be defined as the reduction in volume and/or toxicity of a waste prior to discharge. This definition is generally modified to emphasize that pollution control (i.e. an end-of-pipe treatment) is not considered pollution prevention, and a thorough and comprehensive evaluation of the economic criteria will render this modification a moot point. This is because end-of-pipe treatment will almost always be the most expensive alternative when true lifetime costs are considered. When it is not the more expensive option, it should by all means be considered, since the overall objective is to obtain the "most bang for the buck."

Pollution prevention techniques traditionally fall under one of two categories — "source reduction" or "recycling." Source reduction, in turn, consists of "product changes" or "source control." (See Figure 1).

Product change involves modification of the end product to eliminate a waste. Two recent examples include a major fast food chain's widely publicized substitution of paper for styrofoam wrapping and the replacement of pump sprays for aerosols.

Source control comprises modifications to the production process itself. This may involve the technology, the raw materials, the equipment, and/or changes in operating practices. Changing from traditional spray painting to dry powder painting is an example of technology modification; closing the covers of unused solvent degreasers is an example of using operating practices to achieve source control. Substi-

Pollution Prevention Through Total Quality Management / 349

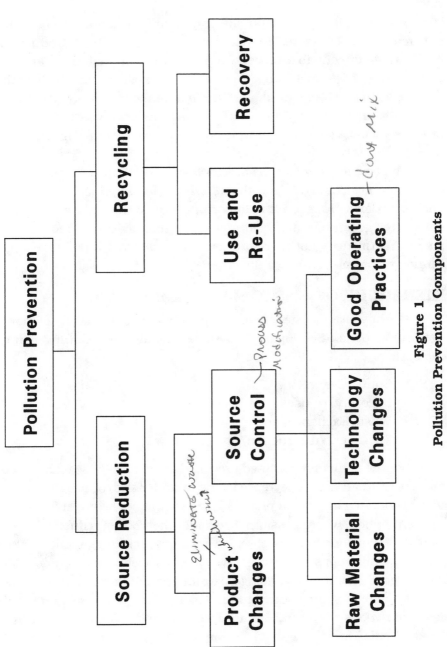

Figure 1
Pollution Prevention Components

tuting aqueous cleaners for chlorinated solvents represents a change of raw materials.

Recycling involves the on-site or off-site utilization, with or without treatment, of a waste. "Re-use" generally refers to the introduction of the waste directly to a process in place of the normal raw material. The smelting of plating sludge in a primary smelter to recover metals (i.e., replacing ore) is an example. "Reclamation" or "recovery" generally involves the separation of a particular component of a waste stream for re-use. The most common example is solvent recovery through distillation.

Treatment comprises the process(es) by which the characteristics of a waste are changed such that the material is reduced in volume and/or toxicity. However, it is important to note that treatment takes place after generation, not before, and is generally conducted "end of pipe." Treatment is not therefore usually considered a pollution prevention technique, although the end result may be the same.

HIERARCHY OF WASTE MANAGEMENT

For any given waste stream, there is a general hierarchy of management options. (See Figure 2). These include the following major categories:

- Source reduction,
- Recovery/re-use,
- Waste Exchange, and
- Treatment/destruction/disposal.

Source reduction, as noted previously, includes those techniques that result in an actual decrease in the quantity and/or toxicity of the waste generated by a given process. This should be the preferred management option, where possible. The major methods of reduction commonly used include product changes, revising operating practices, and process modifications (changes in raw materials, technology, equipment).

A good example of revising operating practices to minimize hazardous waste is source segregation, which refers to the separation of listed hazardous wastes from nonhazardous wastes. Because of RCRA's "mixture rule," any waste that is a combination of listed hazardous and nonhazardous waste is classified in its entirety as a listed hazardous waste. In practice, this means that if a plant combines a small quantity of electroplating rinse water treatment sludge (a listed hazardous waste)

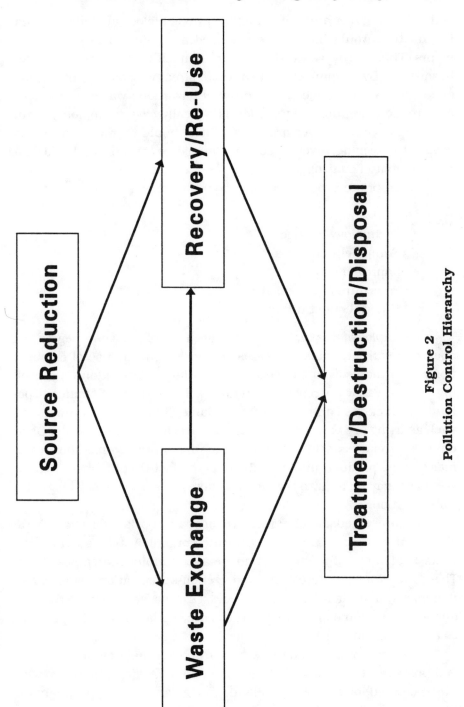

Figure 2
Pollution Control Hierarchy

with a much larger amount of nonhazardous biological sludge, the entire mixture would have to be managed according to RCRA requirements. This is trueeven if the mixture is itself characteristically nonhazardous. By keeping these two waste streams separate, the facility can significantly reduce the quantity of hazardous waste it produces. As a reduction option, segregation is typically easy to implement and very cost-effective. Other practices which should be reviewed include good housekeeping, inventory control, employee training, and spill/leak prevention and control.

Appropriate process modifications might include any of the following:

- Technology changes,
- Equipment changes,
- Improved control,
- Procedure changes, and
- Material substitutions.

Examples of this include the replacement of a solvent- degreasing operation with an aqueous degreasing operation; the use of combustion rather than solvents to clean paint racks; and the addition of cooling coils to vapor degreasers to reduce solvent losses. Significant potential reductions can result from fundamental changes in processes, such as using dry powder paints instead of solvent-based paints. Obviously, a primary consideration in evaluating a process modification is maintaining product quality. This, plus the fact that a substantial investment may be required, generally makes this approach a long-term consideration.

Recovery/reuse techniques are most appropriate for two general classes of wastes: relatively concentrated organic waste, such as solvents, coolants, and waste oils; and metal-containing inorganic wastes. There is a growing interest in the use of off-specification materials either as raw materials for similar processes or as feedstocks to the production of the same material. There is little overlapping of the processes applicable for each.

For waste oil, halogenated solvents, and non-halogenated solvents, four processes have been most commonly identified in conjunction with their potential for recovery. Distillation is by far the most prevalent for solvent recovery because of the relative ease of separation by this means.

Units of various sizes and temperature ranges are readily available for in-plant use. Distillation is also common for waste oil, although generally at a larger scale than is commonly encountered at an individual plant because of the higher temperatures needed. Solvent extraction also has, until recently, been limited to larger scale uses and to only a limited number of organics, particularly phenol. Advances in equipment have expanded this usage somewhat. Carbon adsorption is most commonly utilized for vapor phase solvent recovery and for low concentration aqueous streams. Ultrafiltration, a low pressure membrane separation process, is applicable to the cleaning of coolant oils and the recovery of tramp oil from aqueous coolant mixtures.

The most likely candidates among inorganic wastes for recovery/reuse include the caustic and acidic waste streams resulting from metal finishing operations (e.g., caustic cleaners, phosphate solutions, plating solutions, and pickle liquors). They, of course, have some immediate value as neutralizing agents, but other possibilities include recovery of concentrated metal salts from plating baths and the sale of ferric chloride or sulfatepickle liquors to municipal wastewater treatment plants for phosphate removal. The advantage of this approach is twofold when applicable. Not only is a waste stream recovered for a useful purpose, but the metal content is reduced or eliminated from the wastewater treatment sludge where it would have inevitably ended up.

The processes that can be utilized for the recovery of metals include ion exchange, membrane processes (reverse osmosis, ultrafiltration, electrolysis), and evaporation. All are proven processes, although not necessarily for all applications, and are readily available commercially.

A waste exchange is an administrative technique based upon the premise that "one man's junk is another man's gold." It is essentially a brokerage that handles waste materials that have the potential of being used as raw materials by another company. It then becomes identical to recover/reuse. The feasibility of this operation is largely dependent upon locale, i.e., the distance between a generator and a potential user of a waste material. The potential for waste exchange for any specific waste material is a function of several criteria: the value of the raw material, its purity, the quantity available, and the reliability of generation. The most commonly exchanged wastes include solvents, oil, concentrated acids/alkalis, catalysts, and precious metals.

After attempts are made to reduce waste at the beginning of the process, there is still likely to be a residue that cannot be recovered, reused, exchanged, or otherwise managed. At this point, treatment/destruction/disposal options should be examined.

There is a hierarchy of methods that should be considered. For organic materials, the use of incineration or boilers are an effective, if expensive, means of disposal. For inorganic chemicals and wastes, chemical/physical treatment technologies generally must be utilized to reduce either the volume or toxicity of the wastes. Landfill options should only be considered for those residues that have been rendered essentially inert.

IMPLEMENTING A PROGRAM

Economic Concerns

There are many sound reasons why companies should be interested in developing a pollution prevention program. One of the most persuasive is economics. There is little doubt that in the long term, prevention is always less expensive than responding after the fact. However, these savings may prove to be short term, as well. With respect to hazardous waste, there is only limited waste management capacity, and as that capacity becomes filled, actual disposal costs are naturally going to rise. There are also the long-term "hidden" costs associated with the continued liability for any mismanagement of the wastes (or even proper management in accordance with obsolete rules) in years to come. These costs may be orders of magnitude greater than the original disposal costs. The situation is similar in the areas of water and air emissions primarily because there is an apparent trend to significantly increase permit and other fees on the basis of volume.

There are additional economic pluses, some of which can be substantial. With proper planning, a company can utilize a pollution prevention program to reduce raw material purchases, limit the diverse productive capacity and optimize work in progress. Inventory control, in particular, can result in significant cost savings stemming from two factors. First, less material has to be purchased and stored at any one time; and second, disposal of obsolete materials is minimized because only what is needed is purchased.

Other benefits that can result from an integrated approach to pollution prevention include the following:

- More effective process control,
- Increased product quality, and
- Better means of strategic planning to identify high priority, high return projects or alternatives.

All of these economic benefits can result from the understanding that waste is a quality defect, and that it must be eliminated as programmatically as any other defect would be eliminated in the manufacturing process.

Political Pressures

Simply put, pollution prevention is the politically correct thing to do in the '90s and beyond. This is a natural result of two growing trends in public attitude. The first is a desire to avoid waste and its disposal in any shape, form or manner. This attitude is manifested at one extreme in the continual series of battles between communities or environmental groups, and any companies that want to build or operate a waste disposal facility. At the other end of the spectrum, this trend is illustrated by the "greening" of the marketplace, i.e., the use of materials, that when disposed, are supposed to biodegrade or otherwise just disappear.

The second trend is a general concern for conserving natural resources. The "green market" is also an example of this, as is the current emphasis on recycling, re-using or recovering a plethora of household items once simply tossed away.

Because pollution prevention can directly or indirectly positively respond to both of these trends, a successful program can have substantial public relations value. Although not easily calculable, this value can be seen in better community relations, better sales and better agency relations, and perhaps, better product sales.

REGULATORY

Over the last two decades, but particularly in the last several years, Congress and the U.S. EPA have developed and begun to implement a number of statutes and regulations which directly or indirectly foster pollution prevention. Many states have followed suit. The means of accomplishing this have included both "large stick" and "carrot" approaches.

By far the largest "sticks" for hazardous wastes have been the Resource Conservation and Recovery Act (RCRA) and the Comprehensive Environmental Response, Compensation and Liability Act (CERCLA), or "Superfund"), and their subsequent reauthorizing amendments, the Hazardous and Solid Waste Amendments of 1984 (HSWA), and the Superfund Amendments and Reauthorization Act of 1986 (SARA), respectively. The first decreed what you did with waste; the latter indicated what it would cost if you did not do it right.

Over the past several years, HSWA has mandated a number of regulations that have had a significant impact on how companies manage their hazardous wastes. These regulations have resulted inthe following: land disposal prohibitions, an expanded listing of wastes categorized as hazardous; strict management and design criteria for treatment/disposal facilities; and increased responsibilities for the waste generator. The RCRA is now before Congress for re-authorization; there is no doubt that its regulations will continue, and that its emphasis will be on eliminating rather than controlling waste.

The United States Environmental Protection Agency (EPA) has completed its mandated schedule to review and determine which hazardous wastes will be permitted disposal on land, and under what conditions. Experience indicates that there are problems when generators discover that their waste is prohibited, and/or does not meet the specified treatment criteria. They rush to find an alternative, and their waste disposal becomes costly as a result.

At the same time that the disposal of hazardous wastes is being restricted, the number of wastes identified via listings or characteristics by the EPA is also being expanded. New information on specific wastes (i.e. reducing the applicability of the Bevil exclusion, which exempts certain mining wastes from RCRA standards), and new characterization parameters and methods (e.g., the Toxic Characteristic Leaching Procedure, or TCLP), have resulted in the addition of more waste streams to the hazardous category.

Under HSWA, the federal regulatory requirement for waste minimization sounds simple. Generators must certify on their manifests that they have in place a program "...to reduce the volume and toxicity of wastes generated to the degree...practicable." Generators must also describe in biannual (or annual) reports, the contents and results of their waste minimization programs. The determination of what is practicable will typically be made primarily on the basis of relatively short-term economics by the company itself. To date, this requirement has been relatively toothless.

The EPA is currently reviewing its options for providing increased enforcement and/or encouragement to companies to do more than superficially comply with this requirement. These have included the inception of an Office of Pollution Prevention, the initiation of various research and development programs, and RCRA inspections of waste generators to make sure they are in compliance with the waste minimization requirement.

CERCLA is the statute responsible for the "hidden" costs of waste management referred to earlier. It places retroactive responsibility for problems resulting from the past waste management practices on the generator of the waste, regardless of fault in causing the problem. The costs to investigate, negotiate and remediate these problems have been staggering.

SARA, on the other hand, has provided at least one tool to many generators to assist them in understanding their status with regard to waste. Under the Section 313 ("Form R") program, companies have been forced to perform at least rudimentary mass balances of their facilities to determine what comes in, what leaves and where it goes. This regulation comprehends emissions to all media, not just solid waste.

The EPA and Congress have also embarked on a program of encouragingand/or mandating the concept of pollution prevention. The Pollution Prevention Act of 1990 directs the U.S. EPA to emphasize source reduction in order to prevent pollution. They are to establish standards, consider how to encourage pollution prevention within other statutory programs and assess progress. One means to accomplish the latter was to revise Form R to include pollution prevention information.

In addition to the passage of the Pollution Prevention Act of 1990, pollution prevention incentives are found in the Clean Air Act Amendments of 1990 wherein it is possible to delay compliance with future standards by agreeing and implementing source reduction now. Pollution Prevention Plans are also important components of the two most recent regulations from the Clean Water Act — stormwater permitting and pre-treatment.

The regulatory trends are, therefore, apparent. In the 1990s, generators are going to have to concentrate on prevention and not just proper treatment and disposal. The objectives of an effective waste management program should be to:

- Minimize waste generation,
- Minimize off-site disposal, and
- Minimize the long-term liability associated with such disposal.

The goal of a pollution prevention program is to address these objectives in a cost-effective manner. Reducing the generation of wastes at the source is preferred, both from a regulatory and from the generator's perspective. However, it must be recognized that with most processes and operations, some waste will be produced. Off-site disposal, particularly in landfills, should be considered only in the context of a careful review of all other options. In many instances, a residue will remain that can only be disposed of on land. The objective should then be to ensure that this residual matter is as environmentally harmless as possible.

POTENTIAL OBSTACLES

Given all of the reasons listed above for implementing a pollution prevention program, the fact remains that surprisingly little has been done in this regard. This is because, although almost everyone agrees that pollution prevention is "Mom and apple pie," there are a number of obstacles between the intention and the successful implementation. This is largely due to the fact that in most operations, responsibilities are fragmented. Departments are run with different agendas and objectives, and communications within the facility are often ineffective. Even if the company has a department handling environmental affairs, responsibilities might be divided by media (i.e. air, water, waste). This results in a number of potential barriers to a successful pollution prevention program. These can generally be categorized as follows:

Economic
- There is not enough money to fund the project.
- Existing stocks of raw materials will delay substitutes that may help the program.
- It is not worth the effort by the employees.

Jurisdiction
- Use of a different raw material will have an adverse impact on the process.

- Adequate space, utilities, manpower, etc., are not available.
- Accepting another plant's feedstock will require extensive regulatory difficulties.
- New processes mean more regulation.

Production
- A new procedure will be a bottleneck.
- Production will be stopped for installation.
- The new equipment may not work.
- Product quality will deteriorate.
- Customers will not accept the new product characteristics.
- Attitude
- If it ain't broke, don't fix it!

Any or all of these concerns may be valid, but they can be overcome. To be effective, a pollution prevention program must look at all of the pieces of this puzzle. One way to do this is to use the analytical tools and techniques associated with traditional industrial engineering and quality control, to include environmental factors in the analysis together with the production factors. This, in essence, describes the **integrated manufacturing and environmental (IME)** approach to pollution prevention.

INTEGRATING MANUFACTURING AND ENVIRONMENTAL ISSUES

For many years, factors attributable to environmental issues were not considered significant components of productivity and operating costs; instead, management focused on labor, materials and rework. Today, however, environmentally driven costs, ranging from waste disposal to future Superfund liabilities to the permitting costs of expanding a production operation, must also be considered by plant and corporate management.

An approach to accomplishing this is found in several of the techniques and methods used historically to analyze manufacturing operations for productivity and quality improvement. These methodologies include: functional systems analysis, statistical sampling programs, **design of experiments (DOE),** traditional alternatives analysis, and structured decision-making techniques. When applied to the objective of pollution prevention, the IME approach allows the company to evaluate the entire manufacturing system, not just the components.

SETTING THE STAGE

Before beginning the technical aspects of the IME program, it is essential to set the stage for success. As with any potentially controversial program, good planning and organization can be the difference between achieving success and confirming the predictions of nay-sayers. Experience has demonstrated that four steps are almost invariably present in a successful pollution prevention program:

- Obtain the commitment,
- Establish the goals,
- Identify the champion(s), and
- Staff the team.

The key to a successful pollution prevention program is a company-wide commitment — from the management, the implementors, and the employees. It is absolutely essential that all three of these groups buy into the program. Management is necessary to provide resources, both dollars and people, and symbolic support. Employees must be committed because they are the ones actually implementing the program, and will probably have to change the way they do their jobs. The most elaborate of plans will not succeed unless the employees on the production floor actively participate. Finally, the implementors are needed to keep the process always moving forward, despite the inevitable glitches which occur.

The first step in accomplishing this is to establish the overall goals of the program. This should be developed first by senior management and then filtered down. The senior management-stated goal(s) may start as relatively broad and generic, but as the goal(s) are massaged, they must be made more specific. Whether large or small, these goals must be acceptable, achievable and measurable. The people involved must believe that where they are going is reachable, and that there is a verifiable means to document progress. Once set, these program goals can then be incorporated into individual division, department and/or employee goals.

The second step is to identify one or more individuals to lead the program team. This person must be familiar with the facility, with waste management and production technologies, and with product requirements. He or she must also have a good rapport with both management and employees and be able to bring a diverse group of people together.

Finally, the project team must have input from many departments and specialties, including:

- Production,
- Facilities/maintenance,
- Process engineering,
- Quality control,
- Environmental,
- Research and development,
- Safety/health,
- Marketing,
- Purchasing,
- Material control,
- Legal,
- Finance, and
- Information systems.

Often it is useful, or even necessary, to bring in outside assistance to provide expertise and/or a broader perspective. However, it is essential that key input and direction be provided internally.

BASELINE CHARACTERIZATION

Once the goals and team have been established, the next step is to perform a baseline characterization of the facility. The purpose of this effort is to define the production process(es), as well as ascertain the current regulatory status of the facility. This assessment can be focused on a particular process or on a single medium, but is most effective when used to evaluate the facility as a whole in order not to overlook potential opportunities. It isessential to realistically characterize the operations and establish an accurate baseline, since it is impossible to know how much progress has been made without first knowing the starting point.

In the baseline characterization, both environmental and production data are evaluated. The following types of information should be compiled if possible:

- Process flow charts,
- Piping and instrumentation diagrams,
- Plant layouts,
- Equipment design and operating specifications,

- Labor utilization,
- Labor and utility costs,
- Raw materials usage and costs,
- Effluent and emission data,
- Production rates,
- Standard operating practices,
- Reject rates, and
- Regulatory requirements — current and future.

Much of this information should be relatively available; some may require sampling or other means of direct measurement.

All of this information then needs to be put back together in a "model" which integrates and describes the environmental and operational aspects of the facility or process. This "model" need only be as complicated as is required for a specific circumstance. For example, a complex computer model is neither warranted nor appropriate for a simple plating operation, no matter how large. Conversely, to attempt to assimilate the necessary information for a large multi-operation manufacturing or chemical plant without the aid of at least a simple data software package is not realistic.

In essence, this model should represent a "resource balance" for the facility or process. It will include not only materials, but also energy, labor, and any design constraints.

For simpler operations, this should not be too difficult a problem. A good starting point, when available, is Section 313 of Form R. However, to this must be added the "non-material" aspects of the balance discussed previously.

For smaller or more focused projects, individual processes and functions can be analyzed utilizing several analytical techniques. Proven methods include:

- Statistical sampling programs or charting of historical data, and
- Taguchi and other "Design of Experiments" process diagnostic techniques, which are empirically based methods involving making changes, observing the results, and making further changes until a goal is reached.

For the more complex operations, more exotic modeling tools may be appropriate. One useful analytical technique is "function analysis."

This method, an example of which is Integrated Computer-Aided Manufacturing Definition Methodology (IDEF), permits defining any process by a series of functions or nodes (See Figure 3). Each node can be described in terms of inputs, outputs, the controls or constraints on the function, and the mechanisms of transformation.

In addition to the "resource balance," it is also important toestablish an accurate cost for the product or process. Again, there are any number of techniques available to accomplish this, from simple to complex (see Figure 4). In all, however, it is important to remember that the accuracy of the end result (i.e., the "true" cost) is dependent on the completeness and accuracy of the input criteria. For example, often left out of traditional product cost analyses are costs for items such as: regulatory documentation and compliance (time and materials); waste disposal; disposal liability; and the cost of product rejects (material, labor, lost capacity). (See Figure 5.)

The added value of a baseline cost model is to help identify those operations that are high in cost. Analytical resources can then be focused on those operations or variables that yield the greatest return, environmentally and/or operationally. Among the modeling tools that can be used on more complex efforts is a software package called the Standard Assembly Line Manufacturing Industrial Simulation (SAMIS), which calculates a unit production cost for a process or system. The value can then be used to compare improvement alternatives developed in the next phase.

ALTERNATIVES DEVELOPMENT

Once the baseline characterization model(s) have been completed and validated, then various production improvement and environmental compliance alternatives can be developed and evaluated. Typically, a range of alternatives that partially or completely achieve the previously established pollution prevention goals are identified. Each alternative is then modeled, utilizing the same techniques that were used in the baseline characterization.

The alternatives can be compared in a number of ways. Ideally, the favored approach to decision-making in an uncertain environment is a blend of qualitative and quantitative methods to maximize benefits. Traditional techniques, such as rate of return, minimum capital, unit cost or life cycle cost may be appropriate in many instances. However, a methodology known as structured decision-making can play an im-

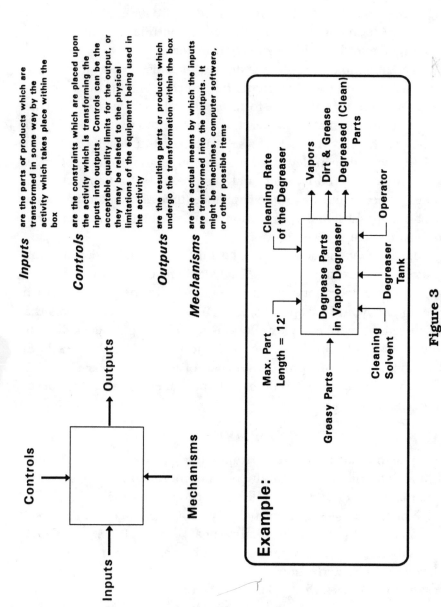

Figure 3
Functional Analysis Diagram

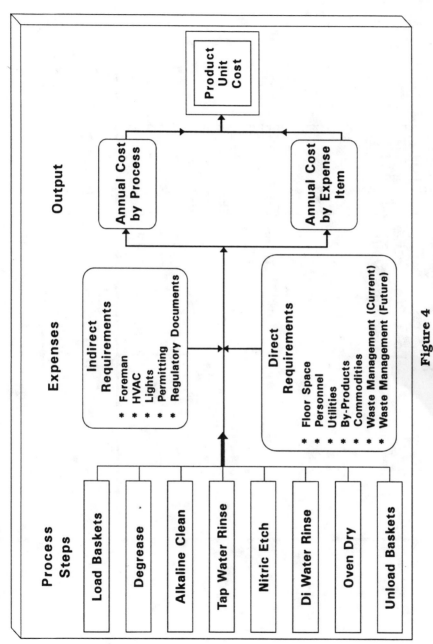

Figure 4
Components of Product Cost

366 / Environmental Science and Technology Handbook

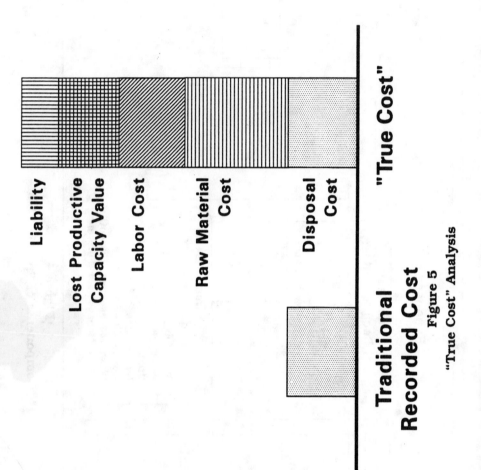

Figure 5
"True Cost" Analysis

portant role, particularly when the process or the alternative are complex, or the selection cannot be made solely on cost factors.

In general, in structured decision-making an alternative's viability depends on its respective environmental and operational improvement potentials. These can be judged from a number of factors deemed appropriate by the user and may include such items as:

Environmental Improvement
- Amount and type of pollutant,
- Potential for disruption or legal action,
- Level of mitigation technology,
- Compliance safety margin provided by control technology, and
- Permitability of technology.

Operational Improvement
- Contribution to increasing inventory turns or reducing lead times (speed),
- Reduction in set up times, product differentiation and mix (flexibility),
- Reduction of scrap and/or rework (quality), and
- Lower operating and capital costs.

An example of these calculations is provided in Figures 6 and 7.

In another example (see Figure 8), three alternatives are presented as noted. The initial cost of each alternative is indicated as proportional to the size of the circle. Alternative No. 1 is the least expensive, provides some environmental improvement, but little operational improvement. This alternative represents a typical "end of pipe" pollution control device, such as a solvent fume scrubber. Alternative No. 3, on the other hand, involves a significantly higher capital expenditure which also significantly improves operations, as well as the environment. An example might be the replacement of solvent painting operations at this facility with a dry powder painting process. Alternative No. 2, while costing just a little more than alternative No. 1, is much better for the environment and improves operations somewhat, as compared to alternative No. 1. This alternative might represent the use of activated carbon adsorbers to recover solvent fumes for re-use in the plant.

If capital cost is the only factor, then the short-sighted selection might be alternative No. 1. However, if initial cost is only a major selection criteria, then alternative No. 2 is the apparent choice. Alternative

(1) Criteria	(2) Weight (0-1)	(3) Baseline Performance	(4) Alternatives Degrade −10pts / No Change 0 / Improve +10pts	(5) Improvement Score [(2) X (4)]
Pollutant	0.2	1,1,1-TCE 2,000 ppm	500 ppm + 7 pts	1.4
Disruption	0.3	Moderate	Low + 5 pts	1.5
Technology	0.1	None	BACT + 9 pts	0.8
Safety Margin	0.1	None	High + 50% + 8 pts	0.9
Permeability	0.3	Yes	High + 8 pts	2.4
Total	1.0		Improvement Index	7.0

Figure 6
Evaluating Potential for Environmental Improvement
Example: Replacing Solvent Dip Tank with Vapor Degreaser

(1) Criteria	(2) Weight (0-1)	(3) Baseline Performance	(4) Alternatives Degrade -10pts / No Change 0 / Improve +10pts	(5) Improvement Score [(2) X (4)]
Cost	0.4	$10/part	$8/part +20% +2pts	0.8
Speed	0.2	14 day cue	5 days +64% +8pts	1.6
Quality	0.2	10/1,000	10/1,000 +0% +0pts	0
Flexibility	0.2	Limited	High +8pts	1.6
Total	1.0		Improvement Index	4.0

Figure 7
Evaluating Potential for Operational Improvement
Example: Replacing Solvent Dip Tank with Vapor Degreaser

370 / Environmental Science and Technology Handbook

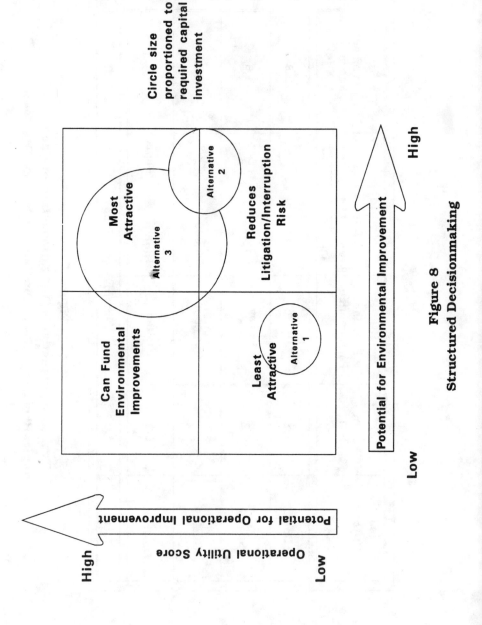

Figure 8
Structured Decisionmaking

No. 3 might be chosen in the situation when other factors (such as public relations, or life cycle costs, or unit costs) override initial capital costs.

With an "end of pipe" solution to an environmental compliance program, no benefit from operational improvement is typically achieved. In addition, negative cash flow is experienced from the purchase and operation of the new control technology equipment. However, a solution combining both operational improvement and environmental improvement can offer an increase in productivity with a return on invested capital.

CASE STUDIES

This approach has been successfully used in both large and small firms, and in industries as diverse as metal finishing to consumer product packaging. Here are some examples of how it has worked.

Plastic Container Manufacturing Facility

The objective of this program was focused on reducing the scrap rate associated with the production of plastic containers. The study had two major components. The first was to identify which of the five primary systems along each of four production lines was the source of the majority of the scrap losses. The second was to identify which variable(s) were causing the losses. Statistical sampling was performed in an attempt to assess any positional (along the line), temporal (time dependent), or cyclical (beginning, middle, or end of a production run) tendencies. This was performed over the course of two weeks, on two shifts, on four production lines for five products, and at five stations on each line. A valuable aspect of the sampling phase was time spent observing the operations, procedures followed, and reactions to problems. The data was plotted and several conclusions drawn about where further effort was to be focused and the range of potentially contributing variables to scrap loss.

Approximately 30 variables were identified as contributors to waste generation. It was impractical to test a range of settings for each variable, so a survey/feedback approach was used to narrow the list of 30 down to eight primary variables that would be the focus of the design of experiments. Another constraint on experimentation was to minimize the amount of down-time experienced by the line on which experiments were conducted. Therefore, close involvement was solicited from production scheduling, plant engineering, maintenance, and operations, as to test sequences, test set points for each variable, etc.

The designed experiments were run with a minimum of downtime on the line, less than one-third of the time that was budgeted. Data were analyzed for both primary and interactive effects of the variables. This analysis of variance showed three variables as primary contributors to waste generation.

A written report was prepared to document what was done, how it was done, and the resulting conclusions. Most importantly, however, oral presentations and discussions were held for management, engineering personnel, and key operators on all four lines on all shifts.

The methodologies used included: statistical sampling, multi-variant analysis, survey-feedback, design of experiments, analysis of variance, and direct operations and maintenance involvement.

The key results included: minimal down-time spent to allow sampling and testing to be carried out; and a recommendation resulting in an immediate 70 percent reduction in waste material generated. Simple payback for this was less than 60 days, including the cost of the study and the value of lost production during the down-time. Although this project did not initially have an environmental goal, the end result certainly was environmentally beneficial, and the company did take some bows for it.

Aircraft Component Manufacturing Complex

This project involved a rather old manufacturing facility that had reached the breaking point with the local air regulatory agency. The company needed to make a decision whether to close, move, or modernize.

The objective of this effort was, to first, quantify the environmental compliance concerns, and then to develop a plan to identify and recommend measures (equipment, facilities, procedures, systems, substitute materials, etc.) which addressed both compliance and factory modernization. The scope of work involved completing an integrated manufacturing and environmental evaluation of this 13,000-employee aircraft component fabrication and assembly complex.

Over 400 production functions were evaluated for environmental concerns and operational improvement potential by adapting a structured system analysis methodology for use by a combined team of environmental, health and safety, manufacturing, and facilities engineers.

The types of operations audited included painting, curing, bonding, composite materials fabrication, foundries, degreasing, metal cleaning (acid and alkaline etch), chemical milling, metal removal, metal cutting, metal joining, and electroplating.

Efficient data analysis, decision making, and prioritization processes were keys to managing this effort. Data were gathered and entered or referenced into a PC-based relational database. The key identifier for all data gathered was a number assigned to each of the more than 400 operating functions. And, because of the relational capabilities of the database, queries were made and reports produced on the basis of regulation, department number, category of emission or discharge, mitigative action strategy, etc. Categories of data gathered included permits, **Material Safety Data Sheet (MSDS)** information, applicable regulations (federal, state and local), production throughput and operating cost data, facility data, such as process tank surface area and paint booth size, and cost estimates for compliance measures.

The methodologies used included: IDEF Function Analysis Methodology; SAMIS Cost Modeling (environmental and production operating costs); a relational database for environmental, operational, and alternative action strategy; cost data; and structured decision-making.

The key results included: a total estimated installed cost of implementing needed control technology that was one-third that of previous internal estimates, and development of a decision model to assist in ranking prospective projects on the basis of potential improvements to both environmental and manufacturing operations. This would allow screening of low or negative project alternatives that may otherwise be equal in their ability to mitigate an environmental, health, or safety problem. In addition, a group of 24 "fast track" projects were identified and defined to the point that further engineering definition was not needed prior to the submission of a capital expenditure request. An operating cost model was also developed to serve as a baseline against which future improvements will be measured.

Metal Product Manufacturing Facility

The objectives of this effort were to establish for a multi-facility metal product manufacturing company the potential air emission limits under the **Maximum Achievable Control Technology (MACT)** provisions of the Clean Air Act Amendments, and then to estimate the cost to implement it. A secondary objective of the project was to identify activities that could be conducted now that would result in significant cost reductions.

After completing a general survey of the facilities with respect to operations, emissions and materials, it was determined to concentrate on one unit process - vapor degreasing with 1,1,1-trichlorethane (TCA) - since it was common to most of the facilities and resulted in a large proportion of the VOC emissions.

It was also decided to focus on one "typical" plant. This facility operated 35 degreasers with individual capacities of five to 300 gallons. These degreasers were used over three shifts, seven days a week. This resulted in over 300 tons of TCA being emitted per year.

Due to the focused nature of this effort, the baseline characterization was performed using a questionnaire, operator interviews, and vendor information and observations. Information was compiled on the following: the design and operating characteristics of each unit and its operators; emission rates and solvent usage; anomalies between recommended and actual operations; and operator and vendor recommendations for improvement.

A list of recommendations ranging from the elimination of certain units from operator training to operational modifications (regulatecrane speed, improve instrumentation, perform regular maintenance) and unit modifications (install covers, increase cooling capacity, disconnect fans, increase freeboard) was compiled. A capital cost of approximately $1 million was estimated to implement all of these changes, which were expected to result in a nearly 50 percent reduction in TCA emissions. The cost was too great. An analysis of alternatives comprised of limiting equipment changes to the most used units indicated that approximately the same TCA reduction could be achieved for one-quarter less initial cost (i.e., about $750,000). Moreover, when the life cycle true cost of the non-emitted TCA was calculated, the payback period was just over one year.

REFERENCES:

1. United States Air Force, Wright-Patterson Air Force Base, *Integrated Computer-Aided Manufacturing (ICAM), Function Modeling Manual (IDEF). UM110231100*, (June, 1981).
2. Jet Propulsion Laboratory, Pasadena, California, *Standard Assembly-Line Manufacturing Industry Simulation (SAMIS) PC User's Guide, SAMIS Release 6.0*, (December, 1985).

3. Vajda, G.F. and Stouch, J.C., *"An Integrated Approach to Waste Minimization"* paper presented at Air & Waste Management Association Conference, Pittsburgh, Pennsylvania, (1990).
4. Stouch, J.C. and Schmidt, P.F., *"An Integrated Approach to Pollution Prevention: From Regulatory Requirements to Shop Floor Benefits"*, HazMat Central, Chicago, Illinois, (1991).
5. Vajda, G.F., *"Overcoming Obstacles to Pollution Prevention"* paper presented HazTech International Conference, Pittsburgh, Pennsylvania (1992).

[pollprev.app]

INDEX

A

Aboveground storage tank, 231-246
    regulation of, 232-233
    Spill Prevention, Control, and Countermeasures Plan, preparation of, 240-242
    standards, 232-233
    system design, 233-240
    underground storage tank, benefits compared, 242-246
Absorbed dose, health risk, 64
Absorption, gaseous air pollutant control, 169-170, 173
Activated carbon adsorption process, waste treatment/disposal, 196
Add-on air pollution control, 161-174
    gaseous pollutant control, 162, 169-174
        absorption, 169-170, 173
        adsorption, 169-170, 173
        catalytic oxidation, 170-172, 174
    particulate control, 161-169
        cyclone, 161-163, 166
        electrostatic precipitator, 162, 165-166, 168-169
        fabric filter, 162, 164, 166-167
        wet scrubber, 162, 166, 168
Administered dose, health risk, 64
Adsorption, 3
    of chemical in water-soil system, 28
    gaseous air pollutant control, 169-170, 173
    suspended sediment, 8
Advection, 2
Air/liquid chemical partitioning, air migration, 5
Air migration control, 4-7
    evaporation into air, chemical spill, 6-7
    partitioning of chemical, between liquid/air, 5
Air pollutant. *See* Air quality
Air quality, 125-153
    ambient air quality evaluation, 135-139
    atmospheric dispersion, 131-132
    chemical analysis, 102-103
    Clean Air Act, 99, 102-103, 141-142, 148, 150-152, 155, 348
        hazardous air pollutant control requirements, 151-152
        operating permit provisions, 150-151
    continuous emission monitoring systems, 130
    depletion, 133
    emission control methods, 133-135
    emission factors, 130
    emission offsets, 147-149
    federal regulations, 143-144
        attainment/nonattainment area, 144
        pollutant type, 143-144
        types of, 143
    indoor, 140-141
    industrial stationary emission pollution source, 128-129
    local regulation, 145
    mass balance, emissions, 130-131
    materials data, 131
    mobile emission pollution source, 127
    non-industrial stationary emission pollution source, 128
    ozone, depletion of, 139-140
    permitting process, 149-150
    pollutant characterization, 155-157
    pollution control technology, 155-175
        add-on pollution control, 161-174
        in-process pollution control, 157-161
    pollution emission rates, stationary emission source, 129-131
    prevention of significant deterioration concept, 145-147
    regulation of, 141-152
    sources of pollution, 127-129
    stack test for emissions, 130
    state regulation, 145
    stratospheric ozone depletion, 139-140
    transformation, 133
    types of pollutants, 126-127
    vendor data, emissions, 131
Air stripping
    with carbon adsorption, groundwater pollution, 335-336
    groundwater pollution, 328-334
Aircraft component manufacturing complex, pollution prevention program, 372-373
Alternatives development, pollution prevention program, 363, 367-371
Ambient air quality evaluation, 135-139
Anticline, structural geology, 276, 278-279
Aquatic organism, toxicity criteria, 88-90

Index / 377

Aquifier, 20
Aquitard, 20
Artesian groundwater, 20-21
Atmosphere, of earth, 252
Atmospheric dispersion, 131-132
Atomic absorption spectrophotometer, chemical analysis, 113
Attainment area, air quality, 144
Attenuation/excavation with off-site disposal, waste treatment/disposal, 204-205

B

Baseline characterization of facility, pollution prevention program, 361-366
Biodegradation, chemical behavior in water-soil system, 31-32
Biological stressor, ecological risk assessment, 76
Biological treatment
   groundwater pollution, 327-328
   waste treatment/disposal, 191-193
     bioreactor, 191-192
     composting, 192-193
     in situ bioremediation, 193
     soil heaping, 192
     solid phase bioremediation, 192
Bioreactor, waste treatment/disposal, 191-192
Bowen's reaction series, 260-261
Buffhide Tank, underground storage tank, 212

C

Calibration, chemical analysis, 120-121
Carbon adsorption
   with air stripping, groundwater pollution, 335-336
   groundwater pollution, 334-335
Carbonate, rock-forming mineral, 258
Carcinogenic health risk, 46-48, 63-64
Carcinogenicity classification scheme, Environmental Protection Agency, 47
Catalytic oxidation, gaseous air pollutant control, 170-172, 174
CERCLA. See Superfund
Chain of custody form, chemical analysis, 116-117
Characterization of facility, baseline, pollution prevention program, 361-366
Chemical
   evaporation into air, 6-7
   partitioning of, between liquid/air, 5
   photolysis of, 9
   in surface water, fate of, 8-9
   transport of, through groundwater advection, 23
   volatilization of, 9
   without toxicity values, 50-51
Chemical analysis
   air quality, 102-103
   atomic absorption spectrophotometer, 113
   calibration, 120-121
   chain of custody form, 116-117
   chemical bond, 104
   chemical compound, 104
   corrosivity, 100-101
   covalent bond, 104
   data validation, 122
   deliverables, 121-122
   documentation, 116-117, 119
   drinking water, 99-100
   duplicate analysis, 120
   elements, 103-104
   environmental chemistry, environmental regulation, 98-99
   EPA Contract Laboratory Protocol, 105
   gas chromatography, 101, 106-111
   graphite furnace atomic absorption, 113-114
   hazardous waste analysis, 100-102
   high pressure liquid chromatography, 111-112
   holding time, 121
   ignitability, 100
   inorganic chemical analysis instrumentation, 113-115
   internal standard area, 120
   ion, 104
   ionic bond, 104
   laboratory operations, 103-104
   mass spectrometry, 101
     with gas chromatography, 110-111
   maximum contaminant level, drinking water, 100
   methods, 104-105
   molecule, 104
   organic analysis instrumentation, 105-106
   quality control, 114, 117-118
   sample preparation, 107
   spiked sample, 118
   standard operating procedure, 115
   storage blanks, 117
   surrogate standard, 118
   synthetic organic chemical, drinking water, 99
   volatile synthetic organic chemical, 99
   water pollution, 102
Chemical behavior in water-soil system, 27-32
   adsorption, 28
   biodegradation, 31-32
   ion exchange, 28-29
   retardation, 27, 30

378 / Environmental Science and Technology Handbook

Chemical bond, 104
Chemical compound, 104
Chemical elements, 103-104
Chemical intake estimation
   ecological risk assessment, 85-86
   health risk, 59
Chemical of concern, 37, 80-82
Chemical of potential concern, selection of, health risk analysis, 44-45
Chemical/physical processes, waste treatment/disposal, 193-200
   dechlorinization, 199
   soil flushing, 197
   soil vapor extraction, 198-199
   soil washing, 196-197
   traditional methods, 193-196
      activated carbon adsorption, 196
      densification, 195
      dewatering, 194-195
      ion exchange, 196
      neutralization, 195
      oxidation/reduction, 195-196
      separation, 193-191
Clean Air Act, 99, 102-103, 141-142, 148, 150-152, 155, 348
   hazardous air pollutant control requirements, 151-152
   operating permit provisions, 150-151
Clean Water Act, 99, 102, 347-348, 357
Colloidal transport in surface water, 7-8
Combustion reaction, in-process air pollution control, 157-160
Commitment, of company, and pollution prevention program, 360
Composting, biological treatment, 192-193
Compound, chemical, 104
Comprehensive Environmental Response, Compensation and Liability Act. *See* Superfund
Conservative transport, defined, 23
Containment/disposal, waste treatment/disposal, 202-204
   Resource Conservation and Recovery Act cap, 202-204
   Resource Conservation and Recovery Act landfill, 204
   soil cover, 202
Contaminants
   groundwater, 306-314
      migration, 324-326
   transport of, in saturated zone, recharge/discharge, 19-20
Continuous emission monitoring systems, air quality, emissions, 130
Coprecipitation of metals, 9
Corrosivity, chemical analysis, 100-101

cost analysis, of total quality management, pollution prevention program, 354-355
Cost analysis, pollution prevention program, 363, 366
Counter-current packed bed air stripping, groundwater pollution, 343
Covalent bond, 104
Cyclone, particulate air pollution control, 161-163, 166

D

Darcy's law, 21-22
   groundwater, 21-22, 290, 293
Data evaluation, health risk, 39-45
   defined, 37
Data validation, chemical analysis, 120
Dechlorinization process, waste treatment/disposal, 199
Decision flowchart, technology, waste treatment/disposal, 184-186
Degradation, 3
Deliverables, chemical analysis, 119
Densification process, waste treatment/disposal, 195
Depletion, air quality, 133
Dermal contact, soil, health risk, 60
Design
   aboveground storage tank, 233-240
   flow, contaminant loading, groundwater pollution, 323-326
   underground storage tank, 211-214
Dewatering process, waste treatment/disposal, 194-195
Diffusion, 2
Discharge, transport of contaminants in saturated zone, 19-20
Dispersion, 2-3
Documentation, chemical analysis, 116-117
Drinking water, chemical analysis, 99-100
Duplicate analysis, chemical analysis, 118

E

Earth
   elements, 252, 254
   geology, 251-254
      atmosphere of earth, 252
      hydrologic cycle, 252
      hydrosphere, 252
      lithosphere, 252
      minerals, 254-258
Ecological endpoint, ecological risk assessment, 82-83
Ecological receptor selection, ecological risk assessment, 79-80
Ecological risk assessment, 75-95

aquatic organism, toxicity criteria, 88-90
biological stressor, 76
characterization of risk, 91-92
chemical intake estimation, 85-86
chemical of concern, 80-82
defined, 75
ecological endpoint, 82-83
ecological receptors, selection of, 79-80
EE. *See* Ecological endpoint
Environmental Protection Agency, current approach of, 76-77
environmental receptor, 79
exposure assessment, 77-87
exposure pathway identification, 78
exposure point concentration, 84-85
fate analysis, 83
field survey, 77
log-probit analysis, 82
physical stressor, 76
polycyclic aromatic hydrocarbon, 81
reasonable maximum exposure, 84
site characterization, 78
species exhibiting toxicological sensitivity, 79
species vital to food web, 79
species with unique life history/feeding habit, 80
terrestrial organism, toxicity criteria, 90
toxicity assessment, 87-91
toxicity quotient, 91-92
defined, 77
toxicity testing, 77
toxicological data available for species, 80
transport analysis, 83
uncertainty analysis, 92-93
upper confidence limit, 81
Electrostatic precipitator, particulate air pollution control, 162, 165-166, 168-169
Elements
chemical, 103-104
earth, 252, 254
Emission control methods, 133-135
Emission factors, air quality, 130
Emission offsets, air quality, 147-149
Endpoints, ecological risk assessment, defined, 75
Environmental chemistry, environmental regulation, 98-99
Environmental improvement, alternatives development, pollution prevention program, 363, 367-371
Environmental issues, manufacturing issues, integration of, pollution prevention program, 359
Environmental process
chemical behavior in water-soil system, 27-32
Darcy's law, 21-22
Henry's law constant, 11-12
hydrogeologic control of migration through groundwater, 32-36
hydrologic cycle, 3-4
migration in air, 4-7
migration through subsurface, 10-26
overview, 1
surface water migration control, 7-9
transport processes, 1-3
universal soil loss equation, 8
Environmental Protection Agency
carcinogenicity classification scheme, 47
Contract Laboratory Protocol, 105
ecological risk assessment, 76-77
establishment of, 99
Environmental receptor, ecological risk assessment, 79
Environmental regulation, environmental chemistry, 98-99
Environmental science/technology
air pollution control technologies, 155-175
air quality, 125-153
ecological risk assessment, 75-95
environmental chemistry and analysis of regulated compounds, 97-121
environmental processes, 1-36
groundwater, 249-345
hazardous waste, 177-208
human health risk assessment, 37-74
solid waste, 177-208
storage tank technology, 209-247
total quality management pollution prevention, 347-375
Environmental transport processes
adsorption, 3
advection, 2
degradation, 3
diffusion, 2
dispersion, 2-3
partitioning, 2
Evaporation into air, chemical spill, 6-7
Exposed population identification, health risk, 55
Exposure assessment
ecological risk assessment, 77-87
health risk, 52-63
defined, 37-38
Exposure pathway
ecological risk assessment, 78
health risk, 55-57
Exposure point concentration

ecological risk assessment, 84-85
estimation of, health risk, 58-59

F

Fabric filter, particulate air pollution control, 162, 164, 166-167
Facility
   alternatives development, pollution prevention program, 363, 367-371
   baseline characterization of, pollution prevention program, 361-366
Failure, underground storage tank, 210
Fate analysis, ecological risk assessment, 83
Fate modeling, health risk, 52-54
Fault, cross section of, 278, 280
Field measure, quality control, chemical analysis, 115-116
Field survey, ecological risk assessment, 77
Fill box, underground storage tank, 216
Flow path determination, 23
Flow velocity, groundwater, 297-304
Food chain model, health risk, 62
Fractured hard rock, hydrogeologic control of migration through groundwater, 34
Freundlich equation, granular activated carbon treatment, 337

G

Galena, rock-forming mineral, 258
Gas chromatography, 101, 106
   chemical analysis, 106-109
   with mass spectrometry, chemical analysis, 109-110
Gaseous pollutant control, add-on air pollution control, 162, 169-174
Geochemistry, defined, 251
Geohydrology, defined, 251
Geologic interpretation, 281-284
Geologic mapping, 281-284
Geologic time, 270, 272-275
Geology, 250-251
   anticline, 276, 278-279
   defined, 250
   earth, 251-254
      atmosphere of earth, 252
      hydrologic cycle, 252
      hydrosphere, 252
      lithosphere, 252
      minerals, 254-258
   fault, 278, 280
   geochemistry, 251
   geohydrology, 251
   geologic interpretation, 281-284
   geologic mapping, 281-284
   geologic time, 270, 272-275
   geomorphology, 251
   geophysics, 251
   groundwater hydrology, 249-316
   historical geology, 250-251
   igneous rock, 259-261
   Law of Cross Cutting Relationships, 272, 275
   Law of Superposition, 272, 275
   metamorphic rock, 268-271
   minerals, 254-258
   mining geology, 251
   monocline, 276, 278-279
   nonconformity, rock relationship, 276-277
   paleontology, 251
   petroleum geology, 251
   petrology, 251
   physical geology, 250
   rock
      Bowen's reaction series, 260-261
      permeability, 264-265
      porosity, 264-265
      types of, 259-260
      weathering, 260-270
   rock-forming mineral, 257-258
   sedimentary rock, 262-268
      transportation, 266-267
   stratigraphy, 251
   structural, 276-281
   syncline, 276, 278-279
Geomorphology, defined, 251
Geophysics, defined, 251
Goal setting, pollution prevention program, 360
Granular activated carbon, groundwater pollution, 337-338
Graphite furnace atomic absorption, chemical analysis, 113-114
Gravel, unconsolidated, hydrogeologic control of migration through groundwater, 32-33
Gravity differential separation, groundwater pollution, 342
Groundwater
   advection, transport of chemicals, 23
   basin, 20
   contamination, 306-314
   flow, Darcy's law, 21-22
   flow velocity, 297-304
   hydrogeologic control of migration through, 32-36
   hydrology, 284-297
   ingestion of, health risk, 61
   modeling, 314-315
   pollution
      air stripping, 328-334
      biological treatment, 327-328
      carbon adsorption, 334-335

Index / 381

carbon adsorption with air stripping, 335-336
contaminant migration, 324-326
control technologies, 317-345
counter-current packed bed air stripping, 343
design flow, contaminant loading, 323-326
evaluation, 318
Freundlich equation, 337
granular activated carbon, 337-338
Henry's constant, 331-332
inorganics removal, 339-340
remedial program development, 319-323
Theis equation, 320, 322-323
remedial technology, 318-319
solids filtration, 343
surge pump, 342-343
surge tank, 342-343
treatment facility, gravity differential separation, 342
ultraviolet oxidation, 336-337
vapor phase granular activated carbon adsorption, 343-344
regulations, 306
treatment, 205-206
velocity calculation, 22-23
Groundwater hydrology, 284-297
Darcy's law, 290, 293
hydraulic conductivity, 289
hydrologic cycle, 284-285
perched water table, 287-288, 290
porosity, formula, 286-287

H

Haase tank, underground storage tank, 212-213
Hard rock, fractured, hydrogeologic control of migration through groundwater, 34
Hazard index value, health risk, 64
Hazard quotient, defined, 64
Hazardous air pollutant control requirements, Clean Air Act, 151-152
Hazardous and Solid Waste Amendments, 356-357
Hazardous waste
  analysis
    chemical analysis, 100-102
    Resource Conservation and Recovery Act, 100-102
    Superfund, 101-102
    Superfund Amendments and Reauthorization Act, 102
    toxic characteristics leaching procedure, 101
  defined, 178-179

treatment/disposal, 177-208. See also Waste treatment/disposal
Health risk
  absorbed dose, 64
  administered dose, 64
  carcinogenic, 46-48, 63-64
  characterization, 63-69
  chemical intake estimation, 59
  chemicals of concern, 37
  chemicals without toxicity values, 50-51
  data evaluation, 39-45
  defined, 37
  Environmental Protection Agency, carcinogenicity classification scheme, 47
  exposed population identification, 55
  exposure assessment, 52-63
    defined, 37-38
  exposure pathway selection, 55-57
  exposure point concentration, estimation of, 58-59
  fate modeling, 52-54
  food chain model, 62
  groundwater, ingestion of, 61
  hazard index value, 64
  hazard quotient, defined, 64
  incidental soil ingestion, 59-60
  intrinsic toxicity, 45
  LOAEL, defined, 49
  log-probit analysis, 42
  NOAEL, defined, 49
  NOEL, defined, 49
  non-carcinogenic, 48-50, 64-65
  octanol-water partition coefficient, 53-54
  polychlorinated biphenyl, 40
  polynuclear aromatic hydrocarbon, 40
  quantitative uncertainty analysis, 67, 69
  reference concentration. See RfC
  RfC, defined, 50
  RfD, defined, 48
  risk characterization, defined, 33
  RME. See Reasonable maximum exposure
  selection of chemicals of potential concern, 44-45
  site information review, 39-41
  soil
    dermal contact, 60
    ingestion, 59-60
    inhalation, 60-61
    sorption coefficient, 54
  statistical evaluation of data, 43-44
  tentatively identified compound, 41
  toxicity assessment, 45-52
    defined, 33
  transport mechanism identification, 52-54

uncertainty analysis, 65-69
uncertainty factor, defined, 49
vadose zone, 54
vapor pressure, 53
water solubility, 53
Henry's law, 11-12, 331-332
High pressure liquid chromatography, chemical analysis, 110-111
Historical geology, 250-251
Holding time, chemical analysis, 119
Hydraulic conductivity, groundwater, 289
Hydraulic gradient, defined, 21
Hydrogeologic control of migration through groundwater, 32-36
   fractured hard rock, 34
   Karst terrain, 35-36
   rock, fractured, 35-36
   unconsolidated sands and gravels, 32-33
Hydrologic cycle, 3-4, 252, 284-285
   residence time, 4
Hydrology. See Groundwater hydrology
Hydrosphere, 252

I

Igneous rock, 259-261
Ignitability, chemical analysis, 100
Immiscible liquid movement, migration through subsurface, 25-26
In-process air pollution control, 157-161
   combustion reaction, 157-160
   process reaction, 161
In situ bioremediation, waste treatment/disposal, 193
Incineration, waste treatment/disposal, 187-189
Indoor air quality, 140-141
Ingestion, groundwater, health risk, 61
Inhalation, soil, health risk, 60-61
Inorganic chemical analysis instrumentation, 111-113
Inorganics removal, groundwater pollution, 339-340
Installation, underground storage tank, 225-227
Integrity testing, underground storage tank, 219-222
Internal quality control, chemical analysis, 117-118
Internal standard area, chemical analysis, 118
Intrinsic permeability, defined, 25
Intrinsic toxicity, health risk, 45
Ion, 104
Ion exchange
   chemical behavior in water-soil system, 28-29
   waste treatment/disposal, 196
Ionic bond, 104

J

Jurisdictional obstacles, in pollution prevention program, 358-359

K

Karst terrain, hydrogeologic control of migration through groundwater, 35-36
Kent-Moore method, underground storage tank leak testing. See Petrotite method

L

Laboratory operations, chemical analysis, 103-104
Laboratory protocol, Environmental Protection Agency, 105
Law of Cross Cutting Relationships, 272, 275
Law of Superposition, 272, 275
Leak detection, underground storage tank, 217-219
Leak Lokator test procedure, underground storage tank, 221
Liquid/air chemical partitioning, air migration, 5
Liquid phase granular activated carbon, groundwater pollution, 337-338
Lithosphere, 252
LOAEL, defined, 49
Local regulation, air quality, 145
Log-probit analysis
   ecological risk assessment, 82
   health risk, 42
Lowest Observed Adverse Effect Level. See LOAEL

M

Manufacturing issues, environmental issues, integration of, pollution prevention program, 359
Mass balance, air quality, e, 130-131
Mass spectrometry, 101
   with gas chromatography, chemical analysis, 109-110
Materials data, air quality, emissions, 131
Materials handling, waste treatment/disposal, 201-202
Maximum contaminant level, drinking water, 100
MCL. See Maximum contaminant level
Mechanical dispersion, migration through subsurface, 24-25
Metal product manufacturing facility, pollution prevention program, 373-374

Metals, waste treatment/disposal, 179, 183
Metamorphic rock, 268-270
  geology, 268-271
Migration through subsurface, 10-26
  aquifier, 20
  aquitard, 20
  artesian groundwater, 20-21
  conservative transport, defined, 23
  Darcy's law, 21-22
  flow path determination, 23
  groundwater basin, 20
  groundwater flow/Darcy's law, 21-22
  groundwater velocity calculation, 22-23
  Henry's law constant, 11-12
  hydraulic gradient, defined, 21
  immiscible liquid movement, 25-26
  intrinsic permeability, defined, 25
  mechanical dispersion, 24-25
  molecular diffusion, 25
  partitioning of volatile organic compound
    between pure liquid and soil gas, 10-11
    between soil gas and soil moisture, 11-13
    between soil moisture and soil solids, 13-14
  porosity, 22
  soil, transport of volatile organics in, 10
  specific yield, defined, 23
  temperature, soil moisture content, effect of on partitioning, 15
  transport of chemicals through groundwater advection, 23
  transport of contaminants in saturated zone, recharge/discharge, 19-20
  transport of volatile organic compound vapor through soil gas, 15-19
Minerals, 254-258
  rock-forming, 257-258
Mining geology, defined, 251
MLE. See Most likely exposure
Mobile emission source, air pollution, 127
Molecular diffusion, migration through subsurface, 25
Molecule, 104
Monitoring, underground storage tank, 217-219
Monocline, structural geology, 276, 278-279
Most likely exposure, ecological risk assessment, 84-85

N

Natural attenuation/excavation with off-site disposal, waste treatment/disposal, 204-205
Neutralization process, waste treatment/disposal, 195
No Observed Adverse Effect Level. See NOAEL
No Observed Effect Level. See NOEL
NOAEL, defined, 49
NOEL, defined, 49
Non-carcinogenic effect, health risk, 48-50
Non-carcinogenic health risk, 64-65
Nonconformity, rock relationship, 276-277

O

Observation well, underground storage tank, 218
Octanol-water partition coefficient, health risk, 53-54
Operating permit provisions, Clean Air Act, 150-151
Operational improvement, alternatives development, pollution prevention program, 363, 367-371
Operations management, underground storage tank, 228-231
Option hierarchy, for waste management, 350-354
Ore mineral, rock-forming mineral, 258
Organic chemical analysis instrumentation, 105-106
Organic compound, volatile. See Volatile organic compound
Overfill protection, underground storage tank, 215-216
Oxidation/reduction process, waste treatment/disposal, 195-196
Oxides, rock-forming mineral, 258
Ozone, depletion of, 139-140

P

PAH. See Polynuclear aromatic hydrocarbon
Paleontology, defined, 251
Particulate control, add-on air pollution control, 161-169
PCB. See Polychlorinated biphenyl
PDS area. See Attainment area
Perched water table, 287-288, 290
Permatank, underground storage tank, 214
Permeability, rock, 264-265
Permitting process, air quality, 149-150
Petroleum geology, defined, 251
Petrology, defined, 251

Petrotite method, underground storage tank, hydrostatic volumetric testing, 221
Physical geology, 250
Physical stressor, ecological risk assessment, 76
Piping, underground storage tank, 214-215
Plasma torch, waste treatment/disposal, 190
Plastic container manufacturing facility, pollution prevention program, 371-372
Political pressure, and pollution prevention program, 355
Pollution
  groundwater
    air stripping, 328-334
    biological treatment, 327-328
    carbon adsorption, 334-335
    carbon adsorption with air stripping, 335-336
    contaminant migration, 324-326
    control technologies, 317-345
    counter-current packed bed air stripping, 343
    design flow, contaminant loading, 323-326
    evaluation, 318
    Freundlich equation, 337
    granular activated carbon, 337-338
    Henry's constant, 331-332
    inorganics removal, 339-340
    remedial program development, 319-323
      Theis equation, 320, 322-323
    remedial technology, 318-319
    solids filtration, 343
    surge pump, 342-343
    surge tank, 342-343
    treatment facility, 340-344
      gravity differential separation, 342
    ultraviolet oxidation, 336-337
    vapor phase granular activated carbon adsorption, 343-344
Pollution Prevention Act, 347-348
Pollution prevention
  program implementation, 354-355
  total quality management, 347-375
    alternatives development, 363, 367-371
    baseline characterization of facility, 361-366
    case studies, 371-374
      aircraft component manufacturing complex, 372-373
      metal product manufacturing facility, 373-374
      plastic container manufacturing facility, 371-372
    commitment, 360
    component chart, 349
    cost analysis, 363, 366
    economic obstacles, 358
    goals of, 360
    jurisdictional obstacles, 358-359
    manufacturing issues, environmental issues, integration of, 359
    option hierarchy, for waste management, 350-354
    potential obstacles, 358-359
    production obstacles, 359
    program implementation, 354-355
      economic considerations, 354-355
      political pressure, 355
      regulation, 355-358
      team leader, 360
      team staffing, 361
Pollution Prevention Act, 347-348
Polychlorinated biphenyl, 40
Polycyclic aromatic hydrocarbon, ecological risk assessment, 81
Polynuclear aromatic hydrocarbon, 40
Porosity
  formula, groundwater, 286-287
  groundwater velocity calculation, 22
  rock, 264-265
Precipitation of metals, 9
Prevention of significant deterioration concept, air quality, 145-147
Process reaction, in-process air pollution control, 161
Process waste stream, waste treatment/disposal, 207
Production obstacles, in pollution prevention program, 359
Pyrolysis, 1, 189-190

Q

Qualitative uncertainty analysis, health risk, 66-68
Quality control
  field measure, chemical analysis, 115-116
  program, chemical analysis laboratory, 114
Quantitative uncertainty analysis, health risk, 67, 69

R

Radioactive waste, treatment/disposal, 205
RCRA. See Resource Conservation and Recovery Act
Reasonable maximum exposure

ecological risk assessment, 84
health risk, 58-59
Recharge, transport of contaminants in saturated zone, 19-20
Recycling, 206
Reference concentration. See RfC
Remedial program development, groundwater pollution, 319-323
Theis equation, 320, 322-323
Residence time, hydrologic cycle, 4
Resource Conservation and Recovery Act, 99-102, 177-178, 209, 347, 356
cap, 202-204
landfill, 204
storage tanks, 209-210
Retardation, chemical behavior in water-soil system, 27, 30
Reuse, for waste treatment, 206
RfC, defined, 50
RfD, defined, 48
Risk characterization, health risk, 63-69
defined, 33
RME. See Reasonable maximum exposure
Rock
Bowen's reaction series, 260-261
fractured, hydrogeologic control of migration through groundwater, 35-36
igneous, 259-261
metamorphic, 268-271
permeability, 264-265
porosity, 264-265
sedimentary, 262-268
types of, 259-260
weathering, 260-270
Rock-forming mineral, 257-258
carbonate, 258
galena, 258
ore mineral, 258
oxides, 258
silicate, 257-258
sphalerite, 258
sulphate, 258

S
Safe Drinking Water Act, 99-100, 106
Sample preparation, in chemical analysis, 107
Sand, unconsolidated, hydrogeologic control of migration through groundwater, 32-33
SARA. See Superfund Amendments and Reauthorization Act
Saturated zone, transport of contaminants in, recharge/discharge, 19-20
Sediment, suspended, settling of, 9
Sedimentary rock, 262-268

transportation, 266-267
Semi-volatile organic compounds, waste treatment/disposal, 179, 181-182
Separation method, waste treatment/disposal, 193-191
Silicate, rock-forming mineral, 257-258
Site characterization, ecological risk assessment, 78
SOC. See Synthetic organic chemical
Soil
chemical behavior in water-soil system, 27-32
cover, waste treatment/disposal, 202
flushing process, waste treatment/disposal, 197
health risk
dermal contact, 60
incidental soil ingestion, 59-60
inhalation, 60-61
heaping, waste treatment/disposal, 192
loss equation, 8
moisture, partitioning of volatile organic compound, 13-14
solids, partitioning of volatile organic compound, 13-14
sorption coefficient, health risk, 54
transport of volatile organics in, 10
vapor extraction process, chemical/physical processes, 198-199
washing process, waste treatment/disposal, 196-197
soil, moisture, temperature, effect of on partitioning, 15
Soil gas
partitioning of volatile organic compound, 10-13
transport of volatile organic compound vapor through
density grandient, 15-16
diffusion coefficient in soil gas, 18-19
gradient concentration, 17-18
pressure gradient, 17
thermal gradient, 16-17
Solid phase bioremediation, biological treatment, 192
Solid waste
defined, 178
treatment/disposal, 177-208. See also Waste treatment/disposal
Solidification/stabilization process, waste treatment/disposal, 199-200
Solids filtration, groundwater pollution, 343
SOP. See Standard operating procedure
Species

exhibiting toxicological sensitivity, ecological risk assessment, 79
with toxicological data available, ecological risk assessment, 80
with unique life history/feeding habit, ecological risk assessment, 80
vital to food web, ecological risk assessment, 79
Specific yield, groundwater velocity, defined, 23
Sphalerite, rock-forming mineral, 258
Spiked sample, chemical analysis, 118
Spill Prevention, Control, and Countermeasures Plan, aboveground storage tank, 240-242
Spill protection, underground storage tank, 215-216
Stack test for emissions, air quality, 130
Standard operating procedure, chemical analysis, 115
Standpipe Test, underground storage tank, 220
State regulation, air quality, 145
Stationary emission source
industrial, air pollution, 128-129
non-industrial, air pollution, 128
pollution emission rates, 129-131
Statistical evaluation of health risk data, 43-44
Statistical inventory reconciliation analysis, underground storage tank, 219
STI-p3 design, underground storage tank, 212
Storage blank, chemical analysis, 120
Storage tank
aboveground, 231-246
regulations, 232-233
Spill Prevention, Control, and Countermeasures Plan, 240-242
standards, 232-233
system design, 233-240
technology, 209-247
underground, 210-231
aboveground, benefits compared, 242-246
Buffhide Tank, 212
design, 211-214
failure, 210
fill box, 216
Haase tank, 212-213
installation, 225-227
Kent-Moore method, leak testing. See Petrotite method
leak detection, 217-219
Leak Lokator test procedure, 221
liquid-tight integrity testing, 219-222
monitoring, 217-219
observation well, 218
operations management, 228-231
overfill protection, 215-216
Permatank, 214
Petrotite method, hydrostatic volumetric testing, 221
piping, 214-215
regulation, 210-211
secondary containment, 215-216
spill protection, 215-216
Standpipe Test, 220
statistical inventory reconciliation analysis, 219
STI-p3 design, 212
system closure, 224-225
system management, 228-231
technical developments, 211
upgrading, 222-224
Stratigraphy, defined, 251
Stratospheric ozone depletion, 139-140
Structural geology, 276-281
Sulphate, rock-forming mineral, 258
Superfund, 101-102, 177-178, 347, 356-357
Hazardous and Solid Waste Amendments, 356
health risk assessment, 37-74
Superfund Amendments and Reauthorization Act, 102, 177-178, 347-348, 357
waste treatment/disposal, 177
Surface water migration control
adsorption on suspended sediment, 8
chemicals in surface water, fate of, 8-9
colloidal transport in surface water, 7-8
coprecipitation of metals, 9
environmental process, universal soil loss equation, 8
partitioning of chemical, into surface water, 7
photolysis of organic chemicals, 9
precipitation of metals, 9
suspended sediment, settling of, 9
volatilization of organic chemicals, 9
Surge pump, groundwater pollution, 342-343
Surge tank, groundwater pollution, 342-343
Surrogate standard, chemical analysis, 118
Suspended sediment
adsorption on, 8
settling of, 9
Syncline, structural geology, 276, 278-279

Synthetic organic chemical, drinking water, 99
System closure, underground storage tank, 224-225
System management, underground storage tank, 228-231

T

Team leader, pollution prevention program, 360
Team staffing, pollution prevention program, 361
Temperature, soil moisture content, effect of on partitioning, 15
Tentatively identified compound, 41
Terrestrial organism, toxicity criteria, 90
Theis equation, remedial program development, groundwater pollution, 320, 322-323
Thermal desorption, waste treatment/disposal, 189
Thermal processes, waste treatment/disposal, 187-191
   incineration, 187-189
   pyrolysis, 189-190
   thermal desorption, 189
   vitrification, 190-191
TIC. See Tentatively identified compound
Time, geologic, 270, 272-275
Total quality management, pollution prevention, 347-375
   alternatives development, 363, 367-371
   baseline characterization of facility, 361-366
   case studies, 371-374
      aircraft component manufacturing complex, 372-373
      metal product manufacturing facility, 373-374
      plastic container manufacturing facility, 371-372
   commitment, 360
   component chart, 349
   cost analysis, 363, 366
   economic obstacles, 358
   goals of, 360
   jurisdictional obstacles, 358-359
   manufacturing issues, environmental issues, integration of, 359
   option hierarchy, for waste management, 350-354
   potential obstacles, 358-359
   production obstacles, 359
   program implementation, 354-355
      economic considerations, 354-355
      political pressure, 355
   regulation, 355-358
   team leader, 360
   team staffing, 361

Toxic characteristics leaching procedure, 101
Toxicity assessment
   ecological risk assessment, 87-91
   health risk, 45-52
   defined, 33
Toxicity criteria
   aquatic organism, 88-90
   terrestrial organism, 90
Toxicity quotient, ecological risk assessment, 91-92
   defined, 77
Toxicity testing, ecological risk assessment, 77
TQ. See Toxicity quotient
TQM, pollution prevention. See Total quality management
Transformation, air quality, 133
Transport analysis, ecological risk assessment, 83
Transport mechanism identification, health risk, 52-54
Treatment, storage, and disposal facilities, 101
Treatment trains, waste treatment/disposal, 201
TSDF. See Treatment, storage, and disposal facilities

U

UF. See Uncertainty factor
Ultraviolet oxidation, groundwater pollution, 336-337
Uncertainty analysis
   ecological risk assessment, 92-93
   health risk, 65-69
   qualitative, health risk, 66-68
Uncertainty factor, defined, 49
Unconsolidated sands and gravels, groundwater, hydrogeologic control of migration through, 32-33
Underground storage tank, 210-231
   aboveground storage tank, benefits compared, 242-246
   Buffhide Tank, 212
   design, 211-214
   fill box, 216
   Haase tank, 212-213
   installation, 225-227
   Kent-Moore method, leak testing. See Petrotite method
   leak detection, 217-219
   Leak Lokator test procedure, 221
   liquid-tight integrity testing, 219-222
   monitoring, 217-219
   observation well, 218
   operations management, 228-231
   overfill protection, 215-216
   Permatank, 214

Petrotite method, hydrostatic volumetric testing, 221
piping, 214-215
regulation, 210-211
secondary containment, 215-216
spill protection, 215-216
Standpipe Test, 220
statistical inventory reconciliation analysis, 219
STI-p3 design, 212
system closure, 224-225
system management, 228-231
tank failure, 210
technical developments, 211
upgrading, 222-224
Universal soil loss equation, 8
Upgrading, underground storage tank, 222-224
Upper confidence limit, ecological risk assessment, 81
U.S. Environmental Protection Agency. See Environmental Protection Agency

V

Vadose zone, health risk, 54
Vapor phase granular activated carbon adsorption, 343-344
Vapor phase granular activated carbon treatment, 338
Vapor pressure, health risk, 53
Velocity, groundwater, calculation of, 22-23
Vendor data, air quality, emissions, 131
VOC. See Volatile synthetic organic chemical
VOL. See Volatile organic compound
Volatile organic compound
  partitioning of
    between pure liquid and soil gas, 10-11
    between soil gas and soil moisture, 11-13
    between soil moisture and soil solids, 13-14
  waste treatment/disposal, 179-180
Volatile organic compound vapor, transport of, through soil gas
  density gradient, 15-16
  diffusion coefficient in soil gas, 18-19
  gradient concentration, 17-18
  pressure gradient, 17
  thermal gradient, 16-17
Volatile synthetic organic chemical, drinking water, 99
Volatilization of organic chemicals, 9

W

Waste management, total quality management, option hierarchy, 350-354
Waste treatment/disposal, 177-208
  biological treatment, 191-193
    bioreactor, 191-192
    composting, 192-193
    in situ bioremediation, 193
    soil heaping, 192
    solid phase bioremediation, 192
  chemical/physical processes, 193-200
    dechlorinization, 199
    soil flushing, 197
    soil vapor extraction, 198-199
    soil washing, 196-197
    solidification/stabilization, 199-200
    traditional methods, 193-196
      activated carbon adsorption, 196
      densification, 195
      dewatering, 194-195
      ion exchange, 196
      neutralization, 195
      oxidation/reduction, 195-196
      separation, 193-191
  containment/disposal, 202-204
    Resource Conservation and Recovery Act cap, 202-204
    Resource Conservation and Recovery Act landfill, 204
    soil cover, 202
  groundwater treatment, 205-206
  hazardous waste defined, 178-179
  materials handling, 201-202
  metals, 179, 183
  natural attenuation/excavation with off-site disposal, 204-205
  plasma torch, 190
  process waste stream, 207
  radioactive waste, 205
  recycling, 206
  Resource Conservation and Recovery Act, 177-178
  reuse, 206
  semi-volatile organic compounds, 179, 181-182
  solid waste, 206-207
    defined, 178
  Superfund Amendments and Reauthorization Act, 177
  technology overview, 178-208
    decision flowchart, 184-186
  thermal processes, 187-191
    incineration, 187-189
    pyrolysis, 189-190
    thermal desorption, 189
    vitrification, 190-191
  treatment trains, 201

vitrification, 190-191
volatile organic compounds, 179-180
Water
　partitioning of chemical into, 7
　solubility, health risk, 53
　surface
　　chemicals in, fate of, 8-9
　　colloidal transport in, 7-8
Water pollution
　Clean Water Act, 99, 102, 347-348, 357
　Safe Drinking Water Act, 99-100, 106
Water-soil system, chemical behavior in, 27-32
Weathering, rock, 260-270
Wet scrubber, particulate air pollution control, 162, 166, 168

Other Related Products by Government Institutes...

BOOKS

For more information on these books and others, please call our Publications Department at (301) 921-2355. Note: prices are subject to change without prior notice.

Book of Lists for Regulated Hazardous Substances, 1993 Edition
Edited by Government Institutes' Staff
Softcover, 345 Pages, May '93, $67 ISBN: 0-86587-337-2

Clean Air Handbook
By F. William Brownell and Lee B. Zeugin
Softcover, 336 pages, Mar '91, $79 ISBN: 0-86587-239-2

Clean Water Handbook
By J. Gordon Arbuckle et al
Softcover, 446 Pages, June '90, $85 ISBN: 0-86587-210-4

Current Environmental Engineering Summaries: 1993 Edition
By Engineering Information Inc.
Softcover, 1,110 Pages, Aug '93, $89 ISBN: 0-86587-346-1

Directory of Environmental Information Sources, 4th Edition
Edited by Thomas F. P. Sullivan
Softcover, 322 Pages, Nov '92, $74 ISBN: 0-86587-326-7

Emergency Planning and Community Right-to-Know Act Handbook, 4th Edition
By J. Gordon Arbuckle et al
Softcover, 192 Pages, Jan '92, $67 ISBN: 0-86587-272-4

Environmental Audits, 6th Edition
By Lawrence Cahill and Raymond Kane
Softcover, 592 Pages, Nov '89, $75 ISBN: 0-86587-776-9

Environmental Engineering Dictionary, 2nd Edition
By C.C. Lee, Ph.D.
Hardcover, 630 Pages, Oct '92, $88 ISBN: 0-86587-328-3

Environmental Law Handbook, 12th Edition
By J. Gordon Arbuckle et al
Hardcover, 550 Pages, Apr '93, $68 ISBN: 0-86587-350-X

Environmental Statutes, 1993 Edition
Hardcover, 1,165 Pages, Mar '93, $59 ISBN: 0-86587-352-6

Environmental Regulatory Glossary, 6th Edition
Edited by Thomas F. P. Sullivan
Softcover, 544 pages, June '93, $65 ISBN: 0-86587-353-4

EPA Engineering Bulletins: *Current Treatment and Site Remediation Technologies*
By U.S. Environmental Protection Agency
Softcover, 178 pages, Aug '93, $55 ISBN: 0-86587-347-X

Greening of American Business: *Making Bottom-Line Sense of Environmental Responsibility*
Edited by Thomas F. P. Sullivan
Softcover, 350 pages, Sep '92, $24.95 ISBN: 0-86587-295-3

Ground Water Handbook, 2nd Edition
By U.S. Environmental Protection Agency
Softcover, 295 Pages, Mar '92, $69 ISBN: 0-86587-279-1

Health and Safety Audits
By John W. Spencer
Softcover, 336 pages, Apr '92, $65 ISBN: 0-86587-297-X

RCRA Hazardous Wastes Handbook, 10th Edition
By Ridgway M. Hall Jr.
Softcover, 464 Pages, Oct '93, $110 ISBN: 0-86587-355-0

State Environmental Law Handbooks
These comprehensive handbooks are written by respected attorneys from each state, with hands-on experience in dealing daily with the maze of state and federal environmental regulations. For more information on available and forthcoming State Environmental Law Handbooks, please call our Publications Department at (301) 921-2355.

Superfund Manual: Legal and Management Strategies, 5th Edition
By Ridgway M. Hall, Jr. et al
Softcover, 468 Pages, May '93, $95 ISBN: 0-86587-344-5

TSCA Handbook, 2nd Edition
By John D. Conner, Jr. et al
Softcover, 490 Pages, Nov '89, $89 ISBN: 0-86587-791-2

Underground Storage Tank Management: A Practical Guide, 4th Edition
By Joyce A. Rizzo
Softcover, 420 Pages, Nov '91, $79 ISBN: 0-86587-271-9

SUBSCRIPTIONS

Environmental, Health & Safety CFR Update Service
*Published Monthly, Code 7200, U.S. $225/year, Outside U.S. $252/year
ISBN: 0-86587-700-9*

Environmental Management Review
Edited by Government Institutes' Staff
*Published Quarterly, Code 6000, U.S. $188/year, Outside U.S. $252/year
ISSN: 1041-8182*

Waste Minimization and Recycling Report
Published monthly, Code 7000, ISSN: 0889-5509 U.S. $198/year; Outside U.S. $252/yr.

COURSES

Our courses reach tens of thousands of professionals each year. Combining the legal, regulatory, technical, management and financial aspects of today's key environmental, health and safety issues — such as environmental laws and regulations, RCRA, CERCLA, UST, TSCA, environmental management, OSHA, pollution prevention, clean air, clean water, and many other topics — we bring together the leading authorities from industry, business and government to shed light on the problems and challenges you face each day. For more information on these and other courses, please call our Education Department at (301) 921-2345.

TRAINING CONSULTING GROUP

These customized training courses ensure effective, relevant, and focused training by tailoring the training program and materials to satisfy the unique requirements of your organization and the specific training needs of your audience. For more information, call our Training Consulting Group at (301) 921-2366 to discuss your environmental training needs.

VIDEOS

For more information on these and other videos, please call our Publications Department at (301) 921-2355. Note: prices are subject to change without prior notice.

Emergency Planning and Community Right-to-Know Act:
What it Means to You
VHS/15 min./Code 115/1990 $98

Environmental Liability
VHS/18 min./Code 136 $495

Essentials of an Environmental Site Assessment
VHS/50 min./Code 114/1989 $198

Construction of RCRA Groundwater Monitoring Wells
VHS/12 min./Code 120/1990 $98

Hazard Communication: Employee Introduction
VHS/19 min./Code 117/1985 $495

Hazwoper Awareness Level Training: Your Role as a First Responder
VHS/15 min./Code 139/1991 $495

Our Environment: The Law and You
VHS/23 min./Code 119/1989 $495

Pollution Prevention: The Bottom Line
VHS/24 min./Code 125/1990 $295

The Chemistry of HazMat
VHS/19 min./Code 159/1992 $495

To receive a free catalog of our books, courses and videos,
please call: (301) 921-2355

or write:
Government Institutes, Inc.
4 Research Place, Suite 200
Rockville, MD 20850